# Gay-Lussac

*Scientist and Bourgeois*

Gay-Lussac in later life.

# Gay-Lussac

## *Scientist and bourgeois*

MAURICE CROSLAND

PROFESSOR OF THE HISTORY OF SCIENCE,
UNIVERSITY OF KENT AT CANTERBURY

CAMBRIDGE UNIVERSITY PRESS

CAMBRIDGE

LONDON · NEW YORK · MELBOURNE

PUBLISHED BY THE PRESS SYNDICATE OF THE UNIVERSITY OF CAMBRIDGE
The Pitt Building, Trumpington Street, Cambridge, United Kingdom

CAMBRIDGE UNIVERSITY PRESS
The Edinburgh Building, Cambridge CB2 2RU, UK
40 West 20th Street, New York NY 10011–4211, USA
477 Williamstown Road, Port Melbourne, VIC 3207, Australia
Ruiz de Alarcón 13, 28014 Madrid, Spain
Dock House, The Waterfront, Cape Town 8001, South Africa

http://www.cambridge.org

© Cambridge University Press 1978

First published 1978
First paperback edition 2004

A catalogue record for this book is available from the British Library

Library of Congress Cataloguing in Publication data
Crosland, Maurice P.
Gay-Lussac, scientist and bourgeois.
Bibliography: p.
Includes index.
1. Gay-Lussac, Joseph Louis, 1778–1850.   2. Chemists
– France – Biography.   I. Title.
QD22.G35C76   540′.92′4   [B]   77-91084

ISBN 0 521 21979 5  hardback
ISBN 0 521 52483 0  paperback

# Contents

v

# Preface

'I should place him [Gay-Lussac] at the head of the
living chemists of France'

Humphry Davy[1]

'one of the first [natural] philosophers of the age'

J. F. Daniell[2]

The name of Gay-Lussac is remembered in many ways. His work on
the density of alcohol–water mixtures is perpetuated in the 'degrees
Gay-Lussac', which in France have come to replace the medieval
'degrees proof' as a means of describing the strength of alcoholic drinks.
The 'Gay-Lussac tower' was the name given to a vital part of the
manufacture of sulphuric acid in recognition of Gay-Lussac's contribution to this industry. The scientist's name is associated with a type of
barometer and a burette. He is also commemorated in the mineral
'Gay-Lussite'[3] and in *Gaylussacia*, the botanical name for the huckle-
berry. Perhaps Gay-Lussac himself would have derived most satisfaction from being remembered above all as the man who formulated
two fundamental laws of nature. If his own scrupulous acknowledge-
ment of unpublished antecedents has meant that his law of the thermal
expansion of gases is now more generally known as 'Charles' law', at
least his discovery of the regularity in the ratio of the volumes of
combining volumes of gases is still appropriately known, and learned
by every elementary student of chemistry as 'Gay-Lussac's law'. Yet he
is so little known as a man that he is listed in the British Library
catalogue – a source of international repute – as Gay-Lussac, *Nicholas
François*, although his Christian names were indisputably *Joseph Louis*.
In his own day the label 'Gaylussacite' (*gaylussacien*) was used by his
enemies to denote a member of the scientific establishment, but such
usage becomes redundant after a man's death.

This biography will help give flesh and bones to one of the names
found in science text-books but, in so doing, it may also make a small
contribution to a fuller understanding of the development of science.
The study of individual cases can do something to correct the picture of

science and scientists given by those who wish to make generalisations about the subject but have no time or inclination to go back to the sources.

The dearth of biographies of important French scientists compared to the relative abundance of biographies of British scientists was commented on at the beginning of this century.[4] Recently the same phenomenon has been commented on again and a historian has remarked that there are not yet enough examples available of detailed lives of scientists. He continues:

French scientists have suffered particularly from this neglect. Biography is not a genre that the French excel in. This is due partly to a preference for more grandiose philosophical themes than mere individuals can provide, and partly to a powerful tradition of *éloges*, raised to a high level by the Academies, which has encouraged subtle essay-writing more than detailed research.[5]

Gay-Lussac certainly obtained his *éloges* and he has also been fêted by French local historians. It is high time that a more serious biography was written of a man who made major contributions to physics and chemistry and who has a claim to be considered as one of the first professional scientists.

But a biography is not a hagiography. Gay-Lussac was a scientist and a man and not a saint. I have tried to place the man in his intellectual, social and national context and have indicated this by the title of the book. The term 'bourgeois' should be understood descriptively rather than pejoratively. It is intended to signify that Gay-Lussac was born into a middle-class French family and always stayed within a middle-class context. This was not so important at the beginning of his career in Napoleonic France, but in the 1830s and 1840s under Louis Philippe Gay-Lussac can be taken as an example of a French citizen in the professional classes or the upper bourgeoisie. His habits of hard work and his sense of duty and responsibility were qualified by the common desire to build up a small fortune. The desire to provide for his family was one of several factors which led Gay-Lussac away from pure science to applied science. Thus in order for us to understand why he chose to do one piece of research rather than another it is not sufficient to look within science. Scientific problems must be seen against the background of Gay-Lussac's employment, his place in the Academy of Sciences, his colleagues, friends and family.

The Academy[6] made a distinction in 1803 between the mathematical sciences and the natural sciences. As Gay-Lussac's work was in physics and chemistry, his research lay at the interface of these two divisions, making some use of mathematics but also having sympathy with more descriptive sciences. Thus his work was more central and perhaps more typical of 'French science' than, say, the work of a mathematician or a botanist. The distinction between mathematical science and natural science was recognised in the Faculty of Science and students for the first degree were required to attend lectures in one branch only. The course on basic physical science, taught by Gay-Lussac, was the only one common to the two programmes.

In describing the work of a scientist it is understandable that one should emphasise his successes, but one should also say something about failures, which can be equally instructive in arriving at an understanding of one's subject and the problems of his time. Any biographer is tempted to pursue relentlessly the details of his subject's life and work. I have exercised some restraint here. Although I have tried to get inside my subject to understand how his mind worked, I have also considered him against the background of his contemporaries. I feel that it is important not only to know what Gay-Lussac did but also how this differed from the work of other scientists of the time.

I have gone out of my way to suggest where Gay-Lussac obtained his ideas and to stress the influence of his mentors. But to do this is not to suggest that his work was mostly derivative. We may wish to know if some of Gay-Lussac's inspiration is found in his predecessors, but we are even more interested in how he took further their ideas and modified them.

I used to think of the biography of a scientist as falling naturally into two parts – life and work. In such an account one would deal first with personal details before examining fairly exhaustively and in chronological order all the scientific work of one's subject. This method of working is clearly easier since the writer is not forced to ask himself continually whether there is any relation between the scientist's life and his work. Indeed it is a convenient format for any author who may not think there is much connection between the two and perhaps merely recounts some biographical information by way of introduction. My approach has been to relate as far as possible the life and the work, believing as I do that every man is a child of his age; his life and his work can best be understood against the backcloth of the culture in which he grew up. Science is the product of scientists, both as individuals and in organised groups and this is the justification for writing the biography of any scientist. I have been selective rather than exhaustive and I have not shirked the responsibility of interpretation and evaluation.

A chronological table has been introduced which fulfils the double purpose of showing what Gay-Lussac did year by year and of providing a reminder of the general historical background. This has left me free to take a thematic rather than a strictly chronological approach to the life and work of the French savant and thus to relate him to his contemporaries and to more general problems. This discussion in one place of particular themes, e.g. scientific rivalry, educational institutions, the search for laws of nature, has certain advantages for the reader with broad interests in the social or philosophical aspects of science and for whom this study of Gay-Lussac might provide fresh data. Such a reader can be selective in his reading of the book and, I hope, easily find material on some aspect of nineteenth-century science which is illustrated in the life and work of Gay-Lussac. A historian with little interest in the details of scientific research might pass over chapters 5 and 6 and still get something from the remainder of the book.

A collection of previously unpublished letters is included in a collection of primary source materials in the Appendix. A few extracts in translation have been given in the text but an appendix allows them to appear in their proper context. The letters have been chosen both because they are representative of different aspects of Gay-Lussac's life and for the information they contain. They provide a series of windows on the world of the scientist, usually seen with his own eyes but sometimes with those of his contemporaries. They cover the period from his arrival in Paris at the age of 17 to after his death. Reasons of space do not allow a larger collection of letters but they have not led to any vital omission.

The idea of writing a biography of Gay-Lussac and some preliminary work towards its realisation was started some time ago. Enquiries which began with descendants of Gay-Lussac took me to France and from Paris to the provinces. Research spread to sources in Britain, Germany, Sweden and the U.S.A. and manuscript materials were examined in many different circumstances ranging from an air-conditioned American library to a converted French chapel, from the office of a *notaire* to the chateau which once belonged to Madame de Pompadour. One of the benefits arising from the curiosity of foreign scholars is a greater appreciation by the owners of the historical value of the material in their keeping and I am glad that Gay-Lussac's library, exposed at the beginning of my research to rain and to rats, has been rescued and saved from further deterioration. The present dispersal and temporary housing of Gay-Lussac's archives precludes complete systematic description. It is to be hoped that the bicentenary celebrations in 1978 will focus attention on the desirability of bringing some of these sources together and cataloguing them.

If I had been looking for an easy piece of research with documents neatly assembled under one roof, I would have considered myself singularly unfortunate in my choice of subject. The task would certainly have been easier if Gay-Lussac's research had been confined to one field or to one institution but also it would have been less worthwhile. A study of a man who contributed to several areas of science, both pure and applied, who was associated with many institutions and who was a public figure, is more likely to reveal the vicissitudes of a career in science than a study of a less versatile figure.

For access to manuscript sources I am indebted to Madame Roger Gay-Lussac, Monsieur Pechdo, Monsieur Decanter, Monsieur Larre, Monseiur Laissus, and to archivists and librarians at the Académie des Sciences, the Archives Nationales, the Ecole Polytechnique, the Service Historique of the French army, the archives of the French artillery, the Royal Institution, the Royal Society and the Wellcome Historical Medical Library. I have used printed books and periodicals at the Bibliothèque Nationale, the British Library, the Institute of Historical Research, London University, the Science Museum Library, the University of Kent and the University of Leeds. Part of chapter 7 was drafted when, as a visiting professor at the University of Pennsylvania, I had access to the Edgar Fahs Smith collection of books on the history of chemistry. My appointment at the University of Kent in 1974 enabled me to go from a few draft sketches to a full-length book. I must thank the Nuffield Foundation for payment of my salary over several years and the standing committee for the Unit for the History of Science under the chairmanship of Professor Graham Martin for encouragement in my research.

I should like to thank various friends and colleagues for private criticism of a first draft of this book. Alec Dolby read and criticised chapter 3, Crosbie Smith looked critically at chapter 6, Graham Smith gave me the benefit of his specialised knowledge in connection with chapter 8, and William Fortescue made some improvements in chapter 10. Raymond Coulon kindly checked the French text in the Appendix. Bill Smeaton read the complete draft and made a large number of detailed criticisms and corrections to the typescript. I should also like to thank Richard Ziemacki and the staff of the Cambridge University Press for all their help. For the typing I owe much to the efficiency and accuracy of Veronica Ansley and Yvonne Procter. Christoph Meinel kindly helped with corrections at the proof stage. Various other acknowledgements are given in endnotes. The author must take responsibility for the contents of the book and the translations provided.

# Chronological table

| | *Life and work of Gay-Lussac* | *French history* |
|---|---|---|
| 1778 | Born at St Léonard (Limousin) | |
| 1789 | | Meeting of Estates General. Fall of Bastille. Abolition of feudal privileges |
| 1790 | | Civil constitution of the clergy |
| 1792 | | France at war. Meeting of Convention. Declaration of a Republic |
| 1793 | | Execution of Louis XVI. Mass mobilisation. Académie des Sciences closed |
| 1794 | | Fall of Robespierre |
| 1795 | Goes to Paris | Ecole Polytechnique. Foundation of National Institute to replace former Academies |
| 1795–9 | | France ruled by Directory |
| 1797 | Enters Ecole Polytechnique | |
| 1799 | | Napoleon Bonaparte seizes power |
| 1799–1804 | | Consulate |
| 1800 | Graduates from Ecole Polytechnique. Enters Ecole des Ponts et Chaussées | |
| 1801 | Joins Berthollet at Arcueil | |
| 1802 | Memoir on thermal expansion of gases ('Charles' law') | Peace of Amiens |
| 1804 | Balloon ascent (7000 m). *Répétiteur* at Ecole Polytechnique | Bonaparte declares himself Emperor |
| 1805 | European tour with Humboldt | |
| 1806 | Return to Paris. Elected to First Class of Institute. Research on specific heats of gases and capillarity | |
| 1807 | Vol. 1 of *Mémoires* of Society of Arcueil | |
| 1808 | Preparation of potassium by chemical method (with Thenard). Preparation of boron (with Thenard). Law of combining volumes of gases | Establishment of University of France including Faculties of Science |
| 1809 | Suspects elementary nature of 'oxymuriatic acid' (chlorine). Professor at Paris Faculty of Science. Marriage | |
| 1810 | Professor at Ecole Polytechnique. Analysis of organic compounds (with Thenard) | Madame de Staël, *De l'Allemagne* |
| 1811 | Publication in book form of collection of memoirs (with Thenard): *Recherches physico-chimiques* | |
| 1812 | | Occupation of Moscow by French troops. Retreat |
| 1813–14 | Research on iodine and hydracids | |
| 1814 | | Abdication of Napoleon. Constitutional Charter of Louis XVIII |

| | |
|---|---|
| 1815 | Research on cyanogen radical |
| 1816 | Editorship of *Annales de chimie et de physique*. Visit to England with Arago |
| 1818 | Memoirs on cold and absolute zero. Estimation of potash. Member of consultative committee to artillery |
| 1819 | Memoir on solubility of salts |
| 1820 | Memoir on estimation of soda (with Welter) |
| 1821 | Memoir on rendering textiles fireproof |
| 1822 | Study of relationship between density and alcohol content of alcoholic liquors |
| 1824 | Instruction on estimating chlorine in bleaching powder. Analysis of fulminating silver (with Liebig). *Instruction* on lightning conductors |
| 1825 | Patent for stearic candles (with Chevreul) |
| 1828 | Estimation of potash. Publication (unauthorised) of his physics and chemistry lectures at Faculty of Science |
| 1829 | Assay master at Bureau de Garantie at Paris Mint |
| 1831 | Elected to Chamber of Deputies |
| 1832 | *Instruction* on the estimation of silver. Professor of Chemistry at Museum of Natural History. First association with Saint-Gobain Company |
| 1835 | Second *Instruction* on estimating chlorine in bleaching powder |
| 1839 | Nominated to Chamber of Peers. Memoir on chemical forces |
| 1840 | Resigns chair at Ecole Polytechnique |
| 1842 | Patent for recycling oxides of nitrogen in manufacture of sulphuric acid ('Gay-Lussac tower') |
| 1844 | Memoir on respiration |
| 1848 | Last published scientific memoir (on aqua regia) |
| 1850 | Death |

| | |
|---|---|
| 1815 | Return of Napoleon ('Hundred Days'). Final defeat at Waterloo. Return of Louis XVIII |
| 1824 | Death of Louis XVIII. Accession of Charles X |
| 1830 | July Revolution. Louis Philippe on throne |
| 1831–2 | Cholera epidemic |
| 1834 | Balzac, *Le Père Goriot* |
| 1840 | Proudhon, *Qu'est-ce que la propriété?* |
| 1842 | Comte, *Cours de philosophie positive* completed |
| 1842–6 | Railway mania |
| 1848 | Abdication of Louis Philippe. Provisional government. Universal suffrage. Constituent Assembly elected. Second Republic |
| 1852 | Napoleon III Second Empire |

# Abbreviations

| | |
|---|---|
| *A.c.* | *Annales de chimie* |
| *A.c.p.* | *Annales de chimie et de physique* |
| *A.C.R.* | Alembic Club Reprints |
| Ch.D. | Chambre des Députés |
| Ch.P. | Chambre des Pairs |
| *C.R.* | *Comptes Rendus hebdomadaires des séances de l'Académie des Sciences* |
| G.L. | Gay-Lussac archives, Limoges |
| *J. de phys.* | *Journal de physique* |
| Liebigiana | Archive series Liebigiana, Bayerische Staatsbibliothek, Munich |
| *M.S.A.* | *Mémoires de physique et de chimie de la Société d'Arcueil* |
| *Phil. Trans.* | *Philosophical Transactions of the Royal Society of London* |
| *P.V. Inst.* | *Procès-verbaux des séances de l'Académie des Sciences* |

# I

# A young provincial in Paris

'I have not chosen a career which will lead me to a
great fortune, but that is not my principal ambition'

Gay-Lussac[1]

## Introduction

By the late eighteenth century science was fairly well established as an
intellectual activity in western Europe. The scientific movement had
reached a zenith with the work of Isaac Newton (d. 1727) who had
applied his mechanics to the whole solar system in his law of universal
gravitation. Among the many followers of Newton in Britain in the
eighteenth century, specially distinguished for their studies of the
nature of matter were Joseph Black, Henry Cavendish and Joseph
Priestley. Each in turn made important contributions to the knowledge
of 'airs' or gases, but the interpretation of the role of gases in the
physical and chemical world had to wait for the Frenchman, Antoine
Laurent Lavoisier (1743–1794). France was not only Britain's political
rival in the eighteenth century but also shared with her supreme
honours in literature, the arts and sciences. With a population of over
twenty million, France had a major advantage over Britain with less
than half that estimated number. But several factors encouraged the
beginnings of an industrial and economic revolution in Britain, while
France, with her government-regulated industry and more rigid social
stratification, carried on the traditional methods of manufacture and
production. One of the few areas where France could claim important
industrial innovation towards the end of the eighteenth century was in
chemical industry and at least some of this advance was due to
Lavoisier and the Royal Academy of Sciences in Paris.

Lavoisier brought about a revolution in chemistry by his new understanding of chemical composition and reaction. He presented a list of
some thirty simple substances or elements which could combine in
certain ways to form compounds. In his system he gave particular
prominence to combination with oxygen (usually by combustion) to
form oxides or, with excess oxygen (he thought), acids. By the early
1780s this oxygen-centred chemistry had begun to win converts. One

of the first of these was a chemist who had come to the subject via medicine, Claude Louis Berthollet (1748–1822), soon to become director of dyeing at the Gobelins dyeworks. Another convert to the new chemistry was Antoine François Fourcroy (1755–1809), a brilliant lecturer who held one of the few official teaching positions in chemistry in Paris before the Revolution. These three were to be joined in 1787 by the Dijon lawyer and chemist Guyton de Morveau (1737–1816) in the reform of chemical language. It seemed sensible when the theory was being reformed to abandon the chaotic and often arbitrary names given to substances in favour of a systematic nomenclature in which chemical compounds were given names reflecting their composition.

All four chemists became involved in the French Revolution, and in the stormy days of 1793–4 their leader Lavoisier was tragically to lose his head. Guyton and Fourcroy both became members of the Revolutionary Convention and like Berthollet, the least politically active, survived the storm. Indeed during the period when France was threatened with foreign invasion all three had applied their chemical knowledge and administrative talents to the organisation and production of supplies and munitions. Fourcroy played an influential part in the educational debates in the Convention and, in the constructive period after 1794, helped to establish several major institutions of higher education, notably the Ecole Polytechnique and several new medical schools. Guyton played a prominent role in the first years of the Ecole Polytechnique and also carried on the journal *Annales de chimie*, which had been founded by Lavoisier. Berthollet became the friend of a brilliant young artillery officer who was soon to alter the shape of France and the map of Europe, Napoleon Bonaparte. Bonaparte took him to Egypt and, when he returned to France in 1799 and seized power, Berthollet was made a member of the newly-established Senate. The Senate was supposed to safeguard the constitution, but duties were minimal and, with a very high salary, membership could be regarded as a sinecure.

Berthollet used his new wealth to buy a country house at Arcueil just outside Paris. There he wrote his *Essai de statique chimique* (1803) in which he formulated a new approach to chemical reaction which was to be taken up later in the nineteenth century. Arcueil was a base for some important scientific research much of which was actually carried out not by Berthollet himself but by various protégés. First and foremost of these was Gay-Lussac. They were soon joined at Arcueil by the great applied mathematician Pierre Simon Laplace (1749–1827), also a friend of Napoleon and Chancellor of the Senate. Laplace was particularly interested during this period in the study of short-range forces of attraction which might explain a whole range of phenomena in physics and chemistry. Although Berthollet flourished under the Napoleonic regime, after the defeat of Napoleon in 1814/15 he went into retirement. As Fourcroy had died in 1809 and Guyton died in 1816 Gay-Lussac became increasingly prominent as the leading French chemist under the restored Bourbon monarchy.

Because there was continuity between Berthollet's conception of chemistry and that of his protégé it may be worthwhile stressing the discontinuity of social support caused by the upheaval of the Revolution.[2] Gay-Lussac was in many ways typical of the new men of science who emerged from post-Revolutionary France. His was the first generation which could receive a full training in science and go on to earn a living as a scientist. Before his time, one was fortunate if one could follow a single course of lectures on some branch of science to supplement what could be learned from books. The great chemists of the generation before Gay-Lussac in France: Lavoisier, Berthollet, Guyton de Morveau, Fourcroy, qualified in medicine or law before turning to chemistry, of which they made virtually a new science. Mathematics had previously been taught at quite a high level in the military academies, but with the foundation of the Ecole Polytechnique young men were able to have mathematical training while remaining civilians.

Of course there were some apparent similarities between the two generations, such as membership of the Academy of Sciences, although closer examination tends to reveal important differences. To understand these it may be useful to review briefly Berthollet's early career. When he came to Paris from Piedmont in 1772 he managed to enter the circle of the Duke of Orleans. Berthollet had taken a degree in medicine and the Duke recommended him for the position of personal physician to Madame de Montesson; this provided him with the necessary leisure to do chemical experiments. His chemical work was favourably received and gained him admission to the Academy of Sciences, thus providing him with credentials as a scientist. But, like every other new member he was elected in the lowest grade (originally known as 'pupil' but then 'assistant'). He could aspire to promotion to 'associate' member when one of the more senior chemists died and eventually to the highest grade of 'pensioner'. After the Revolution, however, the new Institute[3] had no hierarchy of membership. Specialised education and publication could now precede election. This desirable state of affairs was taken too far later in the nineteenth century when the age of election rose steadily so that membership became a final accolade in a scientist's career rather than an honour a scientist could use in his most creative period.

Gay-Lussac was fortunate to be elected at the beginning of his scientific career.

Before the Revolution there was no scientific education at university level with a recognised qualification at the end. Gay-Lussac achieved this through his training at the Ecole Polytechnique. Moreover, his early years after graduation were followed by fruitful research under expert guidance. When Gay-Lussac looked for a livelihood, the days when Lavoisier had earned his living by association with a tax company might have been a century away instead of a mere decade. Berthollet, who depended so much after 1799 on the patronage of Bonaparte, was almost in the position of a royal favourite. This was not to be the pattern of science in the modern world.

For the new generation science was not a leisure-time pursuit for the wealthy bourgeois or nobleman; it was to be a full-time occupation. The scientist had a place in the new state and Condorcet had envisaged him as a civil servant. The scientist was to be paid out of public funds and had a public duty to perform, usually teaching, but occasionally research, as in the Bureau des Longitudes. It is true that there had been a few teaching posts in Paris before the Revolution but the political change brought a national investment in education at all levels; higher education received its full share of attention and resources. The new institutions provided educational opportunities never before available and they also required staff to teach in them. In the acceptance of the principle that education was the responsibility of the state we can see a qualitative change which begins a chapter in the history of the modern world.

Gay-Lussac may be considered as one of the first of the moderns. The analysis we give of his scientific output throughout his life could hardly have been applied before the main method of publication became the scientific paper rather than the book. Gay-Lussac also represents the new era in his many joint publications, a procedure almost unknown before 1800. He was one of the first generation of professional scientists, 'professional' in the sense that, after undergoing a prolonged period of specialised training in the theory and practice of physical science, he was employed to use that training. He made a living out of science and thus helped to establish science as a career to which any young man, in France at least, could aspire. In Britain professionalisation came rather later, but that is another story.

*Early education*

The region of France from which Gay-Lussac came was to become at

4

the time of the Revolution the department of Haute Vienne, after the river Vienne which runs through it and its principal city Limoges before turning north to join the Loire more than 200 kilometres to the south west of Paris. About 22 kilometres to the north of Limoges was the small market town of Saint Léonard en Noblat with its eleventh-century church where the future scientist was baptised. Arthur Young, who travelled through France in 1787, described the country to the north of Limoges as the most beautiful he had seen in that country; thickly wooded with many chestnut trees, an undulating countryside giving picturesque groupings of rock, wood and water.[4] The district was not rich agriculturally, or in any other way, but Turgot, who had been *intendant* of Limoges, had left a memorial to himself in the splendid roads of the region, on which Young commented.

Joseph-Louis Gay-Lussac was born at Saint Léonard on 6 December 1778, one of five children of Antoine Gay-Lussac, advocate and public prosecutor at Saint Léonard, and Leonarde Bouriquet. Joseph-Louis was not the eldest child but he was the eldest son, and family correspondence would suggest that the family hopes were largely pinned on him. He had three sisters, Fanchette, Marguerite and Mariette, and one brother, Pierre, who was to qualify in medicine.

Law and medicine were both represented in the Gay family, which towards the end of the eighteenth century began to use the name Gay-Lussac after their property at the hamlet of Lussac a few miles from Saint Léonard. This distinguished them from other branches of the family which also found some place in history, notably the Gay de Vernon family. Gay-Lussac's contemporary, the military engineer, Simon François Gay de Vernon (1760–1822) was also born at Saint Léonard. The scientist's ancestry can be traced back to members of the Gay family in the seventeenth century who were merchants. Louis, the grandfather of Gay-Lussac, studied medicine at Bordeaux and practised as a physician at St Léonard. Louis' only son Antoine, the father of the scientist, was born in 1744. He was sent to Bordeaux to study law and in 1769 he passed examinations in civil and canon law. In 1775 he became *Procureur du Roi* at St Léonard thus definitely establishing the Gay-Lussac family as one of local importance.

Such a local officer of the crown was not likely to escape the social and political upheaval of the Revolution. The early stages were slow and it seemed then that some reform would be possible under the monarchy. Under political and financial pressure Louis XVI agreed in May 1789 to a calling of the Estates General, a body of representatives of nobility, clergy and the 'third estate', which had not met since 1626. The government as evidence of their sincerity in considering

5

reform invited the people to draw up *cahiers de doléances* or lists of grievances to be discussed by the Estates General. There still exist the suggestions drawn up in the handwriting of Gay-Lussac's father on behalf of the third estate of the region of Saint Léonard. In this preliminary *cahier de doléances*, dated 1 March 1789, it was suggested that the three orders should meet together rather than separately and that each man should have one vote. This document shows that the older Gay-Lussac was opposing the traditional stranglehold of the nobility in the political life of France. By 1793, however, men who had previously appeared as liberals were considered reactionary and in September 1793 the Convention invited the public to denounce as 'suspects' those whose full support of the course of the Revolution was in doubt. Gay-Lussac's father was arrested 'for reasons of aristocracy founded on public opinion'. As he was ill at the time, he was at first placed under house arrest, instead of being sent to the prison in Limoges. A petition of the accused was considered by a local committee on 7 January 1794 and although no definite charges were brought, it was not until after the fall of Robespierre (July 1794) that he was released.

The case of the Gay-Lussac family is typical of many of the bourgeoisie during the Revolution, who after a comfortable existence were reduced to straitened circumstances. As *Procureur du Roi* Gay-Lussac senior had a salary of 2400 livres (just over £100 sterling) but the source of this income dried up when the monarchy was abolished in 1792 and his appointment ceased. The income from his land was reckoned at 800 livres, not much for a family of seven together with servants. The head of the household's first concern was to stay alive but he was also concerned to find some employment, preferably making use of his administrative and legal experience. Under the new regime he had to swallow his pride and on 10 December 1796 he accepted the appointment of bursar at the almshouse of Saint Léonard. He subsequently became the local postmaster. Thus the immediate effect of the Revolution was to depress the fortunes of the Gay-Lussac family and to bring them down in the social scale. Although Joseph-Louis Gay-Lussac at first seemed to his father mindless of such considerations, the humiliation of the family may well have been felt by the eldest son and may have been one factor in his later desire for fame and fortune. The French Revolution, of course, not only brought people down the social scale but also provided opportunities for rising. By his genius, good fortune and hard work Gay-Lussac was to embark on a new career open to talents – that of scientist, professor and consultant.

Before the Revolution education was accepted as the responsibility

6

of the Catholic Church and Gay-Lussac's first education was in the hands of the abbé Bourdeix. In fact there were two brothers Bourdeix at Saint Léonard who were both priests.[5] It seems to have been the elder, Jean-Joseph Bourdeix (b. 1752) who was to be the tutor of Gay-Lussac. He had studied humanities, philosophy and theology at Limoges. He was formally attached to the chapel at Artiges near Saint Léonard, but this position left time for him to make some contribution to education in the Saint Léonard area as a tutor. The Revolution had a considerable effect on the Church in France. In November 1790 all beneficed clergy were required to take an oath to uphold the new political order. Many refused and the following year witnessed the beginning of a mounting campaign of repressive measures against the 'refractory' or non-juring priests. On 6 February 1793 the abbé Bourdeix was arrested and probably later fled the country to return only at the time of the Concordat[6] in 1802, when he opened a school.

In eighteenth-century France boys who were not educated privately by a tutor might have attended a Church school. Although Saint Léonard did not have its own school, there were *collèges* at Limoges and Eymoutiers.[7] Without the Revolution Gay-Lussac would probably have gone to the *collège* at the nearby town of Eymoutiers. A detailed syllabus of examinations for the respective classes in the 1780s was found in the family papers.[8] These include the names and origins of the pupils, several of whom came from Saint Léonard either as *pensionnaires* (boarders) or as *externes* (day pupils). Pupils even came from the city of Limoges to attend the school and one is impressed by the academic classical education offered. Examinations included not only oral translations of texts and recitations but also 'declamation', a useful training for later public speaking. A little science was included in the third class in the form of elementary astronomy, but Latin prose and verse was the staple study with scripture and catechism.

However, political circumstances kept Gay-Lussac for the moment in his home town. As the role of the clergy in education came under challenge we see the beginning of the emergence of the lay teacher. For the next year Gay-Lussac had lessons from two teachers, Courty and Albert, but probably the family did not feel that this local education was at a very high level. The fall of Robespierre in July 1794 marked the beginning of the swing of the pendulum towards moderate opinion. It was in this more constructive period when new educational institutions were being set up that Gay-Lussac senior sent his eldest son to Paris, where he arrived in November 1794 just before his sixteenth birthday.

Gay-Lussac's first months in Paris were of winter and what a winter! There had only been two previous winters as cold in the whole of the eighteenth century. The cold spell lasted from 16 December to 5 February.[9] The cold undermined the morale of everyone. For those with little or no fuel available it was disastrous and it drove back to their homes many of those young students a few years older than Gay-Lussac who had enrolled at the short-lived Ecole Normale and the newly-founded Ecole Polytechnique.

We have no direct evidence about Gay-Lussac's first months in Paris. According to one source he was in the care of the abbé Dumonteil. However, a letter in the family archives of 27 March 1795, signed Daniel Monteil and addressed to Gay-Lussac père, suggests that all had not been well at the beginning.[10] Father Dumonteil spoke of the difficulty of finding him suitable lodgings and of the 'idle and disorganised life' which he had been leading. However the priest had now found a suitable establishment for him where 'food for the body is healthy and sufficient in quantity while that of the spirit and the soul is not neglected'. The family had scraped together enough money to pay for their boy to stay there.

This establishment was a boarding school, called the Pension Savouré, which had been founded about 1729 on the principles of the ancient university of Paris.[11] The education provided was to be based on 'religion and piety without neglecting the progress of studies'. The founder had insisted on the close surveillance of pupils by masters on the principle that, in the absence of a master, the devil might take over. From 1770 to 1803 the school was run by Jean-Baptiste-Louis Savouré, Master of Arts, and was situated in the rue de la Clef near the site of the future Ecole Polytechnique. Gay-Lussac did not stay very long at this school, which seems to have temporarily closed for economic reasons.

Gay-Lussac then went to another private boarding school run by a Monsieur Sencier. Sencier's school had originally been at Nanterre but had moved to Passy near the Bois de Boulogne, then in the country outside Paris. There was a distinct advantage for Gay-Lussac in living in the country in a time of food scarcity. His surviving letters home give us some idea of his life at the school. There is evidence of an enlightened curriculum with, for example, not only Latin but a *modern* language. Gay-Lussac confessed to his father that he had at first been reluctant to undertake the study of English, but Sencier had convinced him of its great utility in the modern world. Drawing too was taught at the school, a subject which should be understood in the post-Revolutionary context of a useful art for engineers and architects. Gay-Lussac sent his father some examples of his architectural drawings and hoped for some encouraging words from home about them.

There would be room for considerable speculation about the exact circumstances which led Gay-Lussac to adopt a scientific career if it were not for the fortunate discovery of a letter written to his father in January 1803 and reviewing his career in Paris.[12] In the first place his father had wanted him to follow in his own footsteps and study law. When the father had subsequently suggested that a grounding in mathematics would also be useful the son had obediently turned to this study and had substituted mathematics for Latin which, perhaps because of the departure of the classics teacher, was no longer available to him at M. Sencier's school. His personal copy of Caesar's *Gallic War*[13] has inside the front cover some long multiplication, as if to symbolise his new concern with mathematics.

## The Ecole Polytechnique

It was the study of mathematics which provided the opportunity for Gay-Lussac to enter the Ecole Polytechnique. Mathematics had received a tremendous stimulus after the Revolution by being made the basis of the entrance examination to the Ecole Polytechnique, the new élite scientific institution. No one therefore could study mathematics, particularly in Paris, without being conscious of the Ecole Polytechnique. We must appreciate the publicity and prestige which this Revolutionary institution had acquired in the first few years of its existence. Although the historian of science is interested in the school as the training ground of many of France's future physical scientists, it was seen at the time in a much broader educational context. The father of Balzac hoped that the future novelist would enter the Ecole Polytechnique and to this end he arranged for him to have additional lessons in mathematics. The writer Stendhal came even nearer to a scientific education. In Grenoble young Stendhal had distinguished himself in mathematics and won a place in the Ecole Polytechnique. Once in Paris, however, Stendhal took fright at the discipline of higher education and used his examination success to escape from the provinces and to taste the life – in all its aspects – of the French capital.

The Ecole Polytechnique was intended primarily for the training of both civil and military engineers although the curriculum was much wider than any purely vocational training. In 1797 the Minister of the Interior, who had overall responsibility for the school, claimed that the curriculum provided 'an education which was complete and yet general enough to serve as a foundation for all possible applications'.[14] The 'applications' here should be understood in the context of the Schools of Public Service or *Ecoles d'application* such as the Artillery School, the School of Mines, the School of Bridges and Highways, etc. After some initial confusion it became the rule that entry to these schools was only available to graduates of the Ecole Polytechnique. This decision had two main effects. It increased the importance of the Polytechnique, giving it a virtual monopoly of recruitment of skilled scientific and technical manpower. It also enabled the Polytechnique to strengthen its tendency to provide a general scientific rather than a strictly vocational training.

The entrance examination to the Ecole Polytechnique was held each year in the last week of October. Candidates were expected to know the rules of arithmetic, algebra, including the solution of equations of the first four powers and the theory of series, geometry, trigonometry, analytic geometry and conic sections. The examiners were instructed to look for promise as much as actual achievement but were obviously able to report more confidently on the latter. From the foundation of the school the examiners were also required to be satisfied as to the 'moral and civic qualities' of the prospective student. Gay-Lussac took the entrance examination in October 1797 and was successful. It has been noted that he was the first student from the department of Haute Vienne to enter the school. This is literally true but it is important to note that he did so from a Paris base. A disproportionately large number of successful candidates took the entrance examination not at one of the provincial centres but in Paris with its special educational facilities. If Gay-Lussac had been an isolated provincial candidate he would have been less likely to have known what was expected from him. He was admitted to the Ecole Polytechnique on 27 December 1797.

His motives in taking the entrance examination for the Ecole Polytechnique were probably of several kinds. First there was the educational one. A new and prestigious institution for higher education was open in Paris and could be entered simply by success in a competitive examination. Secondly one might consider the economic aspect. Gay-Lussac as a student at the Ecole Polytechnique would receive a government grant and no longer be a burden on the family's precarious financial resources. Thirdly there was the problem of conscription. We shall consider the economic and military aspects of Gay-Lussac's life before discussing the details of academic work at the Ecole.

On arrival in Paris, young Gay-Lussac had found himself in a large city with few friends and many problems. Judging from his letters home his greatest problems were economic ones but, even when he had money, the country lad would be swindled by the tradesmen of the metropolis:

I was undoubtedly born to be constantly swindled. The tailor did so well that I never received my trousers. Everyone tells me that the man who made my uniform[15] and the one who sold me the material. . .have robbed me too. Indeed I am robbed everywhere I go. They promise me, they swear to me on their word of honour that it is this or it is that – so well that, credulous creature that I am, I fall in the trap. Yet a woman stationer, from whom I bought for 6 sous a pencil which she originally offered for 16 sous, assured me that she thought I was a married man, because young bachelors did not bargain so much. If, however, she had told me that she could not honestly part with the pencil for less than 16 sous, I would have taken it. But I am not very much further forward if, after not making a mistake over a pencil of 6 sous, I have about a coat costing 70 francs.[16]

One of the innovations of the Revolution was that students accepted for the state schools of higher education were provided with a grant. It is true that the money was sometimes late in being paid. Also it was paid in *assignats*, the new paper money which was being printed in large amounts with a consequent loss in purchasing power. Yet in principle a registered student without private means could scrape by at a minimum subsistence level. At the Polytechnique, although there was no formal means test, enquiries were made about the student's means of support. In the early summer of 1798, monthly payments varied from 15 francs for reasonably well-off students such as Berthollet's son Amédée, to 55 francs for some of the monitors or *chefs de brigade*. Gay-Lussac as a first-year student with virtually no other means of support received 45 francs a month.[17]

In another letter home Gay-Lussac gives a general impression of the struggle he was having to exist in Paris. One tends, however, to judge one's standard of living on a comparative basis and Gay-Lussac seems to have been rather better off than his younger brother, who had recently come to Paris to study medicine:

I have just seen my brother in his little abode, yes I say little abode because it is so tiny that two people could not move in it; but he took it for economic reasons and he has pursued his economy to the highest degree. Alas! Of the 30 francs which he has each month, he uses 7 for his room, he buys a book – one which is particularly necessary – and a pair of shoes. On top of this he has to pay for lighting and laundry and there scarcely remains 6 francs for his food which is reduced to dry bread from morning to night. I am almost in the same situation as he is; I am entitled to about 100 francs[18] at my school and I have used part of what I have been given to procure my most essential books. I thought that I had 60 francs a month, as I told you, but the maximum was fixed at only 45 francs.[19] I have just left a boarding house where I had full board for 24 francs a month because in the way I am paid and the purchases I must make for necessities, I cannot make ends meet, and I have arranged my life in the most economical way possible. One might well describe us as unhappy young men and we would be indeed if our work gave us time to think about our situation. As regards my clothes I can say that I carry them all on me. My brother has given me 24 francs and I am going to buy some trousers because this is what I need most. . .[20]

It would be impossible to ignore the fact that Gay-Lussac received his education in an environment moulded by the forces of Revolution. It might be possible, however, in describing the scientific scene, to overlook the fact that France was at war almost continuously from 1792 to 1814. In August 1793 a general mobilisation was ordered:

All Frenchmen are called by their country to defend liberty. . .From this moment until that when the enemy is driven from the territory of the Republic, every Frenchman is commandeered for the needs of the armies. Young men will go to the front; married men will forge arms and carry food; women will make tents and clothing and work in hospitals; children will turn old linen into bandages; old men will be carried into the squares to rouse the courage of the combatants, to teach hatred of kings and republican unity.[21]

Although everyone was asked to help, it was obviously young men who were expected to contribute the greatest numbers to the Republican army. Conscription was introduced, and, according to one story, which Gay-Lussac believed, it was to be imposed on young men from the age of 16 upwards. Seeking a means of escape, he immediately wrote home asking his parents to send him the certificate of baptism of his younger brother Pierre.[22] However, conscription really began at 18 so he had a temporary reprieve without the need of a subterfuge. Yet in the winter of 1796–7 he would be 18 and might face the prospect of terminating his studies abruptly. An opportunity for legitimate avoidance of conscription presented itself in the Ecole Polytechnique. The Convention and Directory could not have established a unique higher educational institution with students of the highest calibre recruited from all over France and then allowed the conscription laws to siphon them off (almost immediately) into the army. The emergency of 1793–4 having passed, the government could afford to take the long term view[23] that the army would be served by recruiting a substantial proportion of graduates of the Polytechnique who would eventually become officers in the Artillery or Engineering Corps. Gay-Lussac, however, looked to the non-military side of the school. Looking back a few years later, he described the Polytechnique as a haven from conscription which had 'seriously threatened' him.[24]

The educational principles of the Polytechnique were that engineers required two kinds of knowledge. First they should study mathematics to appreciate the form and movement of bodies. But they should also understand the composition of matter and this necessitated the study of physics and chemistry. Although analysis and mechanics were obviously in the syllabus, the Polytechnique made much of engineering drawing, 'descriptive geometry', a subject pioneered by Monge, one of the founders of the school. The fairly prominent position of chemistry in the syllabus seems equally to be accounted for by the many chemists who had drawn up the plans for the school.

There was a basic two-year course for students in which mathematics, physics and chemistry featured in each year. In 1799, when Gay-Lussac was in his second year, the lectures given were as follows (figures in brackets indicate percentage of total time):

| | | |
|---|---|---|
| Mathematical analysis | 48 hours | (8.8%) |
| Mechanics | 72 | (13.3%) |
| Fortification | 54 | (10%) |
| Public works | 54 | (10%) |
| Mining | 27 | (5%) |
| Physics[25] | | |
| Chemistry (technical) | 60 ⎫ | (22.2%) |
| Chemistry (experimental) | 60 ⎭ | |
| Architecture | 45 | (8.3%) |
| Drawing | 120 | (22.2%) |

Although studies were organised on the basis of a two-year course, the actual time spent by the students at the school depended on their particular ambition and their abilities. Students of one year's standing were, for example, eligible for admission to the Gunpowder Service. Future artillery officers were expected to graduate after two years, but military and civil engineers, although examined after two years, were required to spend a third year[26] in the atmosphere of rigorous mathematical training of the Polytechnique before passing on to their respective Schools for public services. Forming a sort of élite among the students, their grants were increased in this third year from 1200 francs to 1500 francs for the year. This arrangement of 1795 was to be modified by the time Gay-Lussac arrived at the school, but it is interesting as showing how from the beginning those students designated for engineering constituted a group which was recognisably academically superior.

In the Republican week of ten days, eight full days were given to work, the fifth day had a class only in the morning and the tenth day (corresponding to the Sabbath) was a day of rest. The working day began at eight when there was a one-hour class. The students were then allowed half an hour for breakfast before some practical work (e.g. drawing). At 1 p.m. there would be another lecture before lunch. The afternoon was free but from 5 p.m. three hours more work was expected from the students before supper at 8 or 9 p.m. Students were therefore expected to attend for nine hours at the school. This timetable was so full that later, when new subjects were to be introduced, it proved impossible to find any free time. The Ecole Polytechnique was remarkable in providing facilities for individual laboratory work in chemistry by the students. These 'manipulations' were supplemented by visits to workshops and factories.

We know something about Gay-Lussac's teachers at the Polytechnique. For chemistry he must have attended the lectures of Fourcroy and Vauquelin in the first year. In the second year students would normally have attended Berthollet's class but, as the latter had gone to Egypt with Bonaparte in 1798–9, this crucial contact was postponed for a year and lectures were given by Chaptal. In his final year Gay-Lussac attended chemistry lectures by Guyton de Morveau and also Berthollet's advanced course, described as a 'cours de perfectionnement'. Gay-Lussac's mathematical education was also affected by the Egyptian expedition, since Monge and Fourier both abandoned their lectures at the Polytechnique in May 1798 in order to follow Bonaparte. Prony, however, remained in Paris as professor of analysis and Hassenfratz was responsible for the course in physics. In 1798 Gay-Lussac's kinsman Simon François Gay de Vernon, a former officer in the corps of military engineers, was appointed to the school as deputy director. He gave a course on fortification but Gay-Lussac probably saw more of him out of class.[27]

Napoleon was to introduce a complete militarisation of the Polytechnique in 1804 but long before then the question of whether the students should be compelled to live at the school had often been discussed. Lack of discipline of students was a frequent complaint. Students still lived in private lodgings. Gay-Lussac moved his lodging several times; one of his addresses was 381, rue de Grenelle, just opposite the Polytechnique. A method of imposing discipline was to select the more responsible students as monitors or *chefs de brigade*. There is a record that Gay-Lussac was appointed a reserve monitor (*suppléant*) in his final year although he resigned half way through the year.

As regards his studies Gay-Lussac presumably made use of the general library of the Ecole Polytechnique. Most of the books needed continually in his studies he probably owned and these were retained in his library in his later life. Fortunately, his descendants did not disperse or sell this valuable collection of books. One of the educational achievements of the Revolution was to sponsor the writing of elementary text-books, but Gay-Lussac's introductory scientific studies came before this programme had been put into effect. We thus find in his library books on mathematics and science which derive from the *ancien régime*. Before the Revolution some of the principal places in which mathematics was taught were the artillery and military engineering schools. Thus the signature of the young Gay-Lussac is found on the flyleaf of the standard *Cours de mathématiques à l'usage du corps royal de l'artillerie* by Bézout (3 vols., 1780), the same work as had been studied by Napoleon Bonaparte when an officer cadet. Gay-Lussac also owned the abbé Bossut's *Cours de mathématiques à l'usage des élèves du corps royal du génie* (2 vols., 1788, 1790). Also in his library was a book written for military cadets on natural history.[28] This is of some interest, since after the section on the animal and vegetable kingdoms, there are some eighty pages on minerals. This may possibly have been the young Gay-Lussac's first encounter with chemistry. The chemistry is presented in the natural history tradition which we may contrast with the more physical approach of the new chemistry of Lavoisier. But Gay-Lussac was not to study the work of Lavoisier until the next decade.[29] Meanwhile an introduction to physics came through Euler's popular *Letters to a German Princess* (French edition, 1787). He also owned a book by the abbé Rochon on mechanics and physics.[30] He was told to buy Lacaille's mathematics text and sent home for this. When the wrong book was sent[31] he bought the book himself, probably: *Leçons élémentaires de mathématiques par M. l'abbé de la Caille...Nouvelle édition augmentée par M. l'abbé Marie* (Paris, 1778).[32]

In the winter of 1797–8 certain changes were proposed at the Ecole Polytechnique which would make for greater economy and higher efficiency. The government had wanted to reduce the number of students to 200 but a figure of 250 was finally agreed to include fifty places for students to study for a third year. As the normal period of study was now fixed at two years students could only stay longer if they fitted into certain categories:

First: Students who had fallen behind in their studies because of illness or other legitimate reasons. Second: Those who, having done the work of the previous two years of instruction wished to devote themselves particularly to the study of a science of their choice and who could in this way, by improving their own knowledge, render a service to the School. Third:

Students recognised as sufficiently up to standard in the exam but who have not been able to enter one of the public services for want of a vacancy. Fourth: Students who, although not wishing to enter one of the public services, wish to increase their knowledge of science or technology, and obtain permission to stay a third year at the School for this purpose.[33]

Thus the terms of reference were sufficiently wide for the authorities of the school to provide a third year of study not only for those students who for personal reasons had not managed to complete the two-year course, but also for many of their more able students as well. Gay-Lussac was one of these favoured students and it was in 1800 in his final year that he worked under Berthollet's direction in the laboratory of the school on the treatment of flax with 'oxymuriatic acid' (chlorine).[34] Thus part of Gay-Lussac's final year was given to research.

The students at the Polytechnique represented a complete cross section of society. A few sons of the former aristocracy mingled with the sons of the new meritocracy. Berthollet and Laplace both sent their sons there and Berthollet's son eventually became a close friend of Gay-Lussac. An exact contemporary of Gay-Lussac was A. C. V. C. Destutt de Tracy, son of the famous *idéologue* philosopher[35] who had taken particular pains over the education of his boy. An analysis of the student population in 1799 claimed that the great majority of the students were of working class or lower middle class origins. In Gay-Lussac's year ('Promotion' of 1797) the most striking feature is the number who failed to graduate (38 out of 108 or slightly more than one third). This may be accounted for by a political purge in 1798.[36]

Of the remaining 70 students who successfully completed the course the largest number went into the artillery (25) and the corps of engineers (14). Various naval careers attracted a further 8 while only 1 entered the infantry. On the civilian side 9 (including Gay-Lussac) elected for the School of Bridges and Highways, and 5 for the School of Mines. It is interesting that 5 should have gone on the staff of various artillery and military schools throughout France. Although Gay-Lussac, who later joined the teaching staff of the Ecole Polytechnique, came to be regarded by that institution as the star student of that year, others also had distinguished careers. Destutt de Tracy became a member of the Chamber of Deputies and was made a viscount. Another graduate at that year, Finot, became prefect of the Cher department and was made a baron.

The Ecole Polytechnique had been established as a multi-purpose institution for higher education in science and the majority of the graduates were able to put their science lectures to some use. It was, however, only the exceptionally able and determined student who could use his scientific education to win fame as a member of the new profession of scientist. In the 'Promotion' of 1798 Poisson was the outstanding student and he was to find his vocation in mathematics and physics. In the year we are concerned with only Gay-Lussac became a scientist in the sense of someone who makes original contributions to the subject by means of research.

## The Ecole des Ponts et Chaussées

The Ecole des Ponts et Chaussées (the School of Bridges and Highways) was a creation of the *ancien régime*. An administrative body for bridges and roads had been set up in 1713 and a related educational centre in 1747. It was only a 'school' in the most general sense of the term since there was no teaching staff; instruction was generally given by older students supplemented by private courses outside. The Revolution established the institution in a rationalised framework of 'schools for public services'; for the first time entry was by competitive examination. The *ancien régime* had established the profession of civil engineer; this career was now thrown open to all young Frenchmen on the basis of ability rather than birth or personal recommendation. By the time Gay-Lussac became involved it had become clearly established that admission was restricted to graduates of the Ecole Polytechnique. The civil engineers were thus produced by a double refining.

We do not know exactly why, among the half dozen *écoles d'application*, Gay-Lussac finally put his name down for the Ecole des Ponts et Chaussées. It was, however, a *civilian* career far from the army and it was also the school which recruited the highest quality graduates. (Fresnel, a proponent of the wave theory of light and Gay-Lussac's junior by ten years, was another graduate of the Ecole Polytechnique who was to go on to the Ponts et Chaussées.) Gay-Lussac could also have gone to the Ecole des Mines, where he would have learned about the chemical analysis of mineral specimens, but it was not to be at any official school that he was going to be trained as a chemist.

In February 1801, shortly after joining up with Berthollet, Gay-Lussac apparently saw engineering as a recognised career to fall back on if he was unable to make a living by the study of pure science. He wrote to his father:

My work in this field [civil engineering]. . .should provide me with a situation which will be my refuge if fortune goes against me.[37]

If we are to accept Gay-Lussac's own account, three years after the event,[38] it was as a student at the Ecole Polytechnique that he fell in love with science for its own sake. His decision to enter for the Ponts et Chaussées was a late one and may very well have been not so much a positive decision as a postponement of a decision about his career. Within the shelter of this *grande école* he could map out his future life. There is, therefore, some doubt whether he really wanted to be an engineer. Looking back in 1803, Gay-Lussac assured his father that he had never wished to become one.[39] Perhaps there would be some comparison with some modern postgraduate students in Britain who register for an education diploma, not because they feel committed to teaching, but because it provides a government grant which enables them to sample one career while exploring other possibilities. Only later is the final choice made.

The school, situated in the rue Grenelle, was a former ducal palace and visitors remarked on the grandeur of the building and its furnishings. The director was Prony (appointed 1798), who had a teaching staff of three. There was a good collection of models of bridges, harbours, etc., and a fine library. There was a total of about fifty students who would expect to graduate in two or three years. They were trained in the theory and practice of the construction of roads, bridges and canals and in quantity surveying. They attended lectures on mathematics, mechanics, architecture and mineralogy and were required to do technical drawing. Additional subjects studied by some students included English and German. The theoretical work was planned to occupy the months from November to May, and from June to October the students concentrated on practical work, learning to superintend the construction of buildings and to estimate costs.

Since his career was to depend almost entirely on Berthollet, it is fortunate that the director of his school, Prony, was a friend of Berthollet. Gay-Lussac reports to his mother:

I have made myself useful to the director, who knows Citizen Berthollet well, so that I need not worry...We can do at home the work that we are given, so I work in my room, although I go fairly frequently to my school.[40]

The Ecole des Ponts et Chaussées, apart from providing Gay-Lussac with an official attachment and further education, provided him with a salary. The archives show that he received a salary of 75 francs a month up to August 1804. However his letter to his father of 15 January 1803 shows that he had recently resigned from the Ponts et Chaussées, much to his father's concern. For the month of November 1803 there is a note to the effect that he was not then attending the school. In January 1804 a marginal note describes him as a laboratory assistant ('aide de laboratoire') at the Ecole Polytechnique but still gives him his usual salary. Most interesting of all is the entry for Gay-Lussac's final months on the payroll of the Ponts et Chaussées. From February 1804 he is said to be 'doing work under the direction of Senator Berthollet related to mortar and cement'. This was the most plausible excuse one could give for a civil engineering student to be drawing his pay while working in the private laboratory of France's leading chemist! However, such a situation could not go on indefinitely and with Gay-Lussac's appointment on 23 September 1804 as *répétiteur* at the Ecole Polytechnique he finally severed his formal connection with civil engineering.

It is impossible to say how seriously Gay-Lussac followed the courses but it would be reasonable to guess that, having begun as a full student, his association became increasingly tenuous so that he became no more than a nominal student. The patronage of Berthollet undoubtedly enabled him to circumvent the normal expectations of student attendance. Although it would be a mistake to regard Gay-Lussac's association with the school as 'wasted' years, probably the only vocational benefits he derived there were some practical ideas useful in chemical engineering.

It was probably soon after Gay-Lussac came under Berthollet's wing at Arcueil that he acquired a book of moral precepts which may have influenced his later life. The book was L.M. Stretch's *The Beauties of History; or Pictures of Virtue and Vice, drawn from real life; designed for the Instruction and Entertainment of Youth.*[41] The author's aim in this book, which went through many editions, was 'to collect. . .records of virtue and to inspire the minds of youth with a noble emulation to rival them'. To this end he had collected many examples from biblical and classical sources to illustrate some thirty virtues and vices listed in alphabetical order and ranging from Affection to Wealth. Ambition is presented as of some potential danger yet, adds Stretch, 'ambition is not a vice but in a vicious mind. In a virtuous mind it is a virtue and will be found to take its colour from the character in which it is mixed'.[42] There is a similar kind of ambiguity in the author's treatment of 'Wealth', but generally a clear moral line is indicated of Stoic virtue, as in Courage, Justice and Magnanimity. Perhaps one of the most interesting entries in view of Gay-Lussac's later life was the comment on industry:

Love labour; if you do not want it for food, you may for physic. . .Action keeps the soul in constant health, but idleness corrupts and rusts the mind.[43]

Courage was not strutting about and uttering oaths, but doing one's duty regardless of danger, and Constancy was executing the demands of reason and conscience with dignity. Much of Gay-Lussac's later career from his balloon ascent to his role as elder statesman of science is reflected in these moral precepts. Despite his early clerical education Christianity does not seem to have played any part in his life. In the eighteenth century the Catholic Church in France was at a low ebb and, with the Revolution, France became much more secularised. Berthollet and other chemists in the Lavoisier circle were influenced by a philosophy of the Enlightenment which replaced faith in a personal god by a humanist philosophy and a belief in progress. Gay-Lussac's belief in the late eighteenth century ideas of Enlightenment was to emerge occasionally, as when in his physics lectures in 1820 he was to speak of 'the progress of Enlightenment'.[44] He presented physics as an ideal science 'to enlighten reason and to deliver it from prejudice'.

Much of Gay-Lussac's later success was no doubt due to his inherent ability, his determination and hard work. Yet he also owed much to the support of friends and patrons. Apart from his family, we have seen how he benefited in his youth from the support of the abbé Dumonteil and also the Sencier family, from whose school he aspired to the Ecole Polytechnique. When he was there it was fortunate for him that the man appointed as deputy governor of the school was a kinsman, Gay de Vernon. Gay-Lussac also claimed to be on good terms with Prony, director of the Ecole des Ponts et Chaussées. Although the patronage of Berthollet was probably the single most important help he received in his career, it was not unique. He was soon to be able to claim close friendship with the explorer Alexander von Humboldt. In Gay-Lussac's later work in applied science the patronage and friendship of artillery officers was helpful and when he finally obtained a key post at the Paris Mint it is worth pointing out that the post depended on the patronage of the Paris prefect Chabrol, a fellow graduate of the Ecole Polytechnique.

# 2

# The apprentice of Arcueil

'At Arcueil...I dined in distinguished company...
There was a lot of very interesting discussion. It is
these gatherings...which are the joy of life'

Gay-Lussac[1]

## Assistant to Berthollet

Gay-Lussac was fortunate in having the resources of the Ecole Poly-technique on which to draw, but he was doubly fortunate in having a second source of support as a semi-permanent guest at Berthollet's country house at Arcueil, a few kilometres to the south of Paris. It was in the autumn of 1801 that Berthollet, General Bonaparte's former companion in Egypt and now a senator, decided to buy another property which he could equip as a centre of research. Complementing his good library there were practical facilities, excellent by the standards of the time, consisting of a chemistry laboratory and a physics labora-tory, well endowed with apparatus. It was in this environment that Gay-Lussac was to work under the supervision of Berthollet.

The role of Berthollet in Gay-Lussac's life merits careful examination. It was without doubt the most important personal influence in the latter's whole life. Berthollet's role was in the first place that of a teacher. But in the case of Berthollet the term 'teacher' has a special meaning. Unlike his colleague, Fourcroy, Berthollet was a poor lecturer and the fact that a person attended his lectures is unlikely to have made any deep impact on him. Gay-Lussac was Berthollet's student in the more personal sense of being initially a research assistant. He learned his chemistry not at Berthollet's feet but at his side, and, if we may be permitted to mix metaphors, it was in this way that the older man's conception of doing chemistry was gradually transfused to him. Berthollet thus became on the cognitive plane his model of the ideal scientist. A second role of Berthollet was as father-substitute and host. Berthollet's friendship gave Gay-Lussac the confidence and stability that one associates with a father-figure. One might also say 'landlord' if this did not imply a commercial relationship. Finally, and most important for Gay-Lussac's career, was the role of Berthollet as patron.

The Ecole Polytechnique explains how Gay-Lussac obtained his scientific and mathematical education but to understand Gay-Lussac's advancement in the period 1800–1815 it is not sufficient to look at his qualifications; one must examine his connections. These three aspects of Berthollet *vis-à-vis* Gay-Lussac are obviously related but it will be convenient to consider them in turn.

As far as Gay-Lussac's scientific research is concerned, the most important aspect of Berthollet was that his work was seen by the younger man as a model. On the personal side, Berthollet was amiable and a good friend but it was as a man of science that Gay-Lussac most closely modelled himself on him. First there was Berthollet's view of what science meant. Of course for Berthollet science was mainly chemistry but although this subject delineation excludes advanced mathematics on the one hand and the biological sciences on the other, it still leaves a very wide range of interpretation. One might try to define 'chemistry' more clearly by asking whether it was 'pure' or 'applied', but this division cuts right across Berthollet's approach. Berthollet was interested in 'applied chemistry' in the tradition of the *Encyclopédie*, namely that science was justified by its applications. In the field of dyeing Berthollet had tried to relate traditional empirical practices to contemporary theory. Only in such a way was progress and enlightenment possible. But Berthollet did not live in the single dimension of applied chemistry. He certainly wanted to help transform the world but he also wanted to understand it. Just at the time when Gay-Lussac made his acquaintance, Berthollet was beginning his enormous task of trying to analyse the factors affecting chemical reaction. His journey to Egypt in which he had observed the large trona[2] deposits had suggested to him that the result of a chemical reaction depends largely on the physical conditions. Thus chemical reactions, far from being simple affairs, required a great deal of study. Berthollet was one of the first to see the problems and his *Essai de statique chimique* constitutes the first sustained attempt at solving them.

Berthollet's great contribution to the theory of chemical reaction was his realisation that the course of a chemical reaction depends not only on the nature of the reactants but their quantities and the physical conditions. This theory was elaborated in a two-volume work intended to view in turn 'all the causes which produce a variation in the results of chemical action'. Berthollet's chemical theory gave prominence to cohesive forces which determined whether a substance was a solid, a liquid or a gas. The physical state of reactants governed the rate of reaction which also affected by the temperature. The modern chemist would be tempted to see Berthollet's work as an anticipation of the

Law of Mass Action of Guldberg and Waage (1864–7) but it differs from it not only by its qualitative approach but by its diffuse presentation. The French chemist J. B. Dumas was later to use it as the basis of his chemical education from the age of 17 to 21 and later described the struggle he had had to understand it.[3] Unfortunately, Gay-Lussac never wrote about the influence of his master's *chef d'oeuvre* on him but we may infer this from a copy of the book in Gay-Lussac's library which was annotated by him. The annotations are interesting not only as direct evidence of Gay-Lussac's study of the book but also for his marked interest in certain topics such as solubility and the theory of acidity, to which he was to make his own contribution. The annotations are often critical. Thus he wrote 'One sees that Berthollet regards affinity as a constant force, which it is not'.[4] One also finds remarks such as 'false',[5] 'inexact',[6] 'inadmissible'[7] and 'to be corrected',[8] some of these comments applying to Berthollet's concept of predisposing affinity.[9] However close Gay-Lussac was to Berthollet, this copy of the *Essai* shows clearly that his critical faculties were not paralysed.

For Berthollet's selection of Gay-Lussac as a promising student from the Ecole Polytechnique we have the testimony of Arago, who was not there at the time but, as a friend of both parties, was probably reliably informed:

Berthollet, who had returned from Egypt with General Bonaparte, asked in 1800 for a student from the Ecole Polytechnique to be his assistant in his laboratory. This privileged student was Gay-Lussac. Berthollet suggested to him a piece of research, the results of which were diametrically opposed to those expected by the illustrious chemist. I dare not state that Berthollet was not a little put out to see himself contradicted in his expectations, but contrary to the attitude of so many other scientists I could name, after his initial unfavourable reaction, the frankness of the young experimenter only increased the esteem of [Berthollet]...'Young man', he said, 'your destiny is to make discoveries; henceforth you will be my table companion. I would like in scientific matters to be like a father to you and this is a title that I am sure I shall one day be proud of'.[10]

It was not merely that Berthollet looked on Gay-Lussac as a son and symbolically bequeathed to him his uniform as a peer; Gay-Lussac looked, and continued to look throughout his life, on Berthollet as a father. In 1841, when having a new house built, he spoke of his plan to have two busts set up in suitable alcoves: one for Berthollet and one for his father.[11]

Berthollet's role as a father was all the more natural as he already had a son about the same age as Gay-Lussac. Amédée Berthollet (1780–1810) had been at the Ecole Polytechnique from 1796 to 1798, thus overlapping with Gay-Lussac. The younger Berthollet had, however, withdrawn from the school without graduating and was seeking a career in applied chemistry. Amédée was Berthollet's only child and he probably welcomed Gay-Lussac as a second son and as a stimulating companion for a son who intellectually may not have lived up fully to his father's hopes and ambitions. Amédée Berthollet and Gay-Lussac worked together to some extent at Arcueil, where they established a model chlorine bleaching works.

There are several references to Gay-Lussac's life with Berthollet in his letters home. Having graduated from the Ecole Polytechnique at the end of November 1800, he spent that winter at Berthollet's house in the rue d'Enfer (now the boulevard Denfert-Rocherau). His mother was apparently delighted that he had gained the protection of such an 'important man'. A letter from him dated 11 February 1801[12] makes it clear that his parents were glad that he was living in Berthollet's house. In his letter he announced that he would be spending some time that year at Arcueil in Berthollet's newly acquired country house.

Gay-Lussac continued his life at Arcueil with the Berthollet family in 1802 and 1803, but living at Arcueil did not have all the advantages of Paris. In a letter of 26 May 1804, Gay-Lussac told his parents that he was going to move back to Berthollet's house in the rue d'Enfer. Also in 1804 came his appointment on the staff of the Ecole Polytechnique. He hoped to move into the school that autumn. However, his written request[13] for accommodation at the school was not met and we find Berthollet in September 1805 writing to the governor of the school repeating the request. In a letter of May 1806[14] Gay-Lussac gives his address as 'chez M. Berthollet, rue d'Enfer No 37' and it was not until later that year that he had a permanent lodging at the Ecole Polytechnique. Although Berthollet was Gay-Lussac's host over this early critical period of the younger man's career, he was not a possessive one. He was using Gay-Lussac as a research assistant but he had no wish to exploit him. His advancement and the furthering of his research demanded that he should have more contact with the scientific life of the capital. However, although Gay-Lussac may not have been resident at Arcueil for more than a part of three years, he belonged to Arcueil in another way for a much longer period. We shall be looking shortly at what it meant to be a member of the 'Society of Arcueil'. The patronage it implied played a large role in Gay-Lussac's early career. First we will turn to examine some of Gay-Lussac's early scientific research, carried out at Arcueil or under Berthollet's wing.

*The thermal expansion of gases as studied by Gay-Lussac and Dalton*

Gay-Lussac's first piece of published research is of considerable importance, not only because of its permanent value in science, not only
because of its obvious importance in a biography of Gay-Lussac, but
also because it involved research carried out almost simultaneously
and independently by the English natural philosopher John Dalton. A
comparative study of the French and British research may provide
some useful evidence of personal differences and even of national
differences.

The paths of Gay-Lussac and John Dalton overlapped at the very
beginning of the young Frenchman's career by the almost simultaneous
publication in 1802 of research on the thermal expansion of gases in
which each independently concluded that all gases expand by the same
proportion for an equivalent rise of temperature at constant pressure.
As in some other instances of 'simultaneous discovery', these memoirs
may be traced to a common origin. In fact the joint memoir of Guyton
de Morveau and Prieur Duvernois,[15] which had been published in the
very first volume of the new French journal, the *Annales de Chimie*,
and which had stimulated Gay-Lussac's criticism, was also the starting
point of Dalton's memoir. Dalton remarks that the French authors'
'conclusions were so extremely discordant with and even contradictory
to those of others, that I could not but suspect some great fallacy in
them'.[16] Dalton and Gay-Lussac agreed that previous work had been
vitiated by the presence of moisture in the gases, the latter being at
pains to point out the major effect this could have. They also agreed in
their caution in taking the mercury thermometer as an absolute measure
of temperature. There were, however, many important differences
between the work of the two men.

Even in Gay-Lussac's first memoir one can see a certain 'professionalism' in thoroughness, precision and recording of results which is
absent from Dalton's paper. Whereas Dalton claimed to have obtained
constantly the same result, Gay-Lussac communicated his actual experimental results, taking the mean reading and recording the deviation.
Gay-Lussac's memoir (twenty pages of Nicholson's *Journal*, where it
appeared in translation, compared with four for Dalton) was a well
planned piece of work and in its presentation was divided into four
sections. Having set out in the first section the aims of his research,
Gay-Lussac gave a historical exposition of previous work. Part III of
the memoir was devoted to a description of his apparatus (with a
drawing) and the last part gave his experimental results and conclusion.

Gay-Lussac made it clear that he was only concerned in his memoir with reporting the behaviour of gases and vapours between the temperatures of 0 and 100°C. This was to be part of a more general research programme on the expansion of gases and vapours and 'the march of the thermometer'. He gave a detailed criticism of the most recent research on the thermal expansion of gases by Guyton de Morveau and Prieur Duvernois, pointing out various sources of error. The gas volume had not been scrupulously reduced to zero but to a few degrees above, Guyton believing that gases expanded very little in the lower temperature range. The most important source of error was the presence of water. The use of gases which were not perfectly dry produced not only error but inconsistency, Gay-Lussac appreciating that rigorous experimentation must be done under strictly controlled conditions. Gay-Lussac made a point of using pure gases and also checked that the mercury in contact with the gases was very pure.

Gay-Lussac divided the gases he examined into two groups and used different apparatus for each. The first batch of gases examined were air, hydrogen, oxygen and nitrogen. The gases were introduced into an inverted flask fitted with a stopcock and the whole was immersed in a water bath and left sufficiently long to reach that temperature. Volumes were calculated by filling the flask with water and weighing, making all necessary temperature and pressure corrections. However in the second part of the experiment with soluble gases, volumes were observed directly into two barometer tubes over mercury, one with air as the standard and the other filled successively with (using modern names) carbon dioxide, hydrogen chloride, sulphur dioxide, nitrous oxide and ammonia. He also applied the same apparatus to study the thermal expansion of diethyl ether, taken as representative of vapours, and concluded that 'all gases and vapours dilate equally at the same degree of heat'.

Dalton's apparatus was typically – as he himself admits – of the simplest kind, consisting of a straight glass tube, sealed at one end and with a mercury index. On the basis of his examination of five gases he was prepared to generalise that 'all elastic fluids expand equally by heat'. But he was no less concerned to establish a second proposition in opposition to Guyton de Morveau, that 'for any given expansion of mercury, the corresponding expansion of air is proportionately something less, the higher the temperature', i.e. the coefficient of expansion decreases with increase of temperature, and he persisted in this conclusion from his hasty experimentation despite theoretical objections which he himself raised.

Thus, contrary to Dalton's own assertion[17] (which has generally been believed) his conclusions were fundamentally different from those of his French contemporary. This difference was pointed out by Dulong and Petit, who expressed it in an alternative form:

The experiments of Mr Gay-Lussac proved that the expansion of gases related to the mercury thermometer is a constant fraction of the volume at a determined temperature [i.e. 0°C]. Mr Dalton, on the contrary, supposed that the increase in volume is for each equal increase of temperature a constant proportion of the volume at the preceding temperature.[18]

Although both claimed constancy, Dalton had guessed at the wrong kind of constancy and his experiments were not sufficiently rigorous to reveal his error.[19] Dalton did not give a coefficient of expansion, but a mean coefficient of 0.00391 per degree C has been calculated from his original data,[20] and thus is considerably less accurate than the result of 0.00375 obtained by Gay-Lussac. (The modern value is 0.00366.)

Gay-Lussac's work was, therefore, superior in its execution, in the selection of gases and the temperature range studied, in the recording of results and in his final presentation. His conclusion that all gases expand equally over the temperature range 0–100°C had a greater experimental basis than Dalton's similar conclusion and his experimental results were considerably more accurate than those of his English contemporary. On the other hand, although Gay-Lussac realised the importance of excluding moisture to obtain consistent results, his experimental result suggests that even he was not sufficiently scrupulous in removing the last traces of moisture from his gases.

Although Gay-Lussac's memoir must rank as one of the classics of science, he does not now receive general credit for his painstaking and definitive work. Credit is popularly assigned to J. A. C. Charles (1746–1823). This is largely because Gay-Lussac mentioned in his memoir that some of his experiments had been carried out fifteen years earlier by Charles. P. G. Tait accepted this statement as recognition of the prior claim of 'Citizen Charles' and called the law after him.[21] However the law should more justly be known as Gay-Lussac's law[22] for several reasons. In the first place credit is normally assigned in science on the basis of publication and Charles did not bother to publish his work. According to Gay-Lussac it was quite by chance that he came to hear of it. Secondly, although Charles had found some gases expand equally, he had also found that soluble gases each had a characteristic rate of expansion *different* from that of other gases. Gay-Lussac was able to show that the coefficient of expansion of the *dry* gas was the *same*. Finally, Charles used crude apparatus which did not measure thermal expansion at constant pressure. Gay-Lussac was shown the apparatus used by Charles and remarked:

I saw that the tube of the barometer was very large in proportion to the capacity of the reservoir, so that the rise of the mercury above 28 in. did not show all the tension the gas had acquired, since for that it would be necessary that its volume in the reservoir had remained constant.

Thus Charles had not measured the absolute rate of expansion and could not legitimately draw conclusions about the thermal expansion of gases.

### Balloon ascents

Two balloon ascents made by Gay-Lussac, the first with Biot on 24 August 1804, and the second alone on 16 September, may be regarded as the first scientific expeditions in a balloon. It was the French who had pioneered ballooning. The historic ascent of a hot-air balloon launched by the Montgolfier brothers in 1783 was followed three months later by a hydrogen balloon sent up by J. A. C. Charles. The first manned free ascent followed in November of the same year. Probably the most famous of the early balloonists was Jean François Pilatre de Rozier, who died in a balloon accident in 1785. In the eighteenth century attention was naturally focussed on distance, both the height of ascent and the distance travelled between the point of departure and the descent. It could not however be taken seriously as a method of transport; it was, rather, an amusement and above all a spectacle. If we exclude the period in the Revolutionary war when balloons were used briefly as military observation posts, the era of Gay-Lussac marks the transition from amusement to professional science. He and Biot were able to take their vehicle more or less for granted and concentrate on carrying out scientific observations. When the editor of the *Journal de physique* reported on these balloon ascents he gave the work considerable prominence because, he said, there were lessons to be learned from these expeditions as opposed to others which were purely frivolous.[23]

Going up in a balloon was, of course, a risky business. It called for a sense of adventure. If Gay-Lussac had been British his sense of adventure and travel might well have taken him in the wake of Captain Cook or on some other naval expedition. Attaining manhood in the Napoleonic wars in a country with a navy in eclipse, he had to make the most of his native country, or at best, a part of western Europe. His balloon ascents of 1804 provide an example of French scientists during this period carrying out new research within French frontiers. If the results came second to the palaeontology of Cuvier and Brongniart in the Paris basin (also begun in 1804), they share with it the substitution of a vertical plane of exploration for the more usual horizontal one.

One of the critics of the Académie des Sciences[24] in the years before the Revolution had remarked that Academicians did not take the risk of actually going up in the new balloons; they preferred to sit comfortably in the Academy and draw their salaries. The same might have been said of some of the senior members of the First Class of the Institute twenty years later. However, in 1804 Gay-Lussac was not yet a member of the official body of science. He was a young unmarried man in search of recognition and he was more prepared to take risks not only because of the advancement it might bring but more immediately because of the adventure of science itself. As an idealist he was even prepared to risk his life in the exploration of the upper atmosphere in the interests of acquiring scientific data.

The immediate precedent for the French balloon ascent of 1804 was an ascent carried out at Hamburg the previous summer by the Flemish illusionist and entrepreneur, Etienne Gaspard Robertson, who had reported a diminution in the intensity of the earth's magnetic field. There was some doubt as to whether Robertson's results could be relied upon and this was part of the motivation for the ascent of 1804. A second motive lay in discovering the state of the atmosphere at high altitudes since this might affect the theory of refraction. This astronomical motive explains the interest which Laplace took in the expedition and Berthollet for his part was particularly interested in the composition of air at high altitudes. From Geneva, de Saussure had made observations of the air in the Alps but the invention of balloons meant that Paris scientists could make independent observations of the upper atmosphere.

Through the intervention of these patrons, Gay-Lussac and Biot were able to acquire the necessary facilities for their balloon ascent from Paris on 24 August 1804. In the history of the evolution of the Arcueil circle into a formal society, this event may be considered one of the milestones, since Berthollet and Laplace were launching their respective protégés on a collaborative scientific exercise which did not fail to attract public attention. The two skilled observers were fully occupied with magnetic, electrical and meteorological observations.[25] They found no difference in the period of oscillation of a magnetic needle and concluded that the intensity of the earth's magnetic field was constant up to 4000 metres. The two young men spent most of their time taking readings of temperature, pressure and humidity and neglected to collect samples of air which had been another of their objectives.

Accordingly a second ascent was planned and this time Gay-Lussac went on his own.[26] With a lighter balloon it was possible to attain a higher altitude and on 16 September 1804 he established a new record, 7016 metres above sea level. Apart from physical and chemical experiments he also made physiological observations, noting the effect of the altitude on his pulse and respiration. He also had a headache but he attributed this to his lack of sleep the previous night, perhaps because of the excitement as well as the preparations. He made further magnetic and meteorological observations although when he tried to measure magnetic declination the cardboard scale of his instrument bent in the excessively dry atmosphere, preventing the free movement of the needle. Gay-Lussac had taken two evacuated flasks with him and opened them when the balloon was over 6000 metres. When the time came for his descent he found that he had been carried north to Rouen. He travelled back to Paris and hastened to the laboratory of the Ecole Polytechnique where he and Thenard analysed the two samples of air. They found that the proportion of oxygen was the same as that in Paris air. In fact it was not so much the scientific results of the ascent (which were largely negative) as the ascent itself which was admired. Gay-Lussac had at great personal risk risen much higher than any other man and his record of 7016 metres remained until 1850, the year of his death, when it was exceeded by a few metres (7049 m).[27] It is interesting that Gay-Lussac's description of his ascent stressed the difficulties of making exact observations, whereas the published account of the previous ascent, written by Biot, adopted a more heroic tone and claimed that the two scientists were quite unperturbed.

Gay-Lussac's new career of science was fraught with a higher degree of danger than is usually appreciated. Gay-Lussac's work on the very reactive new metal potassium was not to be pursued for long without the inevitable explosions and in June 1808 he was temporarily blinded by such an explosion.[28] Although he recovered, his eyes were permanently affected. The work on the electric pile also had its dangers. Gay-Lussac and Thenard reported that the battery gave one of them (they do not say which) such a shock that they were partly paralysed in the arms for 24 hours afterwards.[29] Gay-Lussac worked extensively with poisonous chlorine, he also prepared hydrofluoric acid, and probably suffered considerably from the effects of inhaling small quantities.[30] Iodine was less dangerous but even here he had an accident; when he reacted iodine with concentrated sulphuric acid the acid came out of the flask burning his right hand and foot.[31] The intensely poisonous nature of prussic acid did not prevent him carrying out long experiments with it. Here, as with his work on gunpowder, dangers could be minimised by taking precautions but the possibility of a serious accident could never be excluded. As late as 1844 a flask blew up in his face. Although he was injured in the face and hands, his eyes were protected by the glasses he now wore permanently.

## Election to the Institute

At the height of the Revolutionary turmoil the Royal Academy of Sciences had been closed down as a royalist and élitist institution inconsistent with the ideals of the new Republic. But in the earlier days of 1794–5 voices were raised in support of a general institution of learning embracing science and the arts. In this new National Institute the section concerned with science was not only the largest but was placed first. The 'First Class', as it was called, was thus the official body of science. Membership carried a small honorarium but what was important about it was that it marked recognition of a man's scientific talents. Membership was also the gateway to many other positions which were well remunerated.

Gay-Lussac was fortunate in being elected to the First Class of the National Institute while still in his twenties. He could not have achieved this without the backing and advice of the Arcueil group. One might have expected to find him as a member of the chemistry section but there were no vacancies in this section during the first decade of the nineteenth century and he had to look elsewhere. Gay-Lussac was fortunate in so far as his early memoir on the thermal expansion of gases was more 'physics' than 'chemistry' and, with his mathematical training, he made a suitable candidate.

Election to the Institute depended on dead men's shoes and probably the news that one elderly member of the physics section of the Institute, Brisson, was in failing health was received by some of his colleagues with mixed feelings. Brisson's death on 23 June 1806 meant that sooner or later a vacancy would be officially announced. To become a member of the Institute meant a great deal to an ambitious young man. Gay-Lussac would have to convince first the remaining members of the physics section and then the other members of the First Class that he was a good candidate. Fortunately he had been carrying out magnetic observations earlier that year with Humboldt and their joint memoir was ready to be presented on 8 September 1806. On 15 September Gay-Lussac presented some research on the thermal expansion of gases. Here were two areas of research which were undoubtedly 'physics' and Gay-Lussac decided to explore each further. On 24 November it was announced that a list of candidates would be drawn up at the next meeting. At that meeting Gay-Lussac presented two memoirs. There had obviously been some prior preparation because, at that very same meeting, Haüy was appointed a member of a commission to comment on Gay-Lussac's work. Instead of the usual time lag of several weeks or months Haüy revealed that he had written his report in advance and he proceeded to read this in most favourable terms to his colleagues. The next item on the agenda was the presentation by the physics section of a list of candidates in order of merit. Gay-Lussac's name was at the top of the list. At the next meeting, after further lobbying, he was elected.

## The Arcueil group

Although it was a major step in Gay-Lussac's career to enter the Institute, the Institute tended to be little more than a platform for the presentation of research done elsewhere. In the Napoleonic period Gay-Lussac's base was partly the Ecole Polytechnique and partly Arcueil. Arcueil was a private club with all the disadvantages of one but its advantages and achievement far outweigh criticisms which might be made of its exclusive nature and the orthodoxy expected of its members.[32]

If Berthollet and Gay-Lussac constituted the original nucleus of the Arcueil circle, Laplace and his disciple Biot came to form a second nucleus which was to have many fruitful interactions with the first. Biot, a graduate of the Ecole Polytechnique, had known Laplace since 1799, and Gay-Lussac had begun to carry out experimental work for the great mathematician in 1801. Laplace contributed two extensive footnotes to Berthollet's *Essai de statique chimique* of 1803 and three years later he decided to buy a country house at Arcueil, a property adjoining that of his old friend. The members of the Arcueil circle could now think of themselves as a society and in 1807 they published the first of three volumes of memoirs. The Society was very select with only nine members: the four mentioned above together with the explorer Alexander von Humboldt, the chemist Thenard, the botanist De Candolle, the mineralogist Collet-Descotils and Berthollet's son. Humboldt and Thenard each undertook some important collaborative work with Gay-Lussac more or less on terms of equality. Gay-Lussac thus found himself in an atmosphere guided by the two senior members, Laplace and Berthollet, but also stimulated by colleagues of his own generation. To the names listed above we may add the astronomer and physicist D. F. J. Arago, the chemist Dulong and the short-lived Malus (the discoverer of polarisation of light) who were later to be admitted as neophytes to the charmed circle of Arcueil.[33]

Gay-Lussac had as many as nine different papers published in the *Mémoires* of the Society of Arcueil. These range from major memoirs such as his historic one on the combination of gases by volume to a mere half-page note. These papers were entirely his own work as distinct from a further nine papers published jointly under his own name and that of Thenard. If we were to include these as well as his joint paper with Humboldt we have a total of nearly twenty memoirs and this leaves out a further six papers known to have been read by Gay-Lussac at meetings of the Society of Arcueil but published elsewhere.[34]

Gay-Lussac was therefore not only one of the most longstanding members of the Arcueil group but also one of the most active members in that second phase of its life when it called itself 'the Society of Arcueil'. Nevertheless it would be wrong to think of him as being based at Arcueil during most of the period of the First Empire. From 1804 he lived in Paris, with Arcueil as a pleasant retreat open in spring, summer and autumn and providing the facilities of an exclusive scientific club. Once Gay-Lussac had research facilities at the Ecole Polytechnique, he no longer needed to rely so much on Berthollet's hospitality at Arcueil. If he was frequently to be found there, particularly on Sundays, it was often to discuss his own research and to exchange news. We know however that he continued to take advantage of the experimental facilities at Arcueil. In a paper on the thermal effects of compression and expansion of gases read to the Institute on 15 September 1806 Gay-Lussac acknowledged that his experiments had been performed in the physics laboratory at Arcueil.[35] This research by itself would have earned a permanent place in the history of science for Gay-Lussac if he had done nothing else.[36]

As the Arcueil group became more formalised as a society it established the practice of holding meetings at fortnightly intervals and usually on Sundays. Arcueil was close enough to Paris for the members to be able to walk there. It must have been a delight, particularly in summer, to leave the crowded city for the joys of the countryside, and Berthollet entertained his colleagues well. However, recreation came second to the intellectual content of the meetings. Although memoirs were often read, an important part of each meeting was the discussion, each member hoping to contribute something and perhaps learn from his colleagues. Members would undertake to read specific journals and then present a summary at a meeting. The provision of experimental resources was an important facility which was appreciated by the younger members of the Society, as were the facilities for publication. Arcueil was above all a place for scientific news and criticism. A memoir could be tried out on one's friends on the Sunday before presenting it the following day at the Institute.

Laplace, writing ostensibly about scientific societies in general, cannot have had Arcueil far from his mind when he urged the advantages of scientists working together:

The isolated man of science can dedicate himself without fear to dogmatism; he hears only from afar contradictions of his ideas. But in a scientific society the impact of dogmatic ideas soon results in their destruction, and the desire to win one another over to their point of view establishes necessarily among members the convention of admitting only the results of observation and calculation.[37]

Unfortunately Laplace, in condemning dogmatism in others, could not recognise it in himself. Arcueil became a focus of French Newtonianism in a Laplacian mould as the work of Biot, Malus (and later Poisson) confirms. As Gay-Lussac's main work moved away from physics to chemistry Laplace's Newtonianism became less relevant, but we shall see that even in chemistry Gay-Lussac had sooner or later to come to terms with Laplacian ideas. Laplace and Berthollet both agreed that chemistry and physics were related sciences. That each science could learn from the other was a basic tenet of the Arcueil group. It was open to debate, however, whether chemistry could be reduced to a branch of Newtonian physics.[38]

On one plane Arcueil was a scientific society which published its own memoirs. However, to look on Arcueil simply as a scientific society would be to show a very limited understanding of what it meant in the history of French science. From the very beginning it was a patronage group. Any young man fortunate enough to be invited to join the intimate circle of Berthollet and Laplace was sure not only of the protection of these high priests of Napoleonic science but of their active support in any scientific undertaking. Although they provided valuable facilities, they did not, of course, pay their associates. But they performed a more valuable service to their protégés in providing surer access to salaried posts. When a post was vacant in one of the institutions of higher education in Paris, or even in the Institute itself, it helped enormously to have the support of the pundits of Arcueil.

### On the staff of the Ecole Polytechnique and a European tour

When Gay-Lussac was working as an assistant to Berthollet at Arcueil, his patron sought some appointment for his protégé. Without some inside help it would have been difficult for Gay-Lussac to get on the bottom rung of the ladder, but an opportunity came in December 1802 during Berthollet's presidency of the Conseil de Perfectionnement of the Ecole Polytechnique when Poisson was finally promoted to the position formerly held by Fourier.[39] Poisson had held the position of assistant to the *répétiteurs*[40] of analysis. One could hardly conceive of a more junior appointment. However a precedent existed for such a rank and Gay-Lussac, being no mathematician, was made assistant *répétiteur* of chemistry. This gave him a claim of succession when the post of *répétiteur* fell vacant. In April 1804, Thenard, the *répétiteur* attached to Fourcroy's course, resigned to take up a chair at the Collège de France and Gay-Lussac was appointed in his place (23 September 1804)[41] with the appropriate salary of 1500 francs. Gay-Lussac's establishment at the Ecole Polytechnique was a factor in encouraging the formal resignation of Berthollet from the school in 1805. In any case he had not taken an active part in the work of the school for several years. In his letter of resignation he recommended his 'young friend', Gay-Lussac, to the governor.[42]

The duties of the *répétiteur* were to attend the lectures of the professor and to carry out the oral examinations of the students. He would share with a second *répétiteur* responsibility for practical work by students. Writing in 1806 to a friend who had hoped to see him, Gay-Lussac said:

Today is the day for practical work for the students of the Ecole Polytechnique and I am obliged to be with them.[43]

The Ecole Polytechnique was one of the first institutions to arrange for its students to do individual chemical experiments. Although critics said that the students were sending up large sums of money in smoke, the senior chemists who had advocated such practical work justified this expenditure on pedagogical grounds. By an extension of the original concept of the post, a major responsibility for the chemistry *répétiteur* was to get ready apparatus needed in lecture demonstrations but he was expressly forbidden to do any experiment other than those required by the professor. The *répétiteur* was also a kind of store-keeper, having overall responsibility for ordering and storing chemicals and apparatus. Gay-Lussac complained in 1806 that his salary of 1500 francs was insufficient and certainly incommensurate with his responsibilities and Gay de Vernon made representations to the governor of the school about the low payment of the chemistry *répétiteurs*, whose duties were much more onerous than those of the *répétiteurs* assigned to the more abstract sciences.[44]

Although Gay-Lussac now had a junior post at the Ecole Polytechnique, his life was not to be that of routine or obscurity. About the time of his appointment he had made the acquaintance of Humboldt, who had just returned to France after a long expedition to South, Central and North America. Humboldt was interested in the analysis of the atmosphere and he and Gay-Lussac agreed to collaborate in a series of careful quantitative experiments. These were carried out at the Ecole Polytechnique in November and December 1804. While the two men were working together Humboldt was already planning his next trip – a tour of western Europe during which he intended, among other things, to make systematic observations of the magnetic elements in different places. He asked Gay-Lussac to accompany him not only because of their good personal relations but because the young Frenchman with his training in the physical sciences provided a useful complement to his own training. For Humboldt, a man of independent means, who expected a further income from the publication of his travels, there was no problem in leaving Paris. But Gay-Lussac had only just received a permanent appointment which required his presence in Paris. The problem was to be solved by Gay-Lussac receiving official leave of absence without loss of pay. As one of the earliest examples of sabbatical leave in modern science it is explicable only in terms of the powerful support Gay-Lussac had from friends in high places. Various administrative documents at the Ecole show that Gay-Lussac's absence caused problems which had not been sufficiently worked out before he was granted his leave.

Gay-Lussac's fundamental duty was to help in demonstration for the courses of Fourcroy. Thenard, who had been Fourcroy's *répétiteur* from 1798 to 1804,[45] agreed to stand in for his friend. The question of Gay-Lussac's leave was raised by the governor of the Ecole Polytechnique at a meeting of the professors on 22 February 1805 and it was agreed that the leave should be granted.

Gay-Lussac was able to leave Paris with Humboldt on 12 March 1805. No date had been mentioned in the official documents for Gay-Lussac's return, although it seems that he was expected back early in 1806. An undated letter[46] has been found written by Berthollet on behalf of his protégé asking that Gay-Lussac should have his leave extended. Gay-Lussac was then in Berlin and Berthollet urged that he should be allowed to derive the maximum benefit from his stay there. He would be back in Paris by 15 April 1806.

Gay-Lussac and Humboldt travelled south to Lyons and then crossed the Alps by way of Mont Cenis. On the way they made magnetic observations of inclination and also of the time taken for sixty oscillations of a magnetic needle in a horizontal plane. Altogether on their travels they were to determine the magnetic elements at over forty stations ranging from Naples in the south to Berlin in the north. They did not hurry and it was July before they reached Rome where their host was Humboldt's brother Wilhelm, the Prussian ambassador. Here Gay-Lussac did some research on the famous 'alum of Tolfa'. He found that the sulphur trioxide evolved on heating the alum decomposed further into sulphur dioxide and oxygen, data he was later to find useful. The two travellers then proceeded to Naples where Gay-Lussac was fortunate enough to see an eruption of Vesuvius on 12 August. Humboldt's catholic interests required him to examine several species of electric fish found near Naples and Gay-Lussac helped him in this investigation, a common interest of this period.

On 17 September they left Rome, taking the mountain road to Florence. Gay-Lussac examined a sample of water from the baths of Nocera and observed that the water was so pure that no chemical reagent would give a precipitate. In Milan Gay-Lussac and Humboldt met Volta, whose electric pile was becoming one of the new instruments of nineteenth-century science. They left Italy by way of the St Gothard Pass, going through Zurich, Lucerne, Tübingen, Heidelberg and Göttingen to Berlin, which they reached on 16 November. For Humboldt this was home territory and he saw that his young French friend made contact with men of science in Berlin, particularly the analytical chemist Klaproth (1743–1817) and the physicist Erman (1764–1851), who was later to receive a prize from the French Institute. In Rome Gay-Lussac had worked with the Italian chemist Morichini (1773–1836), so that there resulted from Gay-Lussac's travels not only a general broadening of the young Frenchman's experience but some close contact with men of science in other countries. He acquired an Italian dictionary[47] and a German dictionary.[48] His contact with Humboldt and his stay in Berlin throughout the winter of 1805–6 gave him a good knowledge of German. He was to be asked officially by the Institute to report occasionally on a book in German[49] and he read German scientific periodicals,[50] thus keeping his French colleagues in touch with scientific developments on the other side of the Rhine.

Gay-Lussac's serious concern for the learning of foreign languages is suggested in later correspondence with his son Jules, whom he sent to study in Germany. He also had the idea of teaching one of his children Swedish to help with translations.[51]

The first tangible outcome of the European tour was a paper on terrestrial magnetism read by Gay-Lussac at the Institute on 8 September 1806 on behalf of Humboldt and himself.[52] It was a competent piece of research with tabulation of results obtained and suggestions of possible explanations when the readings obtained did not fit into the general pattern of an increase in the horizontal component of the earth's intensity from north to south and a decrease in the vertical component. They were unable to find any diurnal variation. The memoir has a place in the history of terrestrial magnetism but it was Humboldt rather than Gay-Lussac who was to make this study a lifetime concern.

Probably Gay-Lussac's most famous – although not the most fruitful – research carried out at the Ecole Polytechnique was with a giant voltaic battery. The French Institute had established a prize for work on electricity but it was the news which reached Paris in December 1807 of Davy's decomposition of the alkalis which really fired the imagination of the French savants. Davy had made use of a large voltaic battery at the Royal Institution in London and it was felt in Paris that if a giant battery were constructed there it might produce equally spectacular results. With the approval of Napoleon, 20 000 francs was set aside for the construction of different voltaic piles. By the end of July 1808 a giant pile had been constructed consisting of 600 pairs of plates of copper and zinc each having an area of 900 square centimetres. This prestige research was given to the Ecole Polytechnique and Gay-Lussac and Thenard were entrusted with the planning and execution of experiments. Unfortunately, as even Davy was to find, there was no necessary connection between the size of a voltaic pile and the importance of results obtainable with it. Davy had already taken the cream by his decomposition of the alkalis (November 1807) and the alkaline earths (June 1808). Gay-Lussac and Thenard, after confirming this research, were able to show that the decomposition of an electrolyte was directly proportional to the strength of the current and independent of the size of the electrodes. Their work was published in the *Mémoires* of the Society of Arcueil, in the *Annales de chimie*, and finally in book form.

Having a position with a modest salary, Gay-Lussac was doing important scientific research at the Ecole Polytechnique and Arcueil, but after several years he seemed to be stuck on the career ladder. However, the creation of the new national educational system, the 'University of France', would seem to have provided a number of new posts and we have been fortunate to find a letter[53] written by Humboldt soliciting one of these positions for his friend. Some of the letter is worth quoting:

I take this opportunity of calling the attention of Your Excellency to this friend [Gay-Lussac] who has outstanding merit in the exact sciences combined with modesty, purity of habits, and courage which excites and sustains research. An unfortunate system of accumulating positions and of keeping chairs for those who do not give courses has left M. Gay-Lussac with a place having a salary of 1500 francs. Despite all the efforts of his friends we have not been able to bring about any amelioration in his fortune and there is no immediate prospect open in the years ahead. In this distressing situation I presume to approach Your Excellency once more. M. Delambre, who is as interested as I am in the happiness of M. Gay-Lussac, has also undertaken to pass on to you my solicitude and my wishes. As a member of the Institute could not my friend obtain through the good offices of Your Excellency a position of *Inspecteur des Etudes*?

In fact the creation of the Faculty of Science was to provide Gay-Lussac with a rather better job than Inspector of Studies, which would have involved him in time-consuming administration of little relevance to his science.

Gay-Lussac's friends in the Ecole continued to agitate on his behalf and they created a new position in the establishment for which he was the obvious candidate. A report of February 1809 by the Director of Studies, Gay de Vernon, suggests something of what had been going on behind the scenes. Vernon's report[54] outlined the history of the course of practical chemistry given at the Ecole. This course had been undertaken successfully by Chaptal in the early days of the school. When Chaptal left, the course was entrusted to the *répétiteurs*, Thenard and Desormes. Gay-Lussac and Drappier had then replaced Thenard and Desormes as *répétiteurs*. The course of practical work had been written up by Gay-Lussac and Drappier under the supervision of Berthollet. Vernon argued that practical work by the students fixed the theory of the lectures in their minds, was therefore of special value and should be seen to be so by the students. He therefore recommended that Gay-Lussac and Drappier each be given the title of Professor of Practical Chemistry. Gay-Lussac had taken less than his share in this course up to that time, concentrating on his principal duty of aiding Fourcroy in his course; but

Mr Gay-Lussac intends to devote himself henceforth to the practical course, finding suitable means of ensuring that Mr Fourcroy's course does not suffer by this arrangement...

When this report was considered at a meeting of the Conseil de Perfectionnement in March the governor said frankly that the idea of a new chair at the Ecole was principally to show its recognition of the services of Gay-Lussac, a member of the Institute and a distinguished scientist. The Council should therefore consider whether to give this title to both *répétiteurs*. They agreed that Gay-Lussac only should be given this honorary title but stressed that this was an exceptional case and should not be taken as creating a precedent. A decree of 31 March 1809 signed by the Minister of the Interior formally conferred the title of Professor of Practical Chemistry on him but stressed that he would receive no extra salary.

The creation of this honorary position marks an exception in the history of the *cumul*[55] in France in the nineteenth century. It suggests a certain embarrassment at the Ecole Polytechnique that a member of the Institute, the discoverer of two fundamental gas laws and the French opposite number to Humphry Davy still held only a junior post. Because of his influential friends it was possible to create an honorary position so that from the point of view of title, if nothing else, the anomaly was corrected. From the financial point of view it is obvious that Gay-Lussac was not adequately paid by the Ecole. With a fixed number of salaried positions one could hope only for dead men's shoes. The death of Fourcroy on 16 December 1809 provided the opportunity for which Gay-Lussac's friends had been hoping and on 17 February 1810 an imperial decree nominated Gay-Lussac as Fourcroy's successor to the chemistry chair with a salary of 6000 francs.

### Marriage

But there was another reason why exceptional steps were taken by his friends in 1808 and 1809 to secure for him a position with more than a minimum salary, and this was his impending marriage. As early as 1803[56] his father had plans for him to marry the daughter of a Limousin family which had done him a favour. The young scientist, however, resisted this match and seemed perfectly happy with the company of his male scientific colleagues. He developed a deep friendship with Humboldt when the latter returned to France in the late summer of 1804. In 1805 he wrote of his new friend that 'their hearts felt a great need to see each other often',[57] but four years later Gay-Lussac had found a wife. The lady in question was baptised Geneviève-Marie-Joseph Rojot. She was then 24 and Gay-Lussac 30. In contrast to the wife envisaged by his father, Geneviève Rojot had no money of her own,[58] and was apparently working in a shop when Gay-Lussac met her. Arago[59] tells how Gay-Lussac was struck by the young shop assistant studying a chemistry book under the counter. When they planned to marry Gay-Lussac had only a very modest income. However, the availability of a post at the Faculty of Science in May 1809 provided Gay-Lussac with an additional salary and later that month the wedding took place. The marriage seems to have been a happy one. This is how the new Madame Gay-Lussac described the event to a friend of her husband; in this case she wrote on his behalf because his eyes were giving him trouble after a laboratory accident:

As you express so much interest in what concerns our happiness, I will tell you that it has been settled ('fixé') for life by temporal and spiritual laws on 29 May. I assure you that all these ceremonies, which as you put it so well, will be unable to add to our esteem or our friendship, and which caused me much upset ('beaucoup de tourments') and chagrin, have nevertheless done more than I thought for my peace of mind.[60]

At the end of this letter Gay-Lussac himself added a note to his friend saying that they were happy in their house and adding:

I desire nothing more now than to have a laboratory at home to work there completely at my convenience.

On a more personal note Gay-Lussac added that his wife was expecting a baby the following summer. Gay-Lussac's hopes were realised by the birth on 18 June 1810 of a boy whom they called Jules.

Of his close friends – Thenard, Arago and Humboldt – Gay-Lussac was the first to get married. His collaborative research with Thenard blossomed and resulted in seven publications in 1809 and a further seven joint papers in 1810. Indeed Gay-Lussac's general scientific productivity reached a peak in the early years of married life.

Gay-Lussac's private correspondence shows how much his wife and family meant to him. He was unhappy when separated from his wife. One summer in the early Restoration Madame Gay-Lussac had gone to Saint Léonard, leaving him in Paris. He wrote:

The feast of Saint Louis was very brilliant – according to what everyone said. I spent all day in my laboratory since you know that it is when others are enjoying themselves that I work best. I dined [alone?] for the first time since our marriage.[61]

When separated from his wife he always managed to find time to write a long and affectionate letter:

Who would believe that a man to whom M. Gorin continually cries 'Copy! More Copy!',[62] who has undergone this week all the trouble of examinations, who loses his time in receiving visits and in making them, who has countless other occupations, is tormented, saddened and rejoices in turn in the interval between one letter and the next [from his wife], should devote at least two hours to write to his wife.[63]

The Gay-Lussac household moved several times within Paris in the next twenty years, moving finally to one of the houses attached to the

Muséum d'Histoire Naturelle. It was there that Mary Somerville met Madame Gay-Lussac in 1834 and, twenty-four years after the birth of her first child, estimated her age as 21! Mrs Somerville described her as 'exceedingly pretty and well educated; she read English and German, painted prettily and was a musician'.[64] Five children were born in the marriage. The problems suggested by the support of this large family may be imagined but we will leave until the end of the book the detailed consideration of how the scientist marshalled his resources not only to support his family but to launch his children when the time came on careers of their own.

# 3

# Personal influences and the search for laws

'If one were not animated with the desire to discover laws,
they would often escape the most enlightened attention'
Gay-Lussac[1]

It would be simple to list Gay-Lussac's main scientific achievements. It is less easy, probably more valuable, to investigate how he was able to achieve what he did. Important scientific results have emerged from a variety of contexts – from isolated geniuses as much as scientific societies, from practical men as well as theoreticians, from wild speculation and from severely disciplined reasoning. In different countries and different periods the methods and goals of science have covered a range of possibilities. What did Gay-Lussac see as the job of the scientist? Why should he have followed one line of investigation rather than another? Who influenced him and, if there was such influence, how was it exerted?

For a graduate of the Ecole Polytechnique science meant physical science and he would have sufficient mathematical training to apply this to his research. A graduate would also have some knowledge of chemistry, and although those bent on military or engineering careers may have felt this was an unnecessary component of the syllabus, it was a reflection both of the number of chemists who had shaped the syllabus and more generally of the recent appearance of chemistry as a new science with valuable applications in war and peace. Chemistry, now rid of its last vestiges of Aristotelian theory and alchemy, owed its success to generations of workers in the field. Nevertheless the new chemistry was associated with one man above all: Antoine Laurent Lavoisier.

Men are influenced by books and by institutions but above all by people. I think that Gay-Lussac had three main intellectual mentors: Lavoisier, Berthollet and Laplace. Lavoisier's conception of chemistry, his methods and problems find a strong echo in the work of Gay-Lussac. There is evidence of his influence over a decade constituting the peak

43

of Gay-Lussac's creativity. After about 1820 it became less important since chemistry was moving on to new ground. Berthollet's influence, however, remained with him all his life. Berthollet may not have been good as a lecturer and the way he expressed his ideas in his most important book, the *Essai de statique chimique*, leaves much to be desired. Yet, by working at Arcueil, Gay-Lussac assimilated something which remained in his system throughout his career. Laplace too had a crucial influence on the young Gay-Lussac. Probably without Laplace and the background of the Ecole Polytechnique Gay-Lussac would never have been able to make a contribution to physics. However, one sees the influence of Laplace on Gay-Lussac declining through the Napoleonic period and in his later years there is little to remind us that Gay-Lussac had once been inspired by the man whom some considered as the Newton of his age.

## Lavoisier's influence

Anyone studying the life and work of Gay-Lussac is bound to be struck sooner or later with certain resemblances to his great fellow-countryman Lavoisier. There was no direct contact between them, since Gay-Lussac was only 15 and living in the remote provinces when France's greatest chemist was led to the guillotine set up in the Place de la Révolution in Paris. It was only several years after the Terror that Gay-Lussac began to take an interest in chemistry and then it was by contact with Berthollet and the Ecole Polytechnique. Nevertheless, through Berthollet, Lavoisier's senior convert to the new chemistry, many of the ideas of Lavoisier were passed on to the aspirant chemist. Indeed the friendship and common interests of Lavoisier with Berthollet and Laplace almost amounted to kinship and one might claim for their protégé of the next generation that he belonged to the Lavoisier school by direct descent.

As regards Gay-Lussac's reading, he had copies in his library of both the *Méthode de nomenclature chimique* and the *Traité élémentaire*[2] – Lavoisier's testament addressed to the younger generation of aspirant chemists.

He took Lavoisier's comments seriously enough to mark his copy of the *Traité* in several places. However, Gay-Lussac's copy of the book was of the 1801 edition so it would seem that he only studied this after the beginning of his association with Berthollet.

There are several formal similarities in the position of Lavoisier under the *ancien régime* and Gay-Lussac when he had become an established chemist under the Bourbon Restoration. Both gave their services as technical experts on various commissions for the Académie des Sciences and achieved a certain fame or notoriety as representing 'official science'.[3] Both undertook consultancy work on the quality of gun-powder and had a laboratory at the Arsenal. But these resemblances do not relate to Gay-Lussac's early life. We are more interested in dis-covering whether his research reflected in any way the ideas of Lavoisier.

Of course any chemist beginning research in the early years of the nineteenth century would be conscious of the drastic re-structuring of this science largely due to Lavoisier's oxygen theory and list of elements which constituted the building blocks of the new chemistry. But whereas Davy in England felt compelled to challenge many aspects of the new chemistry, Gay-Lussac, his contemporary in France, was happy to accept the inheritance of Lavoisier. It was hardly an issue in the early years of the nineteenth century that combustion involved oxygen but there were several parts of the oxygen theory which might seem to require modification in the light of new evidence. Davy began his scientific career with a paper attacking Lavoisier's concept of the matter of heat, or 'caloric', and he contined to launch powerful attacks on different aspects of Lavoisier's work. Gay-Lussac's research on the other hand emerged as a continuation of the Lavoisier tradition and only in one instance did the question arise of directly challenging Lavoisier's theory. As we shall see, Gay-Lussac met this typically by a slight adjustment of the theory. He was no heretic.

Gay-Lussac's work on boron provides an example of his work within the Lavoisier tradition. According to Lavoisier, acids are compounds of a radical with oxygen; in many cases the radical was a known element. Thus the respective radicals of sulphuric and carbonic acid are sulphur and carbon. Lavoisier had included the 'boracic radical' in his list of elements but admitted that there was no direct evidence of its existence.

The boracic radical is hitherto unknown; no experiment having as yet been able to decompose the acid; we conclude from analogy with the other acids that oxygen exists in its composition as the acidifying principle.[4]

The isolation by Gay-Lussac and Thenard of appreciable quantities of potassium enabled them to use it to attempt the decomposition of boric acid and in November 1808 they were able to announce the isolation of a new element which they named *bore* (boron).[5] They claimed that they had known of its existence at least since June of that year,

but as we had only decomposed the acid and we had not re-formed the acid from its elements its nature could not be regarded as determined.[6]

In other words not only were they following Lavoisier's theory of the composition of acids, but they also accepted his criterion of experimental proof of composition.[7] Lavoisier had also hinted that the next generation should try to determine the nature of the fluoric radical[8] but the efforts of Gay-Lussac and Thenard were thwarted by the intense reactivity of fluorine, which was not isolated until 1886.

A clear case where Gay-Lussac followed Lavoisier rather than Berthollet was in organic analysis. Berthollet favoured a method of destructive distillation but Gay-Lussac, working again in collaboration with Thenard, preferred to take further an alternative approach which had been used by Lavoisier, that of combustion analysis. In his published memoirs Lavoisier had described how certain inflammable substances such as olive oil and wax could be burned in an atmosphere of oxygen and the carbon content deduced from the amount of carbonic acid gas formed. He had also begun some trials with potassium chlorate as an oxidising agent but this work remained unpublished. In 1810 Gay-Lussac and Thenard announced that they had successfully applied potassium chlorate to the analysis of sixteen vegetable and four animal substances.[9] In theory the use of potassium chlorate was an enormous step forward since it allowed chemists to analyse any organic substance and not simply inflammable ones. In practice potassium chlorate was far too reactive, producing over a short period of time large volumes of gases which broke the apparatus. The two young chemists therefore proposed to make the experiments manageable by dividing the substances into small portions. It was then made into pellets with potassium chlorate and these pellets were dropped one by one through a special tap down into a vertical tube made of thick glass the bottom of which was strongly heated. The gases evolved (carbon dioxide and excess oxygen) passed through a side tube and were collected over mercury. The total carbon dioxide gas evolved during the experiment was measured and the carbon content found. Knowing the weight of the original substance and the amount of oxygen used, the weight of hydrogen could be found by subtraction.

An essential part of Lavoisier's oxygen theory was that acidity is to be explained in terms of the presence of an acidifying principle – oxygen principle. Although this was not formally a quantitative theory, Lavoisier argued that (other things being equal) the more oxygen contained by an acid the stronger the acid; thus sulphuric acid is stronger than sulphurous acid. A problem arose with the quantity of acid necessary to neutralise a base. If the base contained a large proportion of oxygen, would the amount of acid required to neutralise it be less, as might be thought on the basis of Lavoisier's theory? Experiments carried out by Gay-Lussac in the early summer of 1807 suggested that there was no such proportionality.[10] Further experiments showed him that it was not an inverse but a direct proportionality. He had reached this conclusion by 1808,[11] but it is expressed better in his later lectures:

The quantity of acid necessary to produce neutrality increases in proportion to the quantity of oxygen contained in the base.[12]

Although a particular concern with oxygen is a good test of adherence to the Lavoisier school, we can see that quantitative studies of neutralisation are beyond Lavoisier's theory and constitute neither a confirmation nor a refutation. It was with Gay-Lussac's research on hydracids in the years 1813–15 that the testing time came.[13]

Even as late as 1819 there is clear evidence in Gay-Lussac's work of the influence of Lavoisier. In his important memoir on the solubility of salts,[14] Gay-Lussac paid tribute to his famous predecessor. Lavoisier, he said, was the first to give satisfactory explanation of the influence of temperature on solubility; the problem of variable solubility had been neglected since that time. He felt it appropriate to provide a page-long quotation from Lavoisier's *Traité élémentaire.*[15] His own copy of the book had this passage marked and it seems to provide as clear a case as possible of the work of a scientist of one generation continuing the work of a predecessor.

Lavoisier's writings were not generally given the same exaggerated respect as those of Newton had been given in the eighteenth century. Yet Gay-Lussac was involved in a dispute in 1826–7 where one of the points at issue was what Lavoisier had written on the subject of nitrification. Gay-Lussac himself was accused of being under the influence of chemists such as Lavoisier and of accepting ideas not on the basis of experimental evidence but 'on the authority of Lavoisier and of Berthollet'.[16] Gay-Lussac could have ignored this accusation but he chose to take it up and remark that both he and his opponent (Longchamp) were too small (i.e. in comparison with Lavoisier and Berthollet).[17]

## Berthollet's influence

The early influence of Berthollet at the beginning of Gay-Lussac's scientific career has already been discussed in chapter 2. The personal and career aspects were not the least important but here it would be more appropriate to mention briefly some ways in which Gay-Lussac's research was guided by Berthollet. Such influence was not confined to the early years at Arcueil. Even after that circle had dispersed, even long after Berthollet's death, Gay-Lussac could not forget the ideas of his master.

Gay-Lussac's whole conception of chemistry derived from his association with Berthollet. Perhaps if he had worked with another of the senior chemists in Paris at the turn of the nineteenth century he would equally have reflected their approach. Had he been a student of Vauquelin, for example, he would probably have interpreted chemistry as the analysis of mineral substances and the preparation of new compounds. Berthollet, however, did not believe that the chemist's work should be centred on the multiplication of chemical species. Rejecting the old natural history approach, his conception of chemistry, like that of Lavoisier, was more akin to natural philosophy. He was interested in how and why chemical reactions take place. Such problems continued to exercise Gay-Lussac, although there were times (like the hectic years of 1808 and 1809) when, side by side with Thenard, he sought a shortcut to fame by the preparation of new substances in competition with Humphry Davy.

Turning to particular subjects of research, one sees connections with Berthollet's earlier work in such areas as prussic acid, which Gay-Lussac first examined in 1811[18] before continuing it in his brilliant research of 1815 on the cyanogen radical.[19] Gay-Lussac's work in the Restoration on volumetric analysis took further the ideas of his mentor, which had been developed by Berthollet's other associates Descroizilles and Welter. On the whole, however, the origins of Gay-Lussac's research are not to be found simply in Berthollet's work. Gay-Lussac had his own imagination and, living in the intellectual ferment of one of the world's scientific capitals, was exposed to a multiplicity of ideas and possibilities of research.

It was Berthollet's chemical theory rather than specific areas of research which was to dominate Gay-Lussac's thought and emerge from time to time in the interpretation of his experiments. One finds this in his memoir on the combining volumes of gases. His conclusion that gases combined in constant simple volumetric ratios seemed at first to contradict Berthollet's idea that compounds are formed in very variable proportions. Gay-Lussac satisfied himself (and his patron) that gases constituted one of the special cases that Berthollet allowed to explain fixed proportions. If there is a suspicion that in this case Berthollet's personal influence had a restraining effect on Gay-Lussac's genius, the case of chlorine is much clearer. Gay-Lussac and Thenard found new evidence which led them to suspect that the gas was not after all a compound of oxygen but an element.[20] They presented this idea at a meeting of the Society of Arcueil on 26 February 1809 but, they later reported,

it appeared so extraordinary that M. Berthollet prevailed upon us to state it with the greatest reserve.[21]

In fact they presented the elementary nature of chlorine as nothing more than a remote possibility and, as we shall see,[22] it was left to Humphry Davy in the following year to proclaim unequivocally the elementary nature of chlorine.

Berthollet obviously exerted a strong influence in the Arcueil group and its publications and some of this influence might appear to approach coercion. A better test of Gay-Lussac's genuine respect for his mentor is in his work when Berthollet was no longer prominent among the leaders of the French scientific establishment. In 1839, seventeen years after the death of Berthollet, Gay-Lussac brought himself to publish some ideas on chemical reaction involving some criticism of his old friend and master. He admitted:

If my observations are exact, they will weaken considerably the influence which Berthollet has attributed to cohesion in chemical phenomena; but I myself feel too heavily the weight of this illustrious authority to have confidence in my own arguments and not to feel shaken in my new convictions. It is with this sincere feeling of doubt that I shall indicate some applications of the point of view from which I interpret cohesion.[23]

## Laplace's programme and influence

One particular research programme in which Gay-Lussac was involved could be easily overlooked since the work was not published under his name. This was his collaboration with Laplace. Fortunately Laplace gave a full account of the research in one of the supplements to the fourth volume of his *Mécanique céleste* (1805) where, incidentally, he mentions Gay-Lussac no fewer than seventeen times, and in the third edition of his *Système du monde* (1808).

We may recall that it had been the influence of Laplace which had led Lavoisier in 1783 to hope that 'one day the precision of the data [of chemical affinity] might be brought to such a perfection that the mathematician in his study would be able to calculate any phenomenon of chemical combination in the same way, so to speak, as he calculates the movement of the heavenly bodies'.[24] Lavoisier however soon became disillusioned by the complexity of actual chemical reactions so that by 1789 when his text-book was published he had decided that a study of affinity, although not invalid, involved too many variables to

provide a sure basis for chemical theory. But Lavoisier's former collaborator, Laplace, who survived the Revolution, was not to give up so
easily. He was convinced that chemical affinity was only a modification
of gravitational attraction and it was Laplace who carried this Newtonian approach to chemistry into the early nineteenth century. In the
first edition of his *Système du monde* (1796) Laplace expressed his
ambition of bringing under a single general law all the phenomena of
chemistry, physics and astronomy.[25] He mentioned various practical
difficulties and concluded that, in the state of uncertainty prevailing,
the wisest course was to undertake a large number of experiments. But
Laplace was not himself an experimenter and he looked around for
practical help. He made some use of Haüy but the best opportunity
came through Berthollet's young assistant at Arcueil who had already
shown his talents in his work on the thermal expansion of gases.

Of the various phenomena interpreted in terms of short-range forces,
that of capillarity seemed the most amenable to investigation. Laplace
was able to derive formulae for the rise of liquids in capillary tubes and
required experimental confirmation of his theory. The measurements
given by physicists of the rise of water in a capillary tube of a given
diameter differed by as much as 100 per cent largely due to the different extent to which they had wetted the tube. Gay-Lussac was able to
measure capillary rise under controlled conditions for water, alcohol
and mixtures of alcohol and water. Gay-Lussac took care to use
capillary tubes of uniform internal diameter and devised a new method
of measuring this. He introduced mercury into the tube, weighed it
and measured the length. Knowing the density of mercury it was easy
to calculate the diameter. Laplace described this method as one which
introduced 'the precision of astronomical measurements'.[26] From the
author of the *Mécanique céleste* no higher praise was possible. Gay-
Lussac also measured the rise of water between two parallel plates and
investigated the dimensions of large drops of mercury. Laplace was
very satisfied with the agreement of theory with practice and spoke of
having reached 'the true cause of the phenomena'.[27] Moreover he was
now convinced not only of the similarity but of the identity of the
forces governing capillary attraction and those governing chemical
affinity.[28]

One of the experimental investigations undertaken by Gay-Lussac as
part of this programme was the study of the lifting of a disc from a dish
of mercury.[29] This experiment had been carried out previously by
Guyton de Morveau in 1777, using different metals. He had claimed
that, if the metal disc was suspended from one arm of a balance on the
mercury, the weight which had to be added to the other arm of the

balance to lift the disc clear was a measure of its affinity. He had triumphantly produced a series of numbers which were in precisely the same order as that of the displacement series of the metals. Gay-Lussac was able to look critically at this work and show that the total of the successive weights which had to be added to raise the disc depended on the time interval between the adding of the weights. If the weights were added slowly enough it was possible almost to double the total before the disc broke away from the mercury. With this carefully performed experiment, which showed that Guyton's numbers were largely subjective, Gay-Lussac dashed the hopes of the more naive Newtonians who had hoped for an easy solution to the problem of quantifying chemical affinity.

Laplace considered that he had received support for his Newtonian theory of chemical affinity from Berthollet's demonstration that the effective affinity of a reactant depended on the quantity (or mass) present. Moreover, Berthollet showed that, if two acids were present with a base, it was not simply the stronger acid which combined with the base, but rather a partition took place in which each of the acids reacted with the base in a certain proportion, which Berthollet said was that of their respective affinities. Laplace argued that this confirmed the action of short-range forces; contact action alone would not give this result.[30]

Although Laplace argued that there was a general (mechanical) law of chemical affinity, he was prepared to admit that 'the figures of the particles, electricity, heat and other causes, by combining with this general law, modify its effects'.[31] The trouble was that basic data required, such as the figures of the particles of matter and their mutual distances, were unknown, and Laplace admitted that such an approach, although theoretically interesting, was useless in practice.

One aspect of the Newtonian conception of matter did, however, exercise some influence on Gay-Lussac's thought. Although one could not *measure* distances between particles, one could compare them, for example in the three states of matter. Berthollet had argued that in the gaseous state the cohesive force between particles was a minimum and it was with reflections on the implications of this that Gay-Lussac began his famous memoir on the combining volumes of gases in December 1808.

In 1814 in an extensive and crucial note, which has been passed over by historians of science, Gay-Lussac raised more general implications[32] which amounted to a challenge to reductionism; the problem was whether chemistry could be reduced to applied mathematics. It was approached by asking whether the conditions of chemical reaction could be reduced simply, as Laplace had suggested,[33] to considerations of heat. Heat effect could in turn be reduced in purely physical terms to a study of the distances between the constituent particles of reactants. Thus a theoretical approach to chemical reaction was possible. This had been attempted by a number of eighteenth-century Newtonians but it was specially to Laplace, his mentor, that Gay-Lussac referred. Laplace had found that if there was an inverse square law between the particles of matter in the same way as between the planets, then a spherical particle with a radius of a millionth of a metre would have to have a density of more than six million million times that of the mean density of the earth.[34] But this was hardly reasonable. As Gay-Lussac said, 'such a supposition seems exaggerated'. He was, nevertheless, prepared to discuss it. Let us see, he said, whether the decrease in affinity in a body corresponds with the increase of distance between its particles produced by heat. Taking copper as an example and not knowing exactly the force of cohesion in the solid and liquid states, he supposed that it might be at least a thousand times greater in the former state than in the latter. Suppose too, to take an extremely favourable example, that copper increased its volume as much as eight times on melting. But even this exaggerated example would only correspond to a doubling ($\sqrt[3]{8}$) of distance between particles. If the distance was doubled, then according to the inverse square law the attractive force would only be one quarter. This maximum effect bears little relation to the order of magnitude of the estimated minimum change in cohesive force (a factor of a thousand). Thus, although there might be some relation between cohesive force and heat it was not a simple one and other factors must be involved:

The figure, arrangement and inertia of particles can have influence in some chemical phenomena as, for example, the freezing of water and the crystallisation of sodium sulphate; *but there are an infinite number of other cases which are independent of them* [i.e. the above mechanical properties of particles] *as of the separation of the particles*; such is the combination of hydrogen and oxygen which only takes place at red-heat whether the gases are under very high or very low pressures.

With these words, which I have italicised, Gay-Lussac is rejecting the whole reductionist programme as applied to chemical phenomena. He was in fact doing to the simple reductionist programme of Laplace what Berthollet had earlier done to the simple affinity theory. While not willing to condemn it outright, he was suggesting that the factors involved in a chemical reaction were by no means so simple. Gay-Lussac pointed out, for example, that a small amount of electricity could decompose substances which remained unaffected at the highest temperatures. He therefore dismisses the neo-Newtonian programme of Laplace as 'conjecture'.

Although Gay-Lussac did not in the end accept the simple Newtonian ideal of a chemistry of quantified forces, yet at a point in his career when he was most impressionable he spent a considerable time working within Laplace's conception of the natural world. He was fully aware of the important implications Laplace attached to his experiments on capillarity. As late as 1814, when Gay-Lussac was no longer a junior research worker, he took Laplace's arguments seriously enough to argue in detail against them. At least he realised some of the limitations of the Laplacian programme. Four years later, when he was to be a member of an important committee judging a memoir on light by Fresnel which challenged the Laplacian orthodoxy, Gay-Lussac could be quite impartial.

### The influence of the Arcueil circle on the formulation of the law of combining volumes of gases

Even in what might appear as one of Gay-Lussac's greatest feats of independent research, his law of combining volumes of gases, there is considerable evidence of the influence of Laplace and Berthollet. The law may be discussed from two different points of view which have been distinguished by the labels 'context of discovery' and 'context of justification'. This is a distinction based on the fact that the reasons for suggesting a hypothesis may be different from those given for accepting it. The context of discovery of the law was experimental evidence provided by Gay-Lussac and others. This will be considered shortly when discussing his whole attitude to laws. First, however, while on the subject of the influence of Gay-Lussac's mentors, we may consider the context of justification, and I should make it clear that we are interested in justification in the historical context of the Arcueil circle rather than the logical justification. At first sight the law, which states that gases combine together in volumes which bear a simple ratio to one another, seems to contradict Berthollet's views on variable proportions.[35] It was vital for Gay-Lussac, living in the shadow of Berthollet, to be able to produce not only abundant experimental support but also a rational theoretical argument acceptable to his patron to explain why *in the gaseous state* definite combining proportions were possible.

Gay-Lussac began his classic memoir (1809) with a discussion of the differences between cohesive forces in the three states of matter. He pointed out that, although heat had the effect of causing expansion in substances generally, for solids and liquids there was no general law, whereas there was for gases. This was because in solids and liquids the particles were subject to forces of mutual attraction and 'it is only when the attraction is entirely destroyed, as in gases, that bodies under similar conditions obey simple and regular laws'.[36] The first paragraph of the memoir concluded with the hope that

we are perhaps not far removed from the time when we shall be able to submit the bulk of chemical phenomena to calculation.

This paragraph shows clearly the influence of Laplace in its concern with interparticulate forces, its interest in laws and the ideal of reducing chemical phenomena to calculation. But although such ideas may be attributed to Laplace, they were also shared to some extent by Berthollet. It so happened that about October 1808 Berthollet was writing an Introduction to the French translation of Thomas Thomson's text-book of chemistry. In this he tried to give an account of publications connected with chemistry since 1807, when the original English version had been published. He included a description of Laplace's theory of capillarity and other ideas of his colleague on attractive forces.[37] Berthollet, in paraphrasing the ideas of Laplace on attractive forces between the particles of a substance, wrote:

The gaseous state seems to be that in which the particles are already at a great enough distance so that neither the influence of their figures nor their reciprocal attraction has any more sensible effect...[38]

This passage would seem to provide evidence that, in the months when Gay-Lussac was working on his memoir, the question of interparticulate forces was a matter of interest to Berthollet as well as to Laplace. This Arcueil background provides the rationale of Gay-Lussac's conception of matter. Yet no philosophy of matter in itself would explain how the young man of science arrived at his law of combining volumes of gases.

We will consider how Gay-Lussac reached this law in the context of his whole approach to laws of nature.

### The search for laws

Gay-Lussac once remarked: 'If one were not animated with the desire to discover laws they would often escape the most enlightened attention'.[39] It would not be an exaggeration to say that his scientific life, particularly the most creative period, was permeated with a search for laws. His first published research on the thermal expansion of gases resulted in the statement of a law and, even if this subject of research was not entirely his own choice, the success of this work must have given him confidence that the discovery of further laws was within his reach.

In discussing the development of Gay-Lussac's work it might be helpful to distinguish between three different types of law. The first is a law describing a uniformity of behaviour of a group of substances. This is the simplest kind of law and it was where Gay-Lussac started in 1802. Then there was a second, slightly more complex type of law involving proportionality. We shall show that Gay-Lussac may have been thinking in terms of proportional relationships by 1804 and in 1806 was well attuned to this way of thinking. Finally there was a third type of law or perhaps a variation of the second. These were simple ratio laws which Gay-Lussac enunciated in 1808 and 1810. We will consider Gay-Lussac's approach to science in the first decade of the nineteenth century in terms of these three types of law.

First was a law expressing the uniformity of nature – *all* substances of a certain class behave in a certain way when subjected to a particular treatment. Of this kind was the law that *all* gases (regardless of differences in chemical or physical properties) expand uniformly on heating. He was subsequently able to show that the composition of the atmosphere is more or less constant.[40] Yet this was hardly an original discovery but rather the extension of accepted ideas to high altitudes. He also showed, however, in collaboration with Biot, that the magnetic intensity above the earth's surface was the same as on the ground.[41] These conclusions also express a belief in the uniformity of nature albeit in a minor way. Really useful examples of this type of law are, however, comparatively rare and when Gay-Lussac hoped to be able to claim that all gases had the same specific heat (i.e. their temperature was raised by an equal amount when a certain amount of heat was applied) he found that the evidence was against his original hypothesis.

In 1805 he had found in collaboration with Humboldt that the combining proportions of hydrogen and oxygen which react to form water were constant despite variations in the proportions of these gases originally present. The combustion of hydrogen was thus 'of uniform nature' ('de nature uniforme').[42]

A similar confidence in the uniformity of natural phenomena, although on a more limited scale, is shown in his attitude to Volta's eudiometer, the apparatus used to estimate oxygen by sparking with a measured excess of hydrogen, and then noticing the contraction:

Some people had accused this instrument of being inaccurate, of indicating too low a quantity of oxygen in the air; but it seemed to us that if we supposed that some corrections were necessary, we could, by appreciating them as well as *the law of their variations,* make it very exact and very useful.[43]

Thus Gay-Lussac was looking for constancy and regularity even in variations. This was to pursue his search to a second order. If this reference to laws is not merely a figure of speech it shows remarkable confidence that a discerning scientist can find order where others merely found confusion. Chaptal had criticised manufacturing chemists who complained of the arbitrary progress of certain operations: 'Nature. . . obeys invariable laws; and the inanimate substance which we make use of in our manufactures exhibits necessary effects in which the will has no part and consequently in which caprices cannot take place.'[44]

The acceptance of Gay-Lussac's law of thermal expansion of gases undoubtedly gave him confidence and inspired him to look for other uniformities in nature.

In the course of his research with Humboldt on the combination of hydrogen and oxygen in various proportions Gay-Lussac was led to another tentative generalisation. As he confessed a little later:

As one is naturally inclined to generalise, we (or at least I in particular) had retained the opinion that it was very possible that all gases had the same capacity for caloric.[45]

By now, however, Gay-Lussac had found that nature was not everywhere so simple and thus began the second phase of his search for laws. If nature is not uniform in the simplest way, one may at least hope that it varies in a regular way. The second type of law and one which Gay-Lussac was to pursue repeatedly was that of proportionalities. If thermal capacities (unlike coefficients of expansion) are *not* the same for all gases, they are likely to vary in some way related to a basic property of the gas. Such a fundamental property was density. In 1806 Gay-Lussac announced that the specific heat of a gas was inversely proportional to its density.[46] Thus hydrogen, the lightest gas known, had a high specific heat, whereas oxygen and nitrogen, which had approximately the same density, both had a low specific heat.

In 1806 Gay-Lussac had also been working with Laplace on capillarity experiments. From these Laplace drew the following conclusion:

The rise of a fluid which wets completely the sides of a capillary tube is at any temperature directly proportional to the density of the fluid and inversely proportional to the internal diameter of the tube.[47]

It would not be surprising if, after the use made by his distinguished patron of his own experimental work, Gay-Lussac was impressed to think of other general proportional laws.

There is evidence that even in September 1804, when Gay-Lussac made his record solo balloon ascent, he had a propensity towards the simpler type of correlative law. It was known that at high altitudes, e.g. on mountains, the temperature was often lower than in the valleys and it was therefore a reasonable speculation that the decrease of temperature was *directly proportional* to the increase of altitude. Gay-Lussac's readings, however, did not show this.[48] As he expressed it, all they suggested was an 'irregular law'. He rationalised this by suggesting that as some of the readings were taken when the balloon was changing altitude rapidly, the thermometers may not always have had time to register the true temperature of the surrounding atmosphere. He therefore made the bold suggestion that only those temperatures which showed a regular fall should be counted. Not surprisingly this produced 'a more regular law'. The interesting feature here is that Gay-Lussac was almost prepared *despite* the evidence to proclaim a law.

We catch sight of an uncharacteristically reckless Gay-Lussac in a half-page note reporting a paper he read to the Society of Arcueil in June 1807. He thought that he had discovered two general principles (he did not dignify them with the title of law).[49] He thought that the capacity of saturation or equivalent of a body was inversely proportional to its density.[50] He also thought it a general principle that the capacity of saturation of acids and bases was independent of the oxygen they contained. These statements suggest that Gay-Lussac was very prone to make generalisations. The first statement also shows that Gay-Lussac had now come to generalise in terms of proportions. Indeed his second principle appeared the following year, drastically changed and now in the form of a proportionality: the quantity of an acid in a salt is exactly proportional to the weight of oxygen in the corresponding oxide.[51] Later Gay-Lussac tried heroically but vainly to work out a proportional law between the quantity of oxygen in a substance and its acidity.[52] The significance of his failure is discussed elsewhere.[53]

Gay-Lussac's research on the voltaic pile, begun in 1808, illustrates the extent to which he was prepared to investigate possible correlations. It was a reasonable first hypothesis to suggest that the electrical effect of a voltaic pile might be proportional to the strength of the acid used,[54] but Gay-Lussac was prepared if necessary to submit possible correlations to more precise mathematical scrutiny to discover the nature of the proportionality. He then announced that:

The quantities of gas obtained with different solutions of sulphate of soda increase as the *cube roots* of the quantities of the salt they contain.[55]

Although he refers to this correlation as a 'law', his attempts to demonstrate a similar proportionality in the cases of other salts were unsuccessful. Again he suggested that there might be a correlation between the activity of the voltaic pile (measured by the total volume of gas evolved) and the number of plates. The figures he obtained by experiment did not reveal any direct proportionality but Gay-Lussac suggested that the activity varied as the cube root of the number of plates.[56]

Perhaps Gay-Lussac's most interesting testimony on the subject of his search for proportionality laws is contained in his 1816 memoir on the expansion of liquids. Here he tells the reader of his ambitions with usual frankness:

I devoted myself to research on the expansion of liquids, trying to discover some new law and although this work did not fulfil my expectations it has nevertheless confirmed me *in the hope that important correlations will be reached*. Laws are necessarily derived from the observations of a large number of facts; but *if one were not animated with the desire to discover laws, they would often escape the most enlightened attention.*[57]

An interesting apparent exception to Gay-Lussac's search for laws is provided by his collaboration with Humboldt in the winter of 1804–5 when he investigated the combining proportions of oxygen and hydrogen. He failed to recognise the significance of the ratio 2 : 1, although he did remark that 'hydrogen combines with oxygen in double the volume of the latter'.[58] Again in 1807 he remarked of the sulphur dioxide and oxygen obtained by strongly heating copper sulphate:

These two gases are approximately in the ratio by volume of 2:1 but I will return later to the exact determination of this ratio.[59]

It may seem extraordinary that when a man has been shown to have a strong inclination towards the discovery of regularities in experimental data, he should miss such a wonderful opportunity. The answer to this on one level is that what one finds is often related to what one expects. In the case of the balloon ascent Gay-Lussac was expecting to find a correlation. In the second case he was not looking for a law. On another level we may distinguish between different kinds of laws. The law of combining volumes of gases so narrowly missed in 1805 and 1807 and remaining 'undiscovered' for several years was quite different from a proportionality law. In fact it was much more than a numerical regularity – it amounted to an insight into chemical reactions. Looking at reactions in one state (the gaseous) Gay-Lussac detected a regularity hitherto unsuspected. The key to this regularity lay in applying the quantitative approach to volumes rather than weights.

The third type of law fruitfully explored by Gay-Lussac was what we might call a 'ratio law'. It could also be called a law of proportions but it is probably better to avoid this label as it suggests a close connection with the proportionality laws, although it is quite different. There are two examples of this type of law in Gay-Lussac's work, his law of combining volumes of gases of 1808 and his laws of the composition of organic compounds of 1810. The law of combining volumes of gases would have been obvious to Gay-Lussac before 1808 from his own experimental evidence if he had thought at all about simple ratios.

## The law of combining volumes of gases

Any explanation of the development of Gay-Lussac's law of combining volumes must explain why he had found the 2 : 1 relationship of hydrogen and oxygen in January 1805 but did not generalise the law until December 1808. One cannot ignore the interval of four years in the development of an active and ambitious young man on the lookout for new laws. It is not good enough to remark of his 1805 work, as at least one historian has done, that:

Struck by the simplicity of the relations thus found, Gay-Lussac extended his investigations to the volume relations of other gaseous substances. . .and by the end of 1808 he was able to publish results which clearly demonstrated the existence of a simple and general law.[60]

It is my belief that he was not struck by the simple ratio in 1805 because he was not looking for this kind of a law. The reason he waited until 1808 to announce his law was that new evidence and new ideas of that year made him turn back to his earlier work and realise its significance in the general pattern of gaseous reactions. It is true that after he had completed his 1805 research with Humboldt he went on a tour of a year's duration through several countries and was not in a position to concentrate on any prolonged piece of research. As soon as he arrived back in Paris, however, he began an investigation of the specific heats of gases, a piece of research which, he said, arose out of the 1805 collaboration with Humboldt.

The sophisticated modern reader, particularly one who has been taught as a child Gay-Lussac's law of combining volumes of gases, may find it difficult not to regard such a law as obvious. The law as stated by Gay-Lussac was that 'the compounds of gaseous substances with each other are always formed in very simple ratios, so that representing one of the terms by unity, the other is 1, or 2, or at most 3'.[61] Given Dalton's atomic theory, the law may even be deduced from first principles,[62] but such a connection is quite unhistorical. Hindsight can be misleading and it is our concern to examine what experiments and

what news between 1805 and 1808 led a particular scientist to enun-
ciate his wonderfully simple and useful generalisation. The crucial year
was 1808. I think that Gay-Lussac was helped towards the discovery
of the law of combining volumes of gases by his preparation of boron
trifluoride and his study of its reaction with ammonia. But there was
also some news of research in Britain which reinforced his interest in
combining proportions and directed his attention to the idea of simple
ratios. We shall consider each of these factors in turn.

When Gay-Lussac was able to prepare appreciable quantities of
potassium early in 1808 he used it for a variety of purposes, one of
which was to obtain boron trifluoride ('fluoric gas').[63] Struck with the
dense fumes obtained by contact of the new compound with moist air,
he reacted it with ammonia. This led him to compare this reaction with
that of 'muriatic acid' (hydrogen chloride) and ammonia, and carbonic
acid gas and ammonia. In all these cases the volumes of the gases
reacting bore a simple ratio to each other. It is significant that in his
memoir on gases Gay-Lussac at first shows as much interest in the ratio
of the different quantities of alkaline gas required to neutralise a given
volume of acid as in the simple relation between acid and alkali. Gay-
Lussac remarked 'We might even now conclude that gases combine
with each other in very simple ratios; but I shall still give further fresh
proofs.' In other words the experiments on the combination of acid
gases with ammonia, together with the oxygen–hydrogen result, are
really sufficient to suggest the law, but to make his case stronger
Gay-Lussac produced additional (i.e. not fundamental) evidence. This
evidence, which we may interpret in the context of justification, may
be summarised by using modern formulae:

$$NH_3 \ : \quad 100\ N + 300\ H$$
$$SO_3 \ : \quad 100\ SO_2 + 50\ O$$
$$CO_2 \ : \quad 100\ CO + 50\ O \qquad \text{etc.}$$

Experimental work of relevance to the combining volumes of gases
was presented to the Royal Society in January 1808 by Wollaston and
Thomson who were both concerned with simple multiple proportions.
Berthollet referred to this work (published in the first part of *Philosophical Transactions* for the year 1808)[64] in the Introduction he wrote
to the French translation of Thomas Thomson's *System of Chemistry*.[65]
Berthollet probably received this part of the *Philosophical Transactions*
in the late summer.[66] He regarded it as important enough to summarise
and discuss and it would therefore have been known to Gay-Lussac.
Gay-Lussac said in his memoir that any resemblance of his work to
that of Dalton was purely coincidental and we should believe him. He

did not say that the same comment applied to Thomson or Wollaston and circumstantial evidence suggests that it did not. The first evidence he gave of simple ratios for combining volumes of gases was (after oxygen/hydrogen) the cases of acids, particularly carbonic acid gas, combining with ammonia. The point was emphasised of the ratio of volume of 2 : 1 required to produce the carbonate and bicarbonate. Such multiple proportions would be a corollary and hardly introduced at the *beginning* of a logical exposition of the evidence by Gay-Lussac. The reason for its prominence is that Gay-Lussac was not giving a purely logical exposition but one which included an account of his experiments more or less in the order in which he did them. He was not the sort of man to confide in the reader all the circumstances which led him to his conclusion but the broad hint becomes more of a certainty when at the end of his memoir Gay-Lussac referred explicitly to evidence provided by Wollaston and Thomson of ratios 2 : 1 in quantities of acid required to produce different salts.[67]

Thomson had been concerned with oxalates but it seems that Wollaston's experiments were the crucial ones, not only since they included the case of carbonates – the example taken up by Gay-Lussac – but also because they happened to be expressed in volumetric terms. Berthollet had elaborated slightly on Wollaston's description for the French audience:

If 2 grains of super-carbonate of potash freshly precipitated in a paper is passed up a tube filled with mercury and the gas is liberated by muriatic acid, it *will occupy double the volume* of the gas given off from 4 grains of the same salt reduced to the sub-carbonate by exposing it for a little to red heat.[68]

Wollaston had therefore unwittingly provided an example of simple multiple proportions related to the *volumes of gases*. Gay-Lussac was already interested in the volume of gases; he was also interested in the quantities of acid and alkali required for neutralisation; he now took up with enthusiasm the lead of simple proportions. Gay-Lussac in his later lectures contrasted favourably the research in Wollaston's 1808 paper with that of Dalton. He said that Dalton's idea of atoms was no more pure speculation but it had been 'confirmed later by the experiments of Wollaston on the oxalates. Indeed this chemist found that potash, for example, combined with oxalic acid in the ratio of 1 of base with 1, 2 or 4 proportions of acid. This theory has been confirmed since by a very large number of experiments.'[69]

It was no doubt Gay-Lussac's interest in combining proportions as well as his perennial interest in the oxygen content of acids which led him with Thenard in 1810 to present the results of their analysis of vegetable substances in the form of three laws:

*First law* – A vegetable substance is always acid when the oxygen is to the hydrogen in a greater proportion than in water.
*Second law* – A vegetable substance is always resinous, oily, or alcoholic, etc. when the oxygen is in a less proportion to the hydrogen than in water.
*Third law* – Lastly, a vegetable substance is neither acid nor resinous, and is analogous to sugar, gum, starch, sugar of milk, to the ligneous fibre, to the crystallisable principle of manna when the oxygen is [to the hydrogen] in the same proportion as in water.[70]

In the first place these 'laws' are effectively a classification of vegetable substances and from this limited aspect they might have a place in the eighteenth-century natural history approach to chemistry. The important thing about this classification, however, is that it is based not on properties but on *composition*. Gay-Lussac's basic assumption is that substances with similar properties have similar composition and he is able to demonstrate this by rigorous analysis.

## Scientific laws

My study of the development of Gay-Lussac's conception of 'law' in his science has suggested two main stages. After the initial step of a single generalisation: 'All As are Bs', came the simple mathematisation of nature – the proportionality laws and the ratio laws. The view that nature was basically mathematical would have been a common assumption to any graduate of the Ecole Polytechnique. Although Gay-Lussac had followed courses there on analysis, calculus, etc., it was not advanced mathematics which he felt was essential for the physicist (and perhaps in his day it was nearer the truth) but mathematics at a more elementary level. This was what he told his students at the Faculty of Science. In order to understand physical phenomena they would have to understand 'relations ('rapports') and very simple [mathematical] expressions'.[71] He remarked significantly:

To be in a position to discover the laws which link phenomena together, it is important to make use of the very powerful instrument of calculation ('le calcul') by means of which one may more easily grasp the relations between bodies.

In Gay-Lussac's last major chemical memoir published in 1839 he refers to the question of correlation and apparent anomalies in science. Any apparent novelty, he says, can probably be explained in terms of the existing framework of science but if on close examination it really

62

turns out to be some new truth, 'then it is nearly always found to be contrary to those so-called general laws ('ces prétendus lois générales') which had first presented themselves to our mind in so trenchant and decisive a manner'.[72] This rather curious testament may seem a bitter final judgement on the will-o'-the-wisp which had inspired and seduced the young researcher. It represents his memory of his own earlier enthusiasm for establishing new laws but it shows that he now realised that he had pinned too much faith on them, he had regarded them in too absolute a way.

Gay-Lussac's undoubted success in formulating laws of physics and chemistry may be contrasted with what Dulong regarded as his failure in this respect. Having a common background in the Ecole Polytechnique and Arcueil, they both regarded the formulation of laws as what science was about. Despite his success in arriving at the law relating specific heats and atomic weights ('Dulong and Petit's law', 1819), Dulong felt dissatisfied and in 1825, after spending many months investigating refraction in gases, he remarked to Auguste de la Rive:

No general law, nothing but approximate laws. What is wrong? I have no luck at all, whereas Gay-Lussac has only to embark on a subject to find a law.[73]

Was Gay-Lussac lucky, or is it the case here as elsewhere, that 'chance favours only the prepared mind'? Nor did Gay-Lussac achieve success without quarrying deeply into a variety of problems. Perhaps Dulong gave up too easily. While he showed flashes of great originality (sometimes, perhaps, more than Gay-Lussac[74]), he was dogged with bad health. Gay-Lussac was fortunate to have generally good health, he worked hard, and, above all, he had in his youth faith – faith that there were laws of nature waiting to be discovered. These laws were simple enough to be within the grasp of the determined research worker.

## Tables and graphs

Considerable use of tables was made by Gay-Lussac in reporting and summarising his scientific work. They feature significantly in his memoir on the combining volume of gases. Unfortunately the two tables relating to that memoir were printed out of place in the Arcueil *Mémoires*[75] and thus they were missed by many of his contemporaries, translators and historians of science. The first table compares the densities of gases as found by experiment with their densities calculated according to his theory. The close agreement in nearly all cases was a powerful argument in his favour. The second table was probably even more important to the acceptance of his law, since it summarised his discovery that the composition of gases as expressed by volume gave simple ratios, whereas the gravimetric composition as set out in the last columns manifestly did not. The table was therefore an expression of its author's *credo* that the volumetric approach to gaseous composition and reaction was the 'natural' one.

One fairly common purpose for the construction of tables is the didactic one that it makes knowledge easy to assimilate. But Gay-Lussac's use of tables was more a reflection of his conception of scientific knowledge, which for him was normally based on quantitative data obtained by experiment. No one experimental result, however, could mean very much on its own. If it was reinforced by repetition and variation and if an accumulation of results all pointed in the same direction, this was particularly valuable, since it could lead to a generalisation or even to a law. One could always state such a generalisation and then refer to the evidence, but it was a master stroke to present the evidence concisely in such a form that the regularity was seen by the reader. The statement of the evidence in this form *was* the law. I would not wish to suggest, however, that Gay-Lussac's use of tables was simply clever propaganda for his own views; it was something much more basic. It reflected his view of the logical presentation of data. When he set out in tabular form his own views contrasted with those of Davy,[76] he assumed a wonderful impartiality far removed from the rhetoric employed by some other scientists. At the same time he achieved a standard of conciseness and clarity typical of him at his best.

In Gay-Lussac's search for laws he looked for precise and simple correlation. If careful and patient experimentation yielded such a correlation, then one might have a law. But it was inevitable that in the search for relationships some should be found which did not vary in any precise and simple manner and therefore could not be expressed by a law. If no general law were possible, Gay-Lussac had to be content with a table. A table of results might be the first stage in the formulation of a law, but when he was unsuccessful in his search for simple correlations, it was often also the last stage. A good example of this use as an alternative was his apparent hope to find a general law relating the vapour pressure of liquids to their temperature, but, he says,

We recognised that different liquids do not follow the same law in this respect. . .so that it becomes necessary to have a table which gives the [vapour] pressure of each particular liquid.[77]

On at least one occasion an apparent failure to describe a correlation in a simple verbal or mathematical form opened a new chapter in science. Such a case was the solubility of salts. Gay-Lussac's patron, Berthollet, had removed the logical foundation of the traditional eighteenth-century tables of affinity. It is sometimes said that these tables had no replacement as a systematisation of chemistry until the Periodic Table some three-quarters of a century later. Yet in the intervening period Gay-Lussac made a contribution the significance of which should not be overlooked. Despite his desire to apply measurement to chemistry, he had to admit to his students that in general affinities 'could not be measured because they vary according to the circumstances'.[78] Yet if the ambitious eighteenth-century programme of reducing chemical affinity to astronomical attraction had failed, Gay-Lussac was able to proceed with an approach directly related to affinity – solubility. He made a point of distinguishing *physical* cohesion, which was a force expressed in units of weight, from *chemical* cohesion, which was related to solubility. One definition he gave of chemical cohesion was 'resistance to solution'.[79] But solubility was not an absolute property, it depended on the solvent and the temperature. It was Gay-Lussac who first provided a graphical representation to show how the solubility of salts in water varies with temperature (fig. 1). These graphs showed that, whilst the solubility of most salts normally increased fairly regularly with rise of temperature, each salt behaved differently, the anomalous behaviour of sodium sulphate being particularly marked. On the solubility curves of barium and potassium nitrates and potassium chlorate Gay-Lussac commented that:

*They show at once to the eye* that the solubility of each salt, especially nitre, increases at a great rate. It would have been possible to have represented them by algebraic formulas, but their graphical lines have the advantage of giving *immediately* and without calculation, and with almost as great precision, the solubility for all temperatures between which the experiments were made.[80]

However, despite his advocacy of the graphical method of presentation, he stated that he was planning to draw up tables giving the solubility of a greater number of salts at five-degree intervals.

Tables played a prominent part in Gay-Lussac's work in applied science since he did not stop at correlations within pure science but made it his business to provide reliable quantitative information for the artisan. Such data illustrated proportionalities and correlation rather than random variation. The most famous of these tables showed the relationship between the proportion of alcohol in an alcohol–water mixture and density at a given temperature.[81] In his work on volumetric analysis he drew up a table showing what weights of different salts of potassium were equivalent to a given weight of potash.

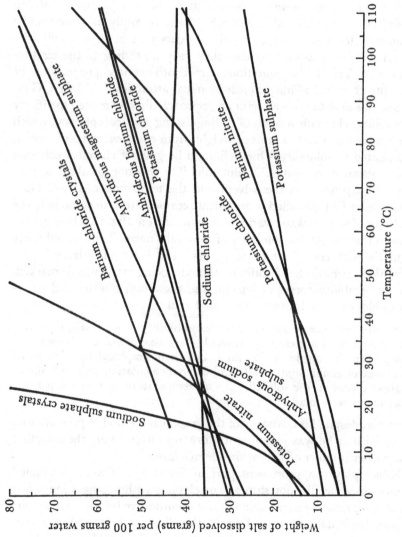

Fig. 1. Gay-Lussac's solubility curves.

Barium chloride crystals

Anhydrous magnesium sulphate

Anhydrous barium chloride

Anhydrous barium chloride

Potassium chloride

Barium nitrate

Potassium sulphate

Sodium chloride

Potassium chloride

Anhydrous sodium sulphate

Potassium nitrate

Sodium sulphate crystals

Temperature (°C)

Weight of salt dissolved (grams) per 100 grams water

## Analogical argument and classification

While some aspects of chemistry could be approached quantitatively, in other cases what was needed was comparison and classification. Generalisation of data was desirable but only at a late stage:

In the natural sciences and especially in chemistry generalisations should result from the detailed knowledge of each fact, they should not precede it. It is really only after having acquired this knowledge that we can be assured whether they have something in common...[82]

This important statement of Gay-Lussac began with a criticism of an *a priori* approach to science. But science could not be simply a knowledge of facts; there must be connections and analogies.

Something of Gay-Lussac's positivism emerged in his attitude to analogies. While scientists would be foolish to ignore analogies, he sounded a warning that they could at best be suggestive, indicating probabilities rather than certain knowledge:

Analogies to which we should not yield too blindly in chemistry (but which ought not to be neglected when founded on a numerous series of phenomena), furnish...some probabilities in favour of the existence of the hydrochlorates.[83]

Analogical argument, therefore, did not have the certainty desirable in science. In his paper on combining volumes of gases he dealt firstly with experimental determinations ('facts') and then went on to extend the discussion to solid and liquid elements. There was, however, a note of warning to the reader:

I shall not discuss more of these determinations, because they are only based on analogies...[84]

It was his work on iodine which gave greatest scope to the application of analogy. Not only did analogy help to elucidate the nature of iodine and its compounds but it also helped in understanding the reactions of elements already known:

The analogy which I have established between chlorine, sulphur and iodine may serve to throw some light on some of the combinations of chlorine...[85]

An important use of analogy was to predict the existence of new compounds and it was in this way that Gay-Lussac discovered a new acid (chloric acid) and a new radical (cyanogen). Reasoning from the analogy between iodine and chlorine and the existence of iodic acid, he postulated the existence and gravimetric composition of a similar oxy-acid of chlorine.[86] On treating barium chlorate carefully with dilute sulphuric acid he successfully prepared chloric acid. A further use of analogy was to consider the series: hydrochloric acid, hydriodic acid, hydrocyanic acid which, according to his volumetric analysis of the respective gases or vapours, he showed could be written respectively as: HCl, HI and HCN. After the recognition of the elementary nature of chlorine and the discovery of the element iodine, Gay-Lussac went in search of a possible 'element', CN. The experimental work involved is described later but the foundation of Gay-Lussac's work on cyanogen in 1815 was undoubtedly his research on iodine in 1813–14.

With the growing number of elements and compounds known to chemists by the early nineteenth century the teaching of chemistry could easily have degenerated into a dull catalogue until the Periodic Table gave inorganic chemistry a new rationale. But between Lavoisier and Mendeleef (or even Lavoisier and Dumas) all was not desert. Gay-Lussac faced squarely the problem of grouping the elements and, although his contributions to the classification of the elements are usually ignored, his concern to compare elements was an important feature of his lecture courses. Gay-Lussac's view on the classification of the elements was that it should be based on reaction with oxygen and with hydrogen.[87] This was as near as possible to a 'natural' classification. He accordingly grouped together:

potassium, sodium, lithium, barium, strontium and calcium
for which he coined the term 'alcaloides', a term later used by chemists for a quite different group of organic compounds. These elements all formed soluble oxides; their oxides were decomposed by chlorine. To a second group of elements he gave the name 'aluminoides':

aluminium, glucinum, yttrium, zirconium, thorium, magnesium
Their oxides were insoluble in water and were not decomposed by chlorine. Gay-Lussac also grouped metals in three classes according to whether their oxides could be reduced to the metal by heat alone, by hydrogen or by carbon. Unfortunately most of these notes are undated but we know that in his final years he placed in separate groups the four halogens; sulphur, selenium and tellurium; and arsenic, phosphorus and antimony.[88] Cannizzaro rightly credited Dumas with a useful early classification of elements but said that this was merely perfecting 'the method of comparison used by Gay-Lussac in a more arid and less attractive form in his courses at the Jardin des Plantes [i.e. the Muséum]'.[89]

*Scientific method*

The principal difficulties of studying scientific method in a historical context are well known: most of what has been written on the subject has not been by practising scientists and when scientists themselves try retrospectively to rationalise their activities what they say they were doing is often not what they actually did. Hence, although one obviously turns to the writings of Gay-Lussac as the basis of any analysis, one must beware of accepting general pronouncements on methodology without supporting evidence. If he did leave some useful advice on scientific method, did he practise what he preached?

To state that Gay-Lussac's scientific method was based on experiment would be fatuous. It was the contribution of the seventeenth century rather than the nineteenth century to make this advance. His denouncement of the *a priori* method is by no means original but his way of saying that scientific theories must be based inductively on experimental data is worth quoting:

In the natural sciences, and particularly in chemistry, generalities must come after the detailed knowledge of each fact and not before it. It is really only after having acquired this knowledge that one may see if the facts have anything in common ('s'ils ont un type commun') and only after that is it permissible to consider them in a general manner.[90]

The presentation of several of Gay-Lussac's memoirs as well as his statement that 'laws are necessarily derived from the observation of a large number of facts'[91] might give the impression that his work provides a good example of induction. But one must distinguish again between the context of discovery and the context of justification. Gay-Lussac's usual process of discovery agrees fairly well with the modern hypothetico-deductive account, but in his presentation of evidence he would collect instances to convince the reader.[92] Thus in the law of combining volumes of gases two instances of simple preparations were probably sufficient for him to guess that other gases might combine in this way and further experiments were carried out to confirm this preliminary hypothesis. In writing his memoir, however, his argument is based on a wider variety of evidence which leads inevitably to the conclusion of simple ratios.

Gay-Lussac's experimental approach was different from many of his contemporaries. Whereas chemists such as Vauquelin were usually content to prepare new chemical species and record their properties, for Gay-Lussac the important thing was to determine *relationships* between phenomena. Ideally relationships were expressed in quantitative terms and might in their most general form express a correlation between a chemical property and some measurable physical or chemical characteristic.

Apparent connections between phenomena could, however, be spurious and any experimental value to be reliable should be the mean of several observations. In case of doubt, experiments should be repeated again and again:

One cannot repeat experiments too often when the problem is one of determining a relationship.[93]

Gay-Lussac's quantitative approach demanded precision, a quality which he was able to bring to most of his work. However, his quest for accuracy was less concerned with myopic scrutiny of instruments than with general principles:

We are convinced that exactitude in experiments is less the outcome of faithful observation of the divisions of an instrument than of exactitude of method.[94]

The importance of laws in physical science was constantly urged by Gay-Lussac and the text of his lecture course in the 1820s provides an opportunity of hearing this view spelled out:

In the study of physics, we see what are called individual facts but which are by no means isolated and which are not independent of each other;[95] on the contrary they are related to each other by laws which the physicist devotes all his attention to discovering ('le physicien met tous ses soins à découvrir'). It is this which is a measure of the true progress of the science.[96]

Gay-Lussac continued in a vein which contrasted his view of science with that of some of his contemporaries. He was not interested in isolated phenomena, or rather he was interested in how any newly discovered phenomenon would fit into the current conceptual scheme. If it did not, then it might be worthy of further investigation:

A fact is not novel if it has an analogue which could have some interest. A fact which does not fit in with a series of known facts is a fact which deserves particular attention. If the mind had to retain all individual facts, it could not manage and science would not exist; but when these facts can be connected by general laws and by theories, when a large number of these facts can be represented by a single one, one can remember them more easily, one can generalise one's ideas, one can compare one general fact with another general fact and discoveries can succeed each other. It is only when laws can be introduced into a science that it assumes the true character of science.

Once more we see the importance Gay-Lussac attached to laws. This chapter has shown that this was not just an ideal mentioned in a lecture, it was a consuming passion which guided the young researcher and led him on his path to fame.

# 4

# Collaboration and rivalry

'A discovery is the product of a previous discovery and in
its turn it will give rise to a further discovery'

Gay-Lussac[1]

'In order to draw any conclusion. . .it is prudent to wait
until more numerous and exact observations have provided a
solid foundation on which we may build a rigorous theory'

Gay-Lussac[2]

The 'chemical revolution' of the late eighteenth century had the effect
not only of expanding the field of scientific endeavour but of focussing
the attention of bright young men of the time on that particular branch
of science.

Within three years of the year of birth of Gay-Lussac (1778) there
were born four other men who were each to establish a considerable
reputation for themselves in chemical science: Amedeo Avogadro
(1776–1856), Jöns Jacob Berzelius (1779–1848), Humphry Davy
(1778–1829) and Louis Jacques Thenard (1777–1857).[3] All came to
have a great respect for Gay-Lussac although most of them had little
direct contact with him. Most remote was Avogadro, on the far side
of the Alps condemned to obscurity and isolation from the main centres
of scientific research and publication. Yet Avogadro's great claim to
fame ('Avogadro's hypothesis') is a direct inference from the Frenchman's law of combining volumes of gases, a connection fully and
explicitly acknowledged by the Italian. The Swede, Berzelius, too was
to derive inspiration from the publications of Gay-Lussac, whom he
was to meet during his stay in Paris in 1818–19. Yet, despite the great
influence which Berzelius exerted on chemistry in post-Napoleonic
Europe, there can be little doubt that Gay-Lussac's work had greater
repercussions for Berzelius than vice versa. For two great contemporaries, holding each other in the highest esteem and working in the
same science, their mutual exchanges were surprisingly slight.[4] That
the Swede had a natural predilection for mineralogical chemistry –
hardly one of Gay-Lussac's passions – is only a partial explanation of
their remoteness from each other.

In the first fifteen years of the nineteenth century two Englishmen, Davy and Dalton, were repeatedly working on problems studied by Gay-Lussac. Sometimes this was done independently, but there were several instances when some research, started on one side of the English Channel, was taken up immediately the first news reached the other side. Finally, there was Gay-Lussac's compatriot, colleague and friend, Thenard. He worked on many of the same problems but he was by no means an identical twin of Gay-Lussac, and differences of interests and approach may help us to understand our subject as much as the contrast with the two Englishmen.

Thenard, who collaborated intensively with Gay-Lussac in the year 1808–9, had more humble social origins than his friend and, although he was made a baron under the Bourbon Restoration, he never completely shook off his peasant origins. However, more instructive than the social contrast with the typically middle-class Gay-Lussac are their different educational backgrounds. Thenard had originally aspired to become a pharmacist, and when he came to Paris he entered the circle of the pharmacist Vauquelin and his friend Fourcroy. One of the principal interests of his mentors was animal chemistry and Thenard's early research in organic chemistry clearly reveals their influence. Gay-Lussac, on the other hand, only entered organic chemistry after he had done important research in other fields and his work is forward-looking, in marked contrast to the traditional animal chemistry of the young Thenard. Gay-Lussac's rigid mathematical training at the Ecole Polytechnique and his close association with Berthollet gave him quite a different view of the purpose of chemistry. Fourcroy was content to carry out and report a succession of analyses in the almost Baconian conviction that science advanced by the multiplication of data. Berthollet, however, insisted that 'chemistry could not be a simple collection of incoherent observations'.[5] He was concerned with understanding chemical reactions rather than preparing new products. Thenard probably considered that the greatest glory to which a chemist could aspire was to discover an exciting new substance, whereas Gay-Lussac, as the protégé of Berthollet and Laplace, thought that his greatest glory was to discover a fundamental principle or law.

Thenard therefore belonged by his original training to what has been called the 'natural history' tradition of chemistry, in which specimens were collected and classified. His association with the Ecole Polytechnique was not as a student but as a member of the teaching staff, thus missing the invaluable mathematical training, in which this famous school was second to none. In his original position as demonstrator to Fourcroy, he stood at the opposite pole to the mathematical approach of Monge and Laplace. Yet when he was drawn into Berthollet's circle in about 1804, he began to imbibe some of the physical approach to chemistry which was characteristic of the Arcueil circle and the influence of Berthollet is to be seen even in his work in organic chemistry. The competence which Thenard had acquired in mineral analysis under the able tutelage of Vauquelin and Fourcroy, remained a valuable skill after he had transferred his allegiance to the Arcueil group. Thus Thenard, although quite different in background to Gay-Lussac, was a most valuable colleague. It could be argued that, in so far as Gay-Lussac became a preparative chemist in the years after 1808, it was largely through the stimulation and help of his friend. Davy, speaking in his later years of the collaboration of Gay-Lussac and Thenard in the period 1808–11, commented that the two French chemists had formed for themselves a single reputation but had no doubt that Gay-Lussac was 'the superior man'.[6] Because of their joint publications over several years, it is impossible to be certain that any particular part of their published research belongs to either exclusively. It is, however, fairly easy to see the hand of Gay-Lussac in much of their quantitative work and especially in certain specific topics, such as gaseous volumes, chemical laws, etc.

## Rivalry with Davy

One cannot discuss the joint work of Gay-Lussac and Thenard for very long without considering Davy. I think that Davy was such a powerful stimulus to the work of Gay-Lussac that it is not only instructive but essential to consider the story in some detail. One might ask what difference did it make to Gay-Lussac that, during the Napoleonic wars a brilliant and imaginative young Englishman, by a series of spectacular discoveries, was threatening to eclipse the French lead in chemistry? Whatever Davy turned his hand to, whether nitrous oxide ('laughing gas') or the alkali metals, it was headline news. Scientists on the continent of Europe might ignore other British chemists: Wollaston, Smithson Tennant, or even Dalton, but Davy demanded attention.

In the standard histories of chemistry any discussion of Davy which has mentioned Gay-Lussac, or discussion of Gay-Lussac which has mentioned Davy, has tended to concentrate on questions of priority. Although we cannot ignore priority disputes, they hardly seem crucial. What is much more important is the effect which one man's work had upon the other. Something must be said of the detailed research of each and what clues provided by the one helped the other in his research. Above and beyond all this, however, I would suggest that at two vital points in Gay-Lussac's career, the winter of 1807–8 and the month of November 1814, news of Davy's work served both as an immediate source of inspiration and as a challenge to the French chemist. It is important here to remember that Davy was the same age as Gay-Lussac and had come to be considered as the leading British chemist. Davy was neither a beginner, whose first research – if noticed at all – might be given condescending praise, nor an elder statesman whose work was deserving the particular respect due to long service or old age. The two occasions referred to above are the news of Davy's electrolytic decomposition of potash and soda in 1807 and Davy's work on iodine seven years later. These each became focal points of the French chemist's research, or, to change the metaphor, springboards for further research. In the hands of Gay-Lussac and Thenard, potassium became a powerful new tool with which they carried out new reactions and analysed gases. Iodine was not only of great interest and value on its own account but because it led Gay-Lussac to further research on non-metals. He looked for analogous properties and I would suggest that his work on cyanogen derives from this. Although he was continually inspired by Berthollet and others of that generation, the case of Humphry Davy demonstrates how great may be the impact of one's peers.

Humphry Davy, born in Cornwall in 1778, had, in one of the first steps in his meteoric rise to fame, acquired something of an international reputation by the publication in 1800 of his researches into the chemical and physiological properties of nitrous oxide. This subject in itself formed a link with the French chemists, since Berthollet had worked on nitrous oxide, which he had prepared in 1785 by heating ammonium nitrate.[7] His views of its composition were corrected by Davy, who, however, expressed nothing but polite admiration for Berthollet, whom he included with Lavoisier, Guyton de Morveau and the British pneumatic chemists as the 'illustrious fathers of chemical philosophy'.[8] Davy referred not only to the work of Berthollet[9] but also to that of Humboldt,[10] since the latter had studied the absorption of nitrous oxide by water. Davy's book is a thoroughgoing study of nitrous oxide which could not fail to have impressed the Arcueil chemists of the talent of its youthful author. Berthollet wrote a long abstract-review of Davy's book,[11] in the course of which he criticised the English chemist's determination of the composition of nitrous oxide. Davy replied and Berthollet reported his reply in the *Annales de chimie*.[12] These exchanges in 1802 must have been known to the young Gay-Lussac. There was a falling off in Davy's research output over the next three or four years, no doubt related to his heavy duties at the Royal

Institution, but enough has been said to show that, when Davy's name came to the fore again in France in 1807, he was no stranger to the Arcueil group. Gay-Lussac for his part, was known to Davy for his very first published research, that on the thermal expansion of gases.[13]

The history of the chemical work of Davy in London and Gay-Lussac and Thenard in Paris in the years 1807–14 is one of rivalry and intense competition. The story might be thought to begin with Davy's famous isolation of potassium, announced in his Bakerian lecture of 19 November 1807. In fact the close relations between the two sides began when the First Class of the Institute decided to award Davy a prize for his Bakerian lecture of 1806. Given the slowness of communications and the time taken for the deliberations of a commission (appointed on 28 September 1807) it was ironically not until 7 December 1807 that the First Class announced the award of the prize for work on galvanic electricity to Davy for his lecture: 'On some chemical agencies of electricity'. The irony arises from the fact that by the time this award was announced, Davy had carried out new research which was more obviously deserving of the prize.

The members of the commission on the prize were: Haüy, Laplace, Hallé, Rumford and Gay-Lussac. Gay-Lussac had not previously served on such a commission and it proved to be no sinecure for him. He was asked to act as secretary and draw up the report. In this he drew attention to Davy's fundamental new theory that the force of affinity which held elements together in a compound could be overcome by electricity, thus providing a powerful new means of decomposing compound bodies into their elements. He seems to have thought that the potential effect of the battery depended on its size and we shall see that a few months later he was hoping by means of a more powerful battery to make new discoveries.

By the time Gay-Lussac came to read this report to a public meeting of the Institute on 4 January 1808,[14] he was able to add a postscript. In this he referred briefly to Marcet's letter of 23 November from London, telling his friends in Paris of Davy's decomposition of potash and soda. This letter gave some details about the remarkably reactive nature of the substance constituting the base of potash and which was to be called 'potassium'.

The above account makes clear how Gay-Lussac came to be involved in the appraisal of Davy's work in the winter of 1807–8. His friend, Thenard, was one of those particularly excited by the new discovery and they decided to collaborate on further research. They succeeded in preparing potassium and sodium by an alternative method, that of heating the appropriate alkali at high temperatures with iron filings.[15]

They thus put to good use Berthollet's views on the dependence of affinity on temperature. The reaction was carried out in a gun barrel and the vapour of the metal was condensed in a receiver attached to the barrel with a cement lute.

Although the makeshift apparatus and the preference for chemical over electrical methods of preparation may seem a step backwards, the advantage of this method was that it enabled the new metal to be prepared in workable quantities and at a comparatively low cost. Whereas Davy had prepared such tiny quantities of potassium and sodium that he could not investigate the new metals with any accuracy, Gay-Lussac and Thenard were able to present their colleagues in the First Class of the Institute with several grams of potassium at the meeting of 7 March 1808. They were thus able to determine the density of potassium with reasonable accuracy. Davy later paid them the compliment of using their method to prepare samples of potassium and sodium for his own experiments.

At the beginning of March 1808, when the French chemists had thus emerged on the Paris stage as possible rivals to Davy, the English chemist had just recovered from a serious illness, which had affected him for several months and had prevented him continuing his research. Davy's brother gave the following account:

upon his recovery, he found the subjects he had been investigating, seized upon by MM. Gay-Lussac and Thenard, rather in the manner and feeling of contending generals intent on conquest, than of philosophical inquirers, members of the common republic of science.[16]

John Davy possibly felt that some rules of chivalry should have been applied, perhaps an agreement not to carry out research while one of the parties was ill. This, however, was not to be, and when an account of this and further research by the two French chemists was published in the *Moniteur* of 27 May and was read by Davy, his brother says that 'it necessarily had the effect of hurrying on his researches'.

To pursue the minutiae of the work carried out on both sides of the English Channel over the next few years would fill a book in itself. It is appropriate here merely to summarize the contributions made by Gay-Lussac with the assistance of Thenard. Particularly worthy of note is their use of potassium as a reagent. Because of its extreme reactivity, the isolation of the new metal placed a powerful new tool at the disposal of chemists. It could be applied to the preparation of new elements and compounds and it was characteristic of Gay-Lussac that he should foresee its possibilities for the quantitative analysis of gases. If the exploitation of the possibilities of the voltaic pile for the preparation of new elements is justly attributed to Davy, it was Gay-Lussac and Thenard who exploited the potentialities of potassium. They began by investigating its reaction with other elements such as sulphur and phosphorus. By heating potassium in hydrogen they obtained the grey solid potassium hydride. Later they successfully decomposed ammonia by heating potassium in a sample of the gas which had been well dried and were thus the first to prepare the amide of a metal. Apart from new compounds they also obtained a new – though hardly spectacular – element, boron. Their description of the preparation of the new substance (to which they gave the name *bore* for the radical of boric or 'boracic' acid) may be worth quoting as an example of their experimental approach:

To decompose boracic acid, place equal parts of the metal [potassium] and very pure vitreous boracic acid in a copper tube to which a bent glass tube is attached. Place the copper tube in a small furnace with the end of the glass tube in a flask of mercury. When the apparatus is ready, heat the copper tube gradually until it becomes faintly red; keep it in this condition for several minutes; then the operation being ended allow it to cool and take out the material. This is what is seen in this experiment:

When the temperature is about 150 degrees, the mixture suddenly glows strongly, which appears in a striking manner if a glass tube is used. So much heat is produced that the glass tube melts slightly and sometimes breaks and the air is almost always driven out of the vessel with force. . . The metal [potassium] is used up decomposing part of the boracic acid; and these two substances are converted by their mutual reaction into an olive grey material which is a mixture of potassium, potassium borate and the radical of boracic acid. Extract this mixture in a tube by pouring water into it and heating slowly, and separate the boracic radical by washing with cold or hot water. That which does not dissolve is the radical itself.[17]

Their success in isolating the radical of boric acid encouraged Gay-Lussac and Thenard to try to find a way to isolate other radicals. They were frustrated in this attempt by practical difficulties, and in the case of fluorine, for example, a succession of attempts by various chemists to isolate the element were not successful until 1886 (Moissan). However they did prepare boron trifluoride ('fluoboric gas') and the striking reaction of this gas with ammonia may have been one of the clues that led Gay-Lussac to his law of combining volumes of gases. His use of potassium to decompose nitric oxide also provided valuable data to confirm this law.

The influence of the intense activity of Gay-Lussac and Thenard on Davy can be seen particularly in his fourth Bakerian lecture, read to the Royal Society in December 1808 with an appendix and extensions in the following February and March.[18] Davy referred repeatedly to an account by the French chemists in the *Moniteur* in May 1808 in which some similar experiments were described, saying that this was the only account of their work available in England. In fact Gay-Lussac and Thenard had published in a French journal of July 1808 experiments, which they said they had carried out on 21 June, in which boron trioxide was heated with potassium.[19] This was the essence of the method of preparing boron and Davy was informed about this work in August 1808 in a private letter from Paris, but he claimed that he had carried out the reaction himself before receiving this information.[20] The first published claim to the discovery of a new element was that by Gay-Lussac and Thenard in the *Moniteur* of 15 and 16 November 1808, a full month before Davy submitted a similar claim to the Royal Society. On the normal criterion of publication, therefore, Gay-Lussac and Thenard would seem to have priority. Among the many uncertainties in the study of this parallel work, one of the clearest things to emerge is how eagerly each side looked to the other for every scrap of information. Davy's research shows signs of haste and gives several indications that he feared anticipation.[21] His own subsequent claims to priority cannot be confirmed with documentary evidence.

From boron Gay-Lussac and Thenard went on to an examination and preparation of fluorine compounds and here Davy was more prepared to give them credit. This is what he later wrote in his text-book:

Concentrated hydrofluoric acid and fluoboric gas were made known by some elaborate researches of Gay-Lussac and Thenard in 1809. My brother, Mr John Davy, in 1810 and 1811 extended the knowledge of the properties of these bodies...[22]

No suggestion was made here that 'extending the knowledge of the properties of...bodies', isolated by other people, constituted interference. It was, of course, perfectly legitimate and, indeed, desirable. One cannot help feeling that after his initial single-handed venture against the 'Napoleonic chemists', Davy had enrolled his brother to balance the odds.

A more important case of the influence of the French chemists on Davy is that of chlorine. Brief histories of chemistry usually report that it was Davy who first recognised its elementary nature. Yet the full story is not quite so simple. At the beginning of the nineteenth century chlorine was known as 'oxygenated muriatic acid', or 'oxymuriatic acid', since it was thought to be a compound of oxygen and muriatic acid. This belief was based on its preparation by heating muriatic (i.e. hydrochloric) acid with a substance such as manganese dioxide which contained an abundance of oxygen. If 'oxymuriatic acid' was dissolved in water, bubbles of oxygen could be collected when the solution was exposed to sunlight. Moreover, many substances burned brilliantly in 'oxymuriatic acid gas' and oxygen was the only known supporter of combustion. It was naturally inferred that it was a compound containing loosely-combined oxygen. It was, therefore, a matter of astonishment to Gay-Lussac and Thenard that when they passed the 'oxymuriatic acid gas' over red hot charcoal, the supposed oxygen present refused to combine with the charcoal, although the latter is one of the best reducing agents. This led them to doubt that the gas contained oxygen and to suggest that it might be an element.

Davy in a paper read to the Royal Society in July 1810 made a brief historical reference to Scheele and Berthollet and then turned to comment on the detailed memoir of Gay-Lussac and Thenard, which had been published in the 1809 volume of the Arcueil *Mémoires* and a copy of which had been sent to him. He remarked:

one of the most singular facts that I have observed on this subject. . .is that charcoal, even when ignited to whiteness in oxymuriatic gas. . .effects no change in [it]. . .*This experiment which I have several times repeated led me to doubt the existence of oxygen in that substance.*[23]

Davy concluded that oxymuriatic acid did not contain oxygen. He suggested that it might be an element and in November 1810 he proposed for it the name *chlorine*.[24] According to Berthollet, Gay-Lussac, although agreeing to modify his original conclusions about the elementary nature of the gas in his paper read to the Institute, was already teaching the new view of its composition in his lectures of 1809.[25] As late as 1813, however, Thenard continued to assert that the gas contained oxygen.[26] I think that the published record shows clearly that Davy deserves credit for presenting chlorine unambiguously as an element but to do so he very properly made full use of the earlier research of the French chemists.

Gay-Lussac and Thenard studied the influence of light on the reaction between chlorine and hydrogen. The change in colour of the gases enabled them to follow the course of combination over several days in hazy sunshine, keeping meanwhile another mixture of the reactants in the dark as a control. In bright sunlight the gases reacted violently, shattering the flasks used. They also prepared new compounds with chlorine, including phosphorus oxychloride, $POCl_3$. Later Gay-Lussac announced the discovery of chloric acid by the action of sulphuric acid on barium chlorate.[27] Davy repeated Gay-Lussac's work and then varied the conditions and reactants, thus obtaining another new compound, chlorine dioxide.[28] On another occasion Davy admitted that his discovery of this new compound originated from 'a statement which. . .[had] been made by M. Gay-Lussac, namely that a peculiar acid, which he has called chloric acid, may be procured from the hyperoxymuriate of baryta by sulphuric acid'.[29]

In their various priority disputes with Davy, Gay-Lussac and Thenard were fortunate in having available the government journal, the *Moniteur*. Although mainly concerned with the publication of official versions of news together with proclamations, decrees, etc., it was prepared to publish items of scientific news at a time when British newspapers were concerned purely with political and social affairs. The *Moniteur* was described by John Davy in the following terms:

For many years this Paper was the Journal of Science as well as of war, presenting an incongruous mixture not uncharacteristic of the individual [Napoleon], of which it was the organ. During this period so little intercourse was there between France and England, that months often elapsed between the publication of a scientific discovery in one country and its being known in the other. And, perhaps, owing to this state connected with war, less delicacy was observed by men of science in the two countries in engaging in researches which they had not themselves originated. . .[30]

While it is true that there was a general feeling in the early nineteenth century that science was international, the men of the time would have been less than human if some national antagonism had not rubbed off on them. A deleted passage in the draft of one of Humphry Davy's lectures delivered in 1810 is quite explicit on this question of national rivalry:

The scientific glory of a country may be considered in some measure, as an indication of its innate strength. The exaltation of Reason must necessarily be connected with the exaltation of the other faculties of the mind; and there is one spirit of enterprize, vigour and conquest in science, arts, and arms.[31]

## The 'discovery' of iodine

Gay-Lussac did important work on iodine. Yet, unfortunately for the historian, the story is a complex one. Perhaps we should begin by dissenting from the popularly accepted view that for every element there is 'a discoverer' and it is the task of the historian of science to identify this person. History is seldom so simple; the person who isolates a new substance often does not appreciate the significance of it, which is left for a succession of later workers. So in the case of iodine it was the saltpetre manufacturer, Bernard Courtois, who noticed in 1811 that the mother liquor of varec produced a violet vapour when treated with sulphuric acid. From the vapour he was able to condense a small quantity of black crystals. However, saltpetre was in great demand for the Napoleonic wars and Courtois was too busy with his livelihood to take this research further.

An extreme view of the discovery of iodine was that of Davy's first biographer, John Ayrton Paris, who considered this a good example of the ignorance of the French chemists.[32] His conclusion was that 'if Davy had not visited Paris, iodine would have remained at the end of the year 1814, as it had been for two preceding years – the unknown X'.[33] Although a spirit of rivalry undoubtedly spurred on both Davy and Gay-Lussac, we shall see that Gay-Lussac began his research quite independently of Davy's visit. The story begins with Courtois, who, being too preoccupied to undertake a full investigation of the new substance, had passed it on to his young friend, Nicolas Clément. Clément in turn had shown the substance to a few friends, including Ampère in the summer of 1813 but had not taken the investigation much further.

It is here that the prestige and procedures of the First Class of the Institute enter the story. The prospect of a vacancy among the corresponding members of the First Class was an incentive for Clément to carry out further experiments, which he did in collaboration with his father-in-law, Charles Bernard Desormes. These they presented to the First Class on 29 November 1813 in a short memoir in which they stated, for example, that the new substance formed 'muriatic acid' with hydrogen.[34] The recognition of this gas as hydrogen iodide was only to come when Davy and Gay-Lussac followed up this research. As far as Clément, however, was concerned, he had reminded the First Class of his existence in the provinces and (he hoped) of his ability. At the very same meeting at which he read the memoir it was agreed to proceed to the election of a corresponding member and when, the following week, a list of candidates was drawn up in the customary order of merit, Clément and Desormes appeared in the third and fourth places.[35] Pride of place at the top of the list, however, was given to their distinguished visitor, Sir Humphry Davy.[36] With such competition poor Clément did not stand a chance of election. Davy was elected by forty-seven votes out of forty-eight, an almost unanimous vote which demonstrates the high regard of the French scientists for Davy's talents.

Gay-Lussac enters the story only at the point where Clément read his memoir to the Institute, i.e. 29 November. The usual practice of the First Class was to appoint a commission of two or three members to examine a memoir presented by a non-member; in this case Gay-Lussac and Thenard were chosen. There is no question of these chemists trespassing on the research of Clément and Desormes. The latter two, as outsiders, had submitted their work for inspection by the academic body and it was the duty of Gay-Lussac and Thenard, as commissioners, to repeat any new or doubtful experiments described in the paper under scrutiny. In this case it meant undertaking an investigation of the properties of iodine. Meanwhile Davy had already had an opportunity of examining the new substance. Clément had previously given some to Ampère, who passed a sample on to Davy on 23 November, almost a week before the First Class became involved.

What now of the relative positions of Gay-Lussac and Davy? When a second worker does research on a subject opened by a first, the question of 'poaching' arises. The rules governing such duplication are not clear but both Davy and Gay-Lussac came to feel that the other man was poaching on his territory. Davy began his experiments on iodine a few days before Gay-Lussac had been shown the substance but it was quite independently of Davy that Gay-Lussac was involved. The main complication in the story of iodine is a moral dimension related to the rules and conventions of hospitality. As a courtesy to a visitor, Clément, or more strictly Ampère, had shown Davy the new substance as they thought it would interest him. Davy later said that Clément had *requested* him to examine it.[37] This is to suggest that Clément and Ampère had asked a favour, whereas it seems much more probable that they had granted one.

According to Davy's first biographer,[38] when Davy arrived in Paris he expressed the wish to be introduced to Ampère 'whom he considered as the only chemist in Paris who had duly appreciated the value of his discoveries'. It seems, moreover, that Davy took no care to conceal this opinion, which naturally caused some surprise, not to say resentment, among the Paris savants. Davy's superior manner also caused some offence.[39] Gay-Lussac for his part was not blameless. Although he called on Davy to present his compliments, his manner was probably cool, in marked contrast to the admiration the English chemist had come to expect and no doubt received from more junior chemists or those outside the establishment. We shall see how, once involved in the excitement of the iodine chase, Gay-Lussac gave little thought to the rights of his British rival but used his privileges, and especially his right of publication in the official *Moniteur*, to maximum effect.

It is time to consider Davy's activities in more detail. He had been in Paris since the end of October. A few days before Clément and Desormes read their paper to the First Class, he had been presented with a sample of the new substance by these chemists then on the periphery of the French scientific establishment. The occasion was described by Davy's assistant, Michael Faraday, in the following terms:

Tuesday 23rd [November 1813] MM. Ampère, Clément and Desormes came this morning to show Sir H. Davy a new substance, discovered about two years ago by M. Courtois, saltpetre manufacturer. The process by which it is obtained is not yet publicly known. . .A very permanent and remarkable property of this substance is that when heated it rises in vapour of a deep violet colour. . .Sir Humphry Davy made various experiments on it with his travelling apparatus, and from them he is inclined to consider it as a compound of chlorine and an unknown body.[40]

Perhaps Davy was lacking in tact when he continued his investigations into this new substance under the very noses of his hosts in Paris. When he later wrote up his research for the Royal Society, Davy felt it necessary to justify it; he said that Courtois had abandoned his research at an early stage and his successor, Clément, had not taken the work very much further.[41] He conceded, however, that Gay-Lussac was then working on the substance and had already presented to the Institute a memoir on the subject on 6 December, i.e. only a few days before Davy wrote his memoir. He excused himself on the grounds first that an investigation of the new substance and its compounds was enough to occupy several people and secondly that it was valuable to have another approach to the subject from a different point of view. In a private letter to his brother he claimed to have Clément's permission to publish.[42]

The real conflict in this situation is masked by consideration of nationalities. The fundamental division was not (as has sometimes been suggested)[43] between a lone Englishman in a hostile country and the French chemists, but between those outside the French scientific establishment and those who represented it. Davy, who considered that he had not been treated with proper deference by the leading French chemists and particularly the Arcueil group, was a natural ally of those like Clément who had not been able to gain recognition for their work. In 1802 Clément had crossed swords with Berthollet, who may have retained a lasting dislike of him.[44] Clément and Desormes had entered for the 1812 prize of the First Class of the Institute on the subject of the specific heats of gases. The competition was judged by a commission of five, including Berthollet, Gay-Lussac and Thenard.[45] The fact that they decided to award the prize not to Clément and Desormes but to Berthollet's protégé Jacques Etienne Bérard and his friend François de la Roche may have been interpreted by the unfortunate party as nepotism. It was as a member of the official body of French science that Gay-Lussac first examined iodine. The normal procedure would have been to report back to the First Class on the merits of the experimental work described. When Davy was known to be working on this, however (and Faraday's diary mentions four different occasions), the usual form became forgotten. The memoir Gay-Lussac was examining was mediocre but it opened up exciting possibilities which he could not resist. It was not so much for France as for official science and for himself that Gay-Lussac was working. It is interesting that at the stage when Gay-Lussac was obviously exceeding his mandate as a member of the Institute commission he, like Davy, sought justification in saying that Clément had asked him to do it.[46]

Davy was next reported by Faraday to be working on iodine on 30 November, i.e. the day after Clément had presented his paper to the Institute. Gay-Lussac was now involved and the old spirit of rivalry returned. Neither of the great chemists made any public announcement about their research for another week and then on Monday 6 December Gay-Lussac read an account of his research to the First Class. We may reasonably suppose that he had been working hard all that week on the new substance, which he now proposed to call *iode*.[47] He reported a number of experiments from which he concluded that *iode* was probably an element. A second significant achievement of this paper was the information that *iode* combined with hydrogen to form an acid for which the French chemist proposed the name *acide hydriodique*. The only thing which calls in question Gay-Lussac's achievement is Davy's later claim[48] that it was he who had told Gay-Lussac of the peculiar nature of the compound of hydrogen and iodine. Gay-Lussac, on the other hand, claimed that *he* had given this information privately to Davy.[49] It is impossible to resolve conclusively these conflicting claims. It may be that each discovered hydriodic acid independently, but there were enough hints in the air for each side to suppose that this was the basis of the other's work. The text of Gay-Lussac's memoir was published in the government newspaper, the *Moniteur*, on Sunday 12 December. By securing publication first, Gay-Lussac was given credit in French eyes for the fundamental research on iodine. Davy's first public statement on the subject was made at a meeting of the First Class on the following day, Monday 13 December. This was in a letter addressed to Cuvier, one of the permanent secretaries of the First Class.[50] Davy stated that his motives in writing the letter were to forestall Gay-Lussac publishing in the *Moniteur* results which the French chemist had learned from him. The letter carried the date of the previous Saturday, 11 December,[51] but Gay-Lussac said that he later overheard Cuvier apologising to Davy for only having read his paper at the end of the Monday meeting *because he had only received it late that very day*.[52] For Gay-Lussac this was evidence that the letter could not have been *sent* until the Monday, i.e. after his own research had been published. Unfortunately for the historian who wishes to settle the matter once and for all there is no independent witness that Davy's work was completed before the Sunday, although Davy would obviously have done most of his experimental investigation before then.

A comparison of Davy's long letter and Gay-Lussac's memoir[53] shows certain similarities. Both Gay-Lussac and Davy stressed the analogy of the new substance with chlorine. Davy's letter gives an interesting preliminary report on 'the new substance' which he thought was 'an undecompounded body' and for which, he said, the name *ione* had been proposed in France. This was the name proposed by Gay-Lussac on 12 December.[54] In Gay-Lussac's memoir the substance was considered as an element and its general chemistry was described together with that of its hydride; proposals were made for naming each. The French chemist's memoir occupied two columns in small type of one of the large pages of the *Moniteur*; it may be considered as one of the most important scientific memoirs ever published in a general newspaper.

Gay-Lussac, however, was only at the beginning of his work on iodine. Davy's decision to report his work to the Institute merely provoked the French chemist into further research. At the next meeting of the First Class on 20 December Gay-Lussac read a second memoir on iodine, in which he announced the discovery of an oxy-acid, the potassium salt of which (potassium iodate) had already been prepared by Davy. Gay-Lussac encouraged further work on iodine in his laboratory by his assistants. On 27 December, just two days before Davy left Paris, J. J. Colin, demonstrator at the Ecole Polytechique, presented to the First Class of the Institute a memoir on some new compounds of iodine.[55] In the following March, Colin and Gaultier de Claubry investigated the reactions between iodine and a number of organic substances.[56] It was in the course of this research that they discovered the blue colour produced by starch and iodine, a reaction which has ever since then been used as a standard test for either starch or iodine.

In 1814 Davy did some further research on iodine but his later achievements were only a pale reflection of his former brilliance. In prolonged and detailed research he was outshone by Gay-Lussac, who presented a full study of the new element in a major memoir read to the Institute on 1 August 1814. When published, its 155 pages filled an entire number of the *Annales de chimie*. In the late nineteenth century Ostwald selected this memoir as a prime example of a classic in science. He said that it was

one of the first and one of the best monographs of all time on a single element and its most important compounds and as such it has served as a model for many later pieces of research.[57]

It was probably the subject matter as much as the treatment of it by Gay-Lussac which caused the normally restrained German editor L. W. Gilbert to suggest that the memoir transports the reader into a sort of fairyland.[58] He could not think of any fairy tale from his childhood which had astonished and enchanted him more by its dimension of apparent magic than Gay-Lussac's study of the chemical history of iodine. When Gilbert wrote this he was obviously not thinking of the careful recording by the French chemist of specific gravity, melting point, boiling point and solubility in water nor of the later quantitative treatment of the compounds of iodine. In order to understand Gilbert's remarks the twentieth-century reader would have to forget his modern sophistication and try to recapture the childlike wonder which might be engendered on first seeing the black crystals of iodine transformed by heat into a beautiful violet vapour which would have no equal among contemporary dyestuffs. He might remember too the rich variety of coloured precipitates which can be produced by iodine compounds. The aesthetic implications of the new element were hardly less than the theoretical.

After the earlier hurried work of Davy and Gay-Lussac, the main value of the French memoir of 1814 is probably its comprehensiveness. John Davy generously summed up the contribution made by the memoir to chemistry by saying:

in less than twelve months, chiefly in consequence of the elaborate and masterly researches of M. Gay-Lussac, the chemical history of iodine was more full and complete than that of most other substances longest known.[59]

Gay-Lussac emphasised throughout his memoir the similarities between iodine and chlorine. Davy too had seen this important analogy, but whereas he also compared iodine to oxygen, the French chemist considered sulphur a more appropriate analogue:

I...place it, in consequence of the experiments which I have made, between sulphur and chlorine; because its affinities are stronger than the former and weaker than the latter of these bodies. Like them it forms two acids, one by combining with hydrogen, and another by combining with oxygen; and most of its combinations have considerable analogy with those formed by these two bodies.[60]

Later in the memoir he returned to the same subject, saying that iodine had less 'energy' than chlorine but more than sulphur.[61]

Gay-Lussac had already used the reaction between iodine and phosphorus to prepare hydrogen iodide. He now made a more detailed study of the gas, which he showed to contain half its volume of hydrogen. He studied its thermal decomposition and solubility in water. He described the preparation of a concentrated solution of the acid and noted its high boiling point. He noted that the acid reacted with a solution of a lead salt to produce a fine orange precipitate, with mercury to produce a red precipitate and with silver a white precipitate insoluble in ammonia. He described the iodides of most of the known metals, noting that there were two iodides of mercury. In the words of his English translator:

In general there ought to be for each metal as many iodurets as there are degrees of oxidation.[62]

He was particularly interested in the reaction with zinc, which enabled him to calculate a value for the equivalent of iodine.[63]

Gay-Lussac then turned to the compounds of iodine with non-metals. He was particularly interested in nitrogen tri-iodide; he did not use this name but noted its composition in volumetric terms to be one of nitrogen and three of iodine. After discussing the reactions of the new element with oxides, he gave further attention to his earlier discovery of iodic acid, which he was now able to show contained five 'proportions' of oxygen.[64] He gave a detailed quantitative treatment of the salts of the acid, noting, as always, crystalline form. He was able to describe the preparation of several new compounds including iodine monochloride and trichloride. By distilling ethyl alcohol with very concentrated hydriodic acid he was able to prepare ethyl iodide for the first time.

### Differences of style and character

There were many differences of temperament between Gay-Lussac and Davy. Both were ambitious young men, keen to make their names in the annals of science and they both worked in a similar area. But whereas Davy acquired fame because of the originality of his work, a study of Gay-Lussac shows that he was often reluctant to commit himself to an original viewpoint and was usually satisfied to collect new data. To go beyond and *interpret* experimental results was a procedure fraught with danger. Thus he concluded a memoir of 1807 by saying:

I only present these conclusions with the greatest reserve, knowing myself how I have still to vary my experiments and how easy it is to err in the interpretation of results.[65]

It was not only characteristic of Gay-Lussac that in nearly every memoir he should stress the provisional nature of his conclusions but also that he should avoid deciding between alternative hypotheses. Someone who makes a practice of sitting on the fence is in danger of losing credit for being on the winning side. So it was in the case of chlorine. Gay-Lussac here presents two points of view:

Oxygenated muratic acid is not decomposed by charcoal, and it might be supposed, from this fact and those which are communicated in this Memoir, that this gas is a simple body. *The phenomena which it presents can be explained well enough on this hypothesis*; we shall not seek to defend it, however, as it appears to us that they are still better explained by regarding oxygenated muriatic acid as a compound body.[66]

Although this last sentence begins in a way typical of Gay-Lussac its conclusion shows a conviction almost alien to him and we know that the wording was changed at a private meeting at Arcueil after the forceful intervention of Berthollet who refused even to consider the possibility of chlorine being an element. It was, therefore, left to Davy to state this less equivocally and receive the credit.

As a second example of Gay-Lussac's caution in presenting his conclusions we may cite his work on iodine. He was almost certain early in December 1813 that this was an element but he refused to state this unequivocally. His conclusion is therefore worded as follows:

All the phenomena of which we have spoken may be explained by supposing iodine is an element and that it forms an acid when it combines with hydrogen; alternatively this acid is a compound of water and an unknown base, and iodine is the same base combined with oxygen. In the light of the preceding facts *the first hypothesis* [i.e. that iodine is an element] *seems to us more probable than the other* and at the same time it serves to give greater probability to the hypothesis according to which oxymuriatic acid is considered an element. If we adopt this hypothesis, the name which would suit the new acid would be hydriodic acid.[67]

Our conclusion may be both personal and methodological. Gay-Lussac's attitude may betray some timidity but it also suggests that he thought that science was necessarily concerned with certain knowledge. Although Auguste Comte had not yet published his work on positivism, one sees in Gay-Lussac, as in several of his French contemporaries, some positivistic features and a total rejection of speculation. The proper role of a scientist was to present his evidence and then show how this could be interpreted. Gay-Lussac was like a judge who had the duty of summarising both the case for the prosecution and the case for the defence. This was then presented to the jury which was to decide the case.[68]

Although Davy was generally more confident in his style than Gay-Lussac there were naturally many instances when on the frontiers of research he was unsure of his ground. His methods of safeguarding his position or covering himself are interesting and sometimes provide a fascinating contrast with those of Gay-Lussac. The French chemist would tend to sit on the fence and gain security at the price of sacrificing possible triumphs. Davy sometimes adopted the tactic of backing several possibilities simultaneously. At the very least he would speculate and later claim credit for those predictions which subsequent experiments had confirmed. Thus he began his Bakerian lecture of 1807[69] by saying that he was now able to provide facts to support ideas on the power of electricity to isolate elements which, he claimed, had been his conclusion to his Bakerian lecture of the prevous year. If we look at the Bakerian lecture for 1806 we find a rather long and diffuse concluding section.[70] The idea mentioned, although brilliant, was no more than one of a number of speculations thrown out at the end. Davy rescues this speculation but we hear no more of some other more bizarre ideas mentioned at the same time. In the same way even to-day Davy tends to be remembered for his (correct) identification of chlorine as an element, while his (incorrect) views on the compound nature of the elements nitrogen, sulphur, phosphorus and carbon have conveniently been forgotten.[71]

Davy scattered seeds in all directions but did not neglect to gather up the best of his crop. Gay-Lussac, on the other hand, sowed more systematically but failed to reap the potential harvest. If Davy was sometimes too bold, Gay-Lussac was often too cautious. Davy always believed himself capable of pointing out new paths. Gay-Lussac, more modest by temperament, did not have the same self-confidence. He felt he could not afford to be wrong and therefore refused to commit himself time and time again.

One wonders why Gay-Lussac should have been so reluctant to make innovatory claims. Part of the answer is, I think, that his training at the Ecole Polytechnique and at Arcueil had impressed on him the authority of the leaders of French science and particularly of Laplace and Berthollet. He felt the weight of this authority long after his mentors were dead.[72] Yet in most of his scientific research there was no question of him contradicting their work. In the end, what Gay-Lussac feared was the responsibility of innovation. Within the framework of established ideas he could work without assuming any great burden. But to make a claim is to commit oneself; it is to assume responsibility and Gay-Lussac may have felt that his responsibility was all the greater as a member of the small professional scientific community based on Paris. An obscure figure in the provinces in France could afford to risk making a mistake. He had nothing to lose and he had much to gain by attracting the attention of the powerful academic politicians in the capital. Also Gay-Lussac, working within a French tradition which respected sound experimental work rather than general theories, knew that he was safe in reporting his carefully executed laboratory work and would be respected for this alone.

The great exception to the generalisation we have made is Gay-Lussac's predilection for general laws of nature. Had he been someone who never went beyond laboratory data he would not have been a great scientist. His claim to our attention in this biography does not rest on his work on potassium, chlorine, etc., but on his study of relations between various physical and chemical phenomena and the enunciation of several basic scientific laws. Nor is there any contradiction here. Gay-Lussac's conception of an order in nature was sufficiently strong to overcome his normal reluctance to go beyond the data.

Another difference between Gay-Lussac and Davy lay in Gay-Lussac's professionalism. Despite Davy's early association with the Royal Institution, he felt it a relief to abandon his employment to live the life of a gentleman. Davy aspired to a social position which in England at this time was incompatible with salaried employment. In 1812 he married a wealthy widow. Berzelius thought that Davy could have been one of the geniuses of all time – if only he had had to study chemistry systematically in his youth. However he was largely self-taught and he contributed only a few brilliant fragments to chemistry. Gay-Lussac, on the other hand, had undergone a specialised scientific education at the Ecole Polytechnique and had then held what in modern terms might be considered as a research assistantship with Berthollet. Gay-Lussac eagerly sought out the salaried posts available in post-revolutionary France for scientists. He thus lost the kind of independence which Davy enjoyed. He lost the freedom to explore any area of the natural world from any point of view which took his fancy. But he gained a certain rigour and discipline lacking in the British dilettante approach.

Gay-Lussac's personality was remarkably different from that of Davy and it is tempting to use the terms 'classical' and 'romantic' in order to bring out two contrasting attitudes to science. Gay-Lussac's love of order and regularity contrasts with Davy's impressionistic and more random incursions into science. Gay-Lussac's measured style and prose of Cartesian clarity contrast with Davy's rhetoric and high-flown poetic imagery. Davy, like many romantic poets, had a short creative life;[73] his important work was confined to the period of his twenties and early thirties. His imagination soared and embraced numerous possibilities a few of which could later be substantiated. Gay-Lussac worked more slowly and steadily for a longer period of time. His mind ran on more predictable lines and he saw science as a taskmaster demanding hard work and sometimes great patience. Thus in 1815 he passed at least 50 000 sparks through a mixture of the vapour of prussic acid and hydrogen.[74] The cult of the ego or vanity of the romantic may be contrasted with Gay-Lussac's self-effacing modesty. Davy's metaphysical intoxication stands in sharp contrast to the positivistic sobriety of the Frenchman. Davy's emphasis on feeling and spontaneity differs from Gay-Lussac's concern with reason and considered action. Davy took a delight in the peculiar properties of new elements. Gay-Lussac was probably less interested than Thenard in this aspect of chemistry. The general laws of nature were more important to Gay-Lussac and we have seen how the quest for such laws dominated his scientific research.

# 5

# The volumetric approach

'Gay-Lussac and the leading French chemists adopt the
language of volumes combining'

John Dalton (1830)[1]

Gay-Lussac's most important pieces of research were either concerned
with gases and their volumes or with preparative chemistry which
involved volatile compounds such as hydrogen iodide and hydrogen
cyanide. His volumetric approach was fundamental, a *leitmotif* which
reappeared throughout his research and particularly in his most creative
years from 1802 until the early 1830s. We shall see that it constituted
at the same time an enormous strength and also a great weakness. On
the one hand Gay-Lussac discovered fundamental volumetric laws and
used them imaginatively in probing the mysteries of chemical composition. However, once committed to volumes as the language of experience, he was insensitive to the possibilities of an alternative gravimetric
viewpoint.

The volumetric approach to matter of Gay-Lussac and his successors
could easily be overlooked today. Yet as one of the principal methods of
investigating the basic problems of chemical composition and reactions,
it influenced much of the chemistry of the first half of the nineteenth
century. When there were so few keys to the understanding of physical
and chemical units, it provided a valuable means of approach and one
which could claim to be solidly based on experimental evidence. In
any sketch of the history of chemistry, writers tend to pass directly from
Dalton and Avogadro (1808, 1811) to the acceptance of a standard
system of atomic weights at the Karlsruhe conference half a century
later. If, however, we are to understand anything of chemical problems
in the years before 1860, the volumetric approach is one which cannot
be ignored. It constituted a major tradition in chemical thought,
beginning with Gay-Lussac and embracing Berzelius, Mitscherlich,
Dumas, Gerhardt, Laurent and Cannizzaro, to mention only some of
the better known names. In England A. W. Hofmann and Odling in
their lectures and text-books emphasised the volumetric aspect of
chemical composition. Even for modern scientists, who may not be

interested in nineteenth-century problems, and who adopt a strictly twentieth-century viewpoint, Gay-Lussac's law of combining volumes of gases, his vapour density determinations and his development of the volumetric analysis of solutions can be appreciated as basic contributions to modern physics and chemistry.

If the reader allows himself to think fancifully for a moment from a modern viewpoint of the basic problems of chemistry in the early nineteenth century, we might suggest a situation in which there were several 'players' doing a jigsaw but each handicapped by missing pieces. Dalton had atoms (and equivalents) but not molecules; nor did he have volumes nor vapour density. Gay-Lussac had volumes, vapour density and equivalents but not atoms and molecules. Avogadro had the concept of atom and molecule (although not clearly expressed); following Gay-Lussac he made use of combining volumes but not of vapour density. Berzelius was able to go further by drawing on both Dalton and Gay-Lussac. Later Gerhardt and finally Cannizzaro were able to bring together all the pieces which, although available earlier, had never all been in the hands of one man who appreciated their value in relation to each other. In this complex story there is no single hero but there are characters of greater and lesser importance and Gay-Lussac, by contributing *two* important elements, volumes and vapour densities, played a major role.[2]

Gay-Lussac's concern with volume may reflect a late strand of French Cartesianism. Descartes had defined matter in terms of extension alone and, although many of his ideas had been displaced by those of Newton, there lingered on in the eighteenth century, particularly on the continent of Europe, ideas of matter different from the 'hard massy atoms' of Newton. Euler, whose popular *Letters to a German Princess* was in Gay-Lussac's library, was prepared to consider the possibility of bodies without mass.[3]

Haüy's *Traité élémentaire de physique*, which was also in Gay-Lussac's library, started off by mentioning extension as a fundamental property of matter. The author remarked that this question had been discussed endlessly by philosophers but what interested physicists was to be able to measure the extension of matter rather than bothering themselves over definitions.[4]

Some idea of Gay-Lussac's views on mass and volume may be obtained from his physics lecture course of the 1820s. He mentioned extension as the first general characteristic of bodies, saying that no one doubted that bodies had dimensions.[5] It came first probably through perception but it was also a fundamental property of matter. When he came to speak of weight (*pesanteur*) he said that, although weight was a general property of matter, it was possible to abstract this property and conceive of it as not an essential property.[6] Yet it was 'very possible' (he did not say 'probable') that matter could not exist without it.

If one reason for Gay-Lussac's concern with volumes was a metaphysical preference for thinking of matter in terms of extension, a more powerful factor in the opening years of the nineteenth century was the influence of Berthollet and Laplace. In his *Statique chimique* Berthollet laid special emphasis on cohesion as a factor explaining why certain reactions took place rather than others. In solution cohesion was exemplified by precipitation, in the gaseous state it took the form of 'condensation'. In other words there was a change in volume (usually a contraction) when gases combined. Gases provided an interesting area for study since in the gaseous state their particles, being a relatively large distance apart, did not exert the mutual attraction found in the liquid and solid states. Berthollet described the situation by saying that 'the quantity of their mass within the sphere of activity is necessarily very small'.[7]

While Berthollet tried to analyse chemical reactions in terms of spheres of influence, the parameter of distance was even more explicit with Laplace. Laplace launched a programme of investigation of the forces operating in the sub-microscopic world, such forces depending on distances between particles of matter. It is true that he had added a hurriedly-written note to Berthollet's book, one of the implications of which was that the force between the particles of a gas was independent of distance.[8] However he quickly realised the un-Newtonian nature of this conclusion and added a further note which re-emphasised the relevance of the distance between particles as a fundamental parameter.[9]

Anyone interested in the distance between particles would be interested in heat. In the laboratory, heat studies could provide valuable information for the mathematician. One could study the amount of 'caloric' contained in a gas and relate this to the repulsive forces between the particles of the gas. Heat was also related to volume changes through thermometry. Before the Revolution Laplace had collaborated with Lavoisier on a study of heat, and it was recognised that changes of temperature could be conveniently measured by studying the concomitant volume changes produced by them in a body, usually a fluid. A study of changes of volume was the key to the objective measurement of temperature. Thus Laplace encouraged Gay-Lussac to undertake research on the air thermometer as a reliable index of temperature. One can therefore see in the context of 'physics' various stimuli directing Gay-Lussac towards a study of volumes.

Gay-Lussac's very first memoir was concerned with the correlation of temperature and volume of gases and a preoccupation with volumes of gases became one of the characteristics of his approach to science. In his balloon ascent Gay-Lussac carried out an analysis of the air at high altitudes and in 1805 he undertook with Humboldt a detailed study of the volumetric composition of air based on the contraction involved in the use of Volta's eudiometer. In 1807 he studied the cooling caused by the expansion of gases, and in 1808, when he and Thenard had prepared potassium by their chemical method, he soon applied it for use as a means of analysing gases volumetrically. All this research pre-dates his famous memoir on combining volumes of gases (1809). In 1810 he was again applying the volumetric approach to organic analysis and in the following year he devised apparatus for measuring the volume of a known weight of liquid – he had thus extended his volumetric view-point from gases to liquids. He was able to put this methodology to good use in the study of hydrogen iodide (1813) and particularly in his study of cyanogen (1815). In 1816 he carried out a careful volumetric study of the oxides of nitrogen and was able to give the compositions of five distinct compounds which correspond accurately to the modern formulae. Finally his volumetric approach to the composition of ethyl alcohol and ether was to suggest a fertile avenue in the labyrinth of organic chemistry.

A good reason for Gay-Lussac's particular concern with volumes was that much of his early research involved gases. A man concerned continually with, say, mineral analyses would tend to turn instinctively to the balance, since the weight of solids is so obvious and basic a characteristic. Yet Dalton, the founder of the atomic theory, was also deeply concerned with gases and he was responsible for a whole new *gravimetric* approach to matter. Working with gases is therefore not a sufficient condition for the volumetric approach, although it might be a necessary condition. What we must do now is to trace the development of Gay-Lussac's views on the importance of volumes.

In Gay-Lussac's first research he studied the thermal expansion of gases and was concerned to compare their change in volume over a range of temperature. This was a programme of research suggested to him by Berthollet and is a reminder of one great inheritance from the last quarter of the eighteenth century – the preparation of a large number of gases and the realisation of the importance of gases in the science of chemistry. Nevertheless, despite the subject of his research, Gay-Lussac made a remark about the greater precision of measurements of weight than of volume.[10] Thus in 1802 Gay-Lussac adopted the common attitude. In other words, far from beginning his career with some innate propensity to volumetric measurements, he studied volumes when he was told to and felt that measurements of volumes were generally inferior in precision to those of weights.

Nevertheless his interest in volumes of gases had been aroused. His careful comparison of the volumes of different gases at different temperatures had enabled him to state a law and gave him great encouragement. Henceforth students of science would learn not only the Boyle–Mariotte law but the Charles–Gay-Lussac law, both fundamental in any study of gases.

Gay-Lussac also became involved in the late eighteenth century tradition of eudiometry. The term 'eudiometer' was introduced by the Italian natural philosopher Landriani for the instrument used by him and his contemporaries to measure the 'goodness' of samples of air. Their eudiometer (depending on the reaction of nitric oxide with the oxygen in the air to form nitrogen dioxide) gave inconsistent and inaccurate results. In 1778, however, Volta suggested a better method of estimating oxygen: the air was mixed with hydrogen in a graduated glass tube fitted with stop-cocks and an electric spark was passed. Priestley tried this method but pointed to certain limitations in the proportions of – to use the modern terms – oxygen and hydrogen used. It is at this point that we can introduce the research of Gay-Lussac and Humboldt.

In the winter of 1804–5 Gay-Lussac collaborated with Humboldt in research on the use of hydrogen as a means of estimating oxygen. This method had not been universally appreciated but Gay-Lussac by careful quantitative work was able to show that, when properly used, it was capable of considerable precision. Thus he came to revise his earlier view of the inferiority of volumetric measurements.

One reason for Gay-Lussac championing volumes as opposed to weights was that he had come to consider it a method of greater intrinsic accuracy. In general this might seem a surprising claim and we must therefore try to understand how Gay-Lussac came to adopt such a position. There was first the special case of the relative volumes of combining gases. In 1805 in his research with Humboldt Gay-Lussac argued that the relative *volumes* (but not the relative weights) of hydrogen and oxygen which combined to form water were independent of water vapour present.[11] Greater precision was therefore possible if one dealt with volumes rather than weights. Having seen the advantage of the volumetric approach in this special case, Gay-Lussac was in a position to look for further advantages of accuracy. The best illustration of this is given in his research on vegetable analysis, not on first sight an obvious field for the volumetric approach. Gay-Lussac and Thenard began their work on organic analysis as a branch of pneumatic chemistry and this is how it remained:

When we had conceived the project of analysing animal and vegetable matters, the first consideration which presented itself to our serious attention was to transform by means of oxygen the vegetable and animal substances into water, carbonic acid and azote [i.e. nitrogen]. It was evident that if we could succeed in operating the transformation so as to collect all the gases, this analysis would be accomplished with very great precision and simplicity.[12]

In the 1820s Gay-Lussac continued to tell his students that in this analysis one should try to deal with gaseous products rather than solids because 'with gaseous products one can obtain a higher degree of accuracy'.[13] In 1810 Gay-Lussac set out clearly his reasons for making such a claim:

The analysis of the air is more exact than any analysis of the salts, and yet it is performed upon two or three hundred times less matter than the latter. This is because in the former, where we judge of weights and volumes which are very considerable, the errors which we may commit are perhaps 1000 or 1200 times less perceptible than in the latter, where we are deprived of this resource. Now as we transform into gas the substances which we analyse, we bring our analyses not only to the certainty of the common mineral analyses, but to that of the most precise mineral analyses; more particularly as we collect at least a litre of gas, and as we find even in our way of proceeding the proof of an extreme exactitude and [only] of the most trifling errors.[14]

The basic justification therefore for Gay-Lussac's confidence in volumes was the greater volume : mass ratio of gases compared with other states of matter, which meant that even with small quantities accurate results would be obtained. Another justification followed from his law of combining volumes of gases. Any actual experimental value of the reacting volume of a gas or vapour would approximate to a whole-number ratio. Thus a number could be 'rounded off' to the nearest integer, particularly if it was very close to it. By this method, assuming the absolute accuracy of Gay-Lussac's law of combining volumes of gases, even an experimenter of only moderate ability would be able to find the *precise* composition of a compound even if the values he obtained by successive experiments varied by several per cent.

Furthermore by 1810 Gay-Lussac had been convinced of the superiority of the volumetric aproach for reasons even more fundamental than that of accuracy. He was convinced by his law of 1808 that volumetric measurements were more 'natural' than gravimetric measurements. Whereas reacting volumes of gases bore a simple ratio to each other, reacting weights did not:

It is very important to observe that, in considering weights, there is no simple and finite relation between the elements of any one compound.[15]

It was true that if two elements reacted together in more than one proportion, there was a simple ratio between the weights of element A which combined with a fixed weight of B. Gay-Lussac was not over-awed by this however, since, in addition to the more fundamental law of simple reacting volumes, he found that a law of multiple proportions also applied to the volumes of gases such as nitrogen and oxygen which combined in various proportions. He even went so far as to suggest that when measurements are made of the quantity of acid required to neutralise a given quantity of alkali (and vice versa), these should be volumetric measurements, at least if the reactants could be produced in the gaseous state:

These ratios by volume are not observed with solid or liquid substances, nor when we consider weights, and they form a new proof that it is only in the gaseous state that substances are in the same circumstances and obey regular laws.[16]

Two points emerge from this. First Gay-Lussac's concern with volumes went hand in hand with an interest in the gaseous state. He did not pretend at this stage that the volumes of liquids or solids could have any special significance. Second Gay-Lussac's volumetric *credo* was not a matter of blind faith. It was founded on a sound theoretical basis. Indeed Gay-Lussac began his historic memoir by explaining why certain fundamental regularities may be found in the gaseous state, while there were no corresponding laws for solids or liquids:

Substances, whether in the solid, liquid, or gaseous state, possess properties which are independent of the force of cohesion; but they also possess others which appear to be modified by this force (so variable in its intensity), and which no longer follow any regular law. The same pressure applied to all solid or liquid substances would produce a diminution of volume differing in each case, while it would be equal for all elastic fluids. Similarly, heat expands all substances; but the dilations of liquids and solids have hitherto presented no regularity, and it is only those of elastic fluids which are equal and independent of the nature of each gas. The attraction of the molecules in solids and liquids is, therefore, the cause which modifies their special properties; and it appears that it is only when the attraction is entirely destroyed, as in gases, that bodies under similar conditions obey simple and regular laws.[17]

These views were fully in accord with Berthollet's ideas set out in the

*Essai de statique chimique.* Indeed the modern reader might be misled by the prominence given to this statement into thinking that Gay-Lussac had been looking continually for laws relating to gases since his first success of 1802. Evidence presented elsewhere[18] suggests that this is unlikely. However from 1808 onwards Gay-Lussac had a programme. He made a special study of volumes, not only because they were the units of composition but also because he considered them the natural units of reaction. Thus in his research on iodine of 1814 he remarked:

We shall compare the bodies according to their volumes in the elastic state and not according to their ponderable quantities which have much less influence on the combination.[19]

This serves as a reminder that his attitude to volumes was an extension of the importance attached by Berthollet to condensation in a chemical reaction.

Gay-Lussac's concern with volumes is nowhere illustrated better than in his famous law of combining volumes of gases. Historians of chemistry have been in the habit of jumping from the main statement of Gay-Lussac's law of combining volumes of gases to Avogadro's idea of molecules which was based on it. Anyone who looks at the conclusion to Gay-Lussac's historic memoir, however, may see that for him the implications were in quite a different direction:

It is remarkable to see that ammonia gas neutralises exactly its own volume of gaseous acids; and it is probable that if all acids and alkalis were in the elastic state, they would all combine in equal volumes to produce neutral salts. The capacity of saturation of acids and alkalis measured by volume would then be the same and *this might perhaps be the true manner of determining it.* The apparent contraction of volume suffered by gases on combination is also very simply related to the volume of one of them.[20]

Gay-Lussac was therefore returning to his old interest in equivalents. Already in the late eighteenth century chemists such as Kirwan had experimented on combining weights of acids and alkalis. Gay-Lussac had done such experiments in 1807, looking in vain for a general pattern. But Gay-Lussac was interested not only in the combination of acids and bases but in all chemical combination. A typical case was the combination of 1 volume of oxygen with 2 volumes of hydrogen to produce (above 100°C) 2 volumes of water vapour. This contraction of one third could be measured directly. Alternatively it could be deduced from a comparison of the *densities* of the reactants and products. Going one stage further, if the densities were known, the probable contraction could be deduced.

By the time Gay-Lussac began his investigations of iodine in 1813 he had long been convinced of the primacy of volumes rather than weights. The question arose again then because of the very high density of iodine.[21] There was a great difference in weight between the two constituents of hydrogen iodide, the iodine weighing more than a hundred times the hydrogen. Gay-Lussac had an immediate answer to this apparent imbalance. It was that bodies usually combine 'according to the volume of their vapours'.[22] Considered volumetrically the hydrogen and iodine were equal partners and this seemed much more reasonable. Thus the high weight ratio I : H served to confirm Gay-Lussac's hypothesis of the primary importance of volumes. The aim in natural history had been to find a natural classification. In chemistry Gay-Lussac believed he had found the way to express chemical combinations in natural terms. Thus what began as a methodology developed into an ontology.

It may well be that Gay-Lussac's loyalty to Lavoisier's theory of acidity helped to confirm his belief in volumes as 'natural' units. His great predecessor had committed himself to the oxygen theory of acidity a few years *before* he had established the composition of water, which showed not only that water was an oxide, but also that it contained a high proportion of oxygen. According to Lavoisier's analysis, water contained approximately 85 per cent of oxygen by weight. It was surprising, to say the least, that a compound which contained such a high proportion of oxygen should not be acidic. This may have been one of the factors encouraging Gay-Lussac to reject a gravimetric approach in favour of a volumetric one since, using the language of volumes, the composition of water could be expressed as two volumes of hydrogen to one volume of oxygen. Expressed this way, the acidifying effect of the oxygen could be seen to be submerged by the hydrogen.

## Reacting volumes and chemical composition

One of the major problems of physical science has been to understand in what ways physical measurements of gross matter could reveal the secrets of composition. One learns little about chemical composition from weighing solids and liquids but Gay-Lussac was to show how fruitful the measurement of the density of gases and vapours could become. The rather extravagant claim has been made that Gay-Lussac 'built the first bridge between physics and chemistry, two sciences which had been separated till then'.[23] If we accept the claim this 'bridge' depended on the law of combining volumes which could be stated in the form that there was a relation between the densities of gases and composition.

Gay-Lussac's law of combining volumes of gases relates primarily to the ratio of the volumes of the reactants. Further experiments convinced Gay-Lussac that the law could be extended to cover the volume of the products:

Not only, however, do gases combine in very simple proportions, as we have just seen, but the apparent contraction of volume which they experience on combination has also a simple relation to the volume of the gases, or at least to that of one of them.[24]

As an example of this he gave:

100 volumes carbonic oxide + 50 volumes oxygen
= 100 volumes carbonic acid

The extension of the law to cover the volume of a gaseous product is the second aspect of Gay-Lussac's law.

There was no necessity for the product to be a gas for one to be able to provide useful information about its composition. If the *constituents* were gaseous, then they should combine in simple ratios by volume. In one of the tables[25] which he drew up in conjunction with his memoir he included several salts such as ammonium chloride and ammonium carbonate, the composition of which could be expressed simply in terms of the volumes of their constituents.

The implication of Gay-Lussac's work for knowledge of chemical composition has not received sufficient attention. Although his law is primarily a statement of chemical reactions, it was the implication for composition which was to make it of fundamental importance. Only the most elementary notions of chemical composition were available before Gay-Lussac's time. Lavoisier had helped to establish the beginnings of quantitative analysis by calculating the (approximate) gravimetric composition of water and of some oxy-acids. Any new knowledge of reacting proportions was potentially a valuable tool in solving the intractable problems of composition. In some cases, such as that of carbonic acid gas, the composition could be inferred directly from experiment. In other cases, such as carbonic oxide (carbon monoxide), the question of composition could not be solved without a leap in the dark.

Let us consider first the composition of the more straightforward compound gases: steam, ammonia and carbonic acid gas. Whereas, according to Dalton's arbitrary law of simplicity, water was represented by (to use Berzelius' atomic symbols) HO and ammonia NH, Gay-Lussac was quite clear that water contains *two* parts by volume of hydrogen to every part of oxygen, and ammonia contains *three* parts of hydrogen for every part of nitrogen. Gay-Lussac was able to support his ideas on composition of compound gases by evidence from densities. Thus if

100 vols. carbonic oxide gas + 50 vols. oxygen
= 100 vols. carbonic acid gas

then not only would it be possible to predict that the density of carbonic acid gas would be greater than that of carbonic oxide, but it could be said that its density was an additive property of its constituents in the ratio of $1 : \frac{1}{2}$. Conversely, if the density of carbonic acid is known, it should be possible to predict that of carbonic oxide:

| | | |
|---|---|---|
| Density of carbonic acid gas | = | 1.5196 |
| Minus half the density of oxygen [density = 1.1036] | = | 0.5518 |
| Theoretical density of carbonic oxide | = | 0.9578 |

This compared very well with Cruickshank's experimental value for the density of 0.9569 (air = 1).[26] Another success in the application of this method was with ammonia. The theoretical value for the density of ammonia on the basis of a composition of 3 volumes of hydrogen for 1 of nitrogen and a contraction of 50 per cent was 0.594. This compared fairly well with the experimental value of 0.596.[27]

Having used the densities as experimental evidence for his ideas, Gay-Lussac was faced with the problem of the density of carbonic oxide gas. It was difficult to understand how a gas which contains both carbon and oxygen should be *lighter* than oxygen itself. With the hindsight of Avogadro's hypothesis a schoolboy studying science today could solve this problem on the assumption of a diatomic molecule for oxygen. This imaginative leap was not taken by the French chemist, and indeed when it was suggested it was rejected or ignored by the majority of chemists for half a century.

We return to the problem faced by Gay-Lussac in 1808. He knew that 'oxygen gas doubles its volume on forming carbonic oxide gas with carbon'.[28] Thus 50 volumes of oxygen produced 100 volumes of carbonic oxide. Such an apparent *expansion* would account for the lower density of the product. Yet the formation of nearly all other compound gases known was accompanied by a *contraction*.[29] Gay-Lussac solved this dilemma by considering the other reactant, the solid carbon, to be in a gaseous state. He took the minimum proportional volume required to produce a contraction, 100 volumes of 'gaseous carbon' and supposed that this combined with 50 volumes of oxygen to form 100 volumes of carbonic acid gas.

We must now consider fully Gay-Lussac's extension of his gas law to non-gaseous elements. Avogadro is sometimes criticised for his reckless extension of his law to include solids and this is given as one of the reasons why his ideas were not generally accepted for half a century. Berzelius too is given prominent treatment in a standard history of nineteenth-century chemistry[30] for his general theory based on volumes. In fact the idea originates with Gay-Lussac and both Avogadro and Berzelius were extending an idea found in the French chemist's 1809 paper. Gay-Lussac wrote:

The observation that the gaseous combustibles combine with oxygen in the simple ratios of 1 to 1, 1 to 2, 1 to $\frac{1}{2}$, can lead us to determine the density of the vapours of combustible substances, or at least to approximate closely to that determination. For if we suppose all combustible substances to be in the gaseous state, *a specified volume* of each would absorb an equal volume of oxygen, *or twice as much, or else half*; and as we know the proportion of oxygen taken up by each combustible substance in the solid or liquid state, it is sufficient to convert the oxygen into volumes and also the combustible, under the condition that its vapour shall be equal to the volume of oxygen, or else double or half its value. For example, mercury is susceptible of two degrees of oxidation, and we may compare the first one to nitrous oxide. Now, according to MM. Fourcroy and Thenard, 100 parts of mercury absorb 4.16, which reduced to gas would occupy a space of 8.20. These 100 parts of mercury reduced to vapour should therefore occupy twice the space, viz., 16.40. We thence conclude that the density of mercury vapour is 12.01 greater than that of oxygen, and that the metal on passing from the liquid to the gaseous state assumes a volume 961 times as great.[31]

Gay-Lussac had thus extended his consideration of solid non-metallic elements to the case of a metal, albeit a volatile one. The basic assumption of the validity of speaking of 'carbon gas', 'mercury vapour', etc., on the same basis as the ordinary gases may have been influenced by a passage in Lavoisier's *Traité*, where he says that if the earth were moved much nearer to the sun a whole range of new substances, even mercury, would become gases in the fullest sense.[32] The words which we have italicised in the above passage reveal a second assumption in such arguments. Gay-Lussac could have his volumes in simple ratios but he could never be sure which simple ratios to choose. Analogies provided some guide but these could always be disputed. Gay-Lussac assumed, for example, that red mercuric oxide is composed of one volume of mercury combined with one volume of oxygen.

Gay-Lussac's 'contraction hypothesis' may seem as arbitrary as Dalton's rule of simplicity in the atomic constitution of compounds. It could be argued that it was superior in so far as it was based on sound analogy and was subject to experimental verification.

## Vapour densities

If the history of science consisted of 'discoveries' then Gay-Lussac could be credited with having 'discovered' vapour density. In his research the vapour density of an element or compound became for the first time a fundamental datum which was capable of determination either by calculation or experimentally. If a volume of gas or vapour is a natural unit, then the vapour density is the weight of a natural unit. Thus in Gay-Lussac's scheme densities could be used to find equivalents and he was able to do valuable work in chemistry without using the concept of atomic weight. The use of densities to compare substances is not in itself a very difficult step. It had, for example, occurred to Dalton but he had rejected the whole method after considering the paradoxical case of water vapour having a *lower* density than that of oxygen despite the fact that water contained oxygen. In so far as Gay-Lussac was concerned with densities he had not rejected the gravimetric approach entirely, since the density was the weight of a given volume. But both could be measured *directly* in the laboratory which was more than could be said for Dalton's atoms.

In his very first memoir Gay-Lussac, having found certain simple regularities in gases, tried to extend his research to other states of matter. The easiest extension was with volatile liquids. Accordingly he measured the thermal expansion of ether vapour and concluded that it was the state of matter rather than the individual properties of a substance which governed its expansion. We have already seen that in those cases where, for practical reasons, he did not examine vapour densities directly, he used indirect methods to arrive at their value.

In his research on iodine carried out in the first half of 1814 Gay-Lussac made a special study of hydrogen iodide. He calculated its theoretical vapour density according to the volumetric composition he had established of equal volumes of hydrogen and iodine vapour. He wrote:

I must put the reader in mind in the first place that from the experiments of M. Thenard and myself, one volume of chlorine by combining with one volume of hydrogen produces exactly two volumes of hydrochloric gas; hence it follows that the density of this last gas is the mean of that of chlorine and hydrogen. But since a volume of chlorine takes one volume of hydrogen, its ratio to oxygen in bulk will be that of two to one, and its ratio in weight may be immediately deduced from this. We do not know the density of the vapour of iodine; but from the experiments to be stated below, I have found that the ratio of oxygen to iodine is 1 to 15.621. Now the density of a demivolume of oxygen being 0.55179, 0.55179 × 15.621 = 8.6195 will represent the density of iodine under the volume taken for unity. If to this density we add that of hydrogen, 0.07321, and take half the sum, we have 4.4288 for the density of hydriodic gas.[33]

The reference to the combining ratio of iodine and oxygen as 15.621 requires explanation and indeed it is of interest in its own right since it provides an excellent example of how Gay-Lussac and many other chemists in the years following the publication of the atomic theory were able to do excellent work on the basis of combining proportions or equivalents. Gay-Lussac explained the conditions under which he was able to react iodine with zinc and reported:

In three experiments which differed little from each other, and of which I have taken the mean, I found that 100 iodine combined with 22.225 zinc. But 26.225 zinc combine with 6.402 oxygen, which saturate 0.849 hydrogen; consequently the ratio of oxygen to iodine is 6.402 to 100, or 10 to 156.21, and the ratio of hydrogen to iodine is 0.849 to 100 or 1.3268 to 156.21. Thus if we represent the number for oxygen with Dr Wollaston by 10, the number for iodine will be 156.21.[34]

However, we must return to the theoretical density of hydrogen iodide of 4.4288. In Gay-Lussac's first experiment, which he was honest enough to report, he obtained a value of 4.602 (air = 1). He repeated this determination with greater care and obtained a value of 4.443 which gives an agreement to within 0.5 per cent. He was ready with an explanation of why the experimental value should be slightly higher. It was that, despite the precaution he had taken to dry the gas by passing it through a tube cooled to −4°C, there were slight visible traces of moisture. Thus Gay-Lussac could feel satisfied that experiment had vindicated his theoretical method and he had begun to establish vapour density as part of the fundamental data of any volatile new compound. By the roundabout method described above he had successfully overcome the fact that one of the constituents of his compound (iodine) was not gaseous at ordinary temperatures.

An example of the analytical power of Gay-Lussac's volumetric approach is provided by his quantitative analysis of the spontaneously explosive compound, nitrogen iodide. Obviously no direct quantitative experiment was feasible. He therefore approached the problem in an indirect manner:

We have seen that the ratio of hydrogen to iodine is 1.3268 to 156.21 and as ammonia is composed of: Hydrogen 18.4756

Nitrogen 81.5244

it follows that the ratio of nitrogen to iodine is that of 5.8544 to 156.21; and such is the ratio of the elements of the fulminating compound. If we reduce these elements to volumes by dividing 5.8544 by 0.96913, the density of nitrogen, and 156.21 by 8.6195, the density of the vapour of iodine, we find that the proportion in volume of the elements is one nitrogen to three iodine.[35]

Or, to use the symbols then being introduced by Berzelius, the compound was $NI_3$ or nitrogen tri-iodide. Gay-Lussac was able to confirm this analysis by a more direct volumetric method, namely the observation that iodine vapour and hydrogen combine in equal volumes and that in ammonia 1 volume of nitrogen is combined with 3 of hydrogen.

Gay-Lussac also used a volumetric method in his analysis of prussic acid. This is not a gas but a volatile liquid, and he accordingly took advantage of the hot days of August 1815 to carry out the analysis in a Volta eudiometer. He mixed 100 volumes of the vapour with excess of heated oxygen and passed a spark between the platinum electrodes. There was a bluish-white flame and a contraction of 78.5 volumes due to the formation of water of negligible volume. When caustic potash was introduced there was a further contraction (101.0 volumes) due to the absorption of carbonic acid gas. The remaining gas was nitrogen together with unused oxygen. The proportion of the latter was obtained by further sparking with hydrogen thus showing that 46.0 volumes was nitrogen. The failure of these figures to produce exact ratios was partly explained by the formation of a little nitric acid during the detonations. Thus the figures could be rounded off to 75, 100 and 50 respectively. A contraction of 75 volumes represents the combination of 50 volumes of hydrogen.[36] Also 50 volumes of nitrogen is formed. One hundred volumes of carbon dioxide resulted – Gay-Lussac supposed – from 100 volumes of carbon. He therefore concluded that prussic acid had the composition:

One volume of the vapour of carbon
Half a volume of hydrogen
Half a volume of nitrogen

It was necessary to confirm that prussic acid contained only these three elements and, in view of the approximations used, in the ratio suggested. This is where he was able to use his method of densities of compounds based on the densities of their constituents.

| Density of prussic acid vapour | | Density of vapour of carbon | $\frac{1}{2}$ Density of hydrogen | $\frac{1}{2}$ Density of nitrogen |
|---|---|---|---|---|
| | = | + | + |

and

|  | | Density of | | |
|---|---|---|---|---|
| Density of carbon vapour | = | carbonic acid gas | − | Density of oxygen |
|  | = | 1.5196 | − | 1.1036 |
|  | = | 0.4160 | | |
| ½ Density of hydrogen | = | 0.0366 | | |
| ½ Density of nitrogen | = | 0.4845 | | |
| Total | = | 0.9371 | | |

(By experiment, density of
    prussic acid vapour      = 0.9476)

There was, therefore, agreement with the theoretical value to within 1 per cent.

Since however, this was an acid it might be thought to contain oxygen. When prussic acid vapour was passed through a red hot porcelain tube containing iron wire, the iron was not oxidised. Finally, to check on the proportion of carbon in prussic acid, Gay-Lussac passed the vapour over heated copper oxide.[37] Drops of water were formed together with carbonic acid gas and nitrogen in the proportion of 2 : 1 by volume. He deduced from this that there was twice as much carbon as nitrogen in prussic acid, thus confirming the composition he had given.

*Organic chemistry and the practical determination of vapour densities*

Strangely Berzelius had considered that a disadvantage of the volumetric approach was that it could not be applied to organic compounds.[38] The development of organic chemistry was to show how mistaken he was. Gay-Lussac himself applied his volumetric method to the elucidation of the composition of ethyl alcohol and ether in 1814; this, however, depended also on other considerations. It was his isolation of ethyl iodide in 1815 which opened up the field of alkyl halides which, because of their low boiling points, are particularly suitable for the volumetric approach.

A practical method of determining the vapour densities of volatile liquids was a valuable contribution to the expanding science of organic chemistry. One of the problems of chemists in the nineteenth century was to integrate the science of (non-volatile) mineral compounds with that of the volatile organic compounds.[39] With additional problems such as the dissociation or association of several inorganic compounds, it was only too easy for chemists like Dumas to be perplexed. But this was not before Dumas himself had used and improved Gay-Lussac's method of determination of vapour densities.

If Gay-Lussac was to extend his studies of gases to include liquids it was necessary for him to find a method of determining vapour densities. In his classic memoir on combining volumes of gases he had drawn on a wide range of values for the densities of *gases* but no data were available for liquids. Gay-Lussac therefore turned his attention to this second state of matter and devised apparatus to determine the vapour density of volatile liquids. It was perhaps typical of him that he should *measure the volume* of a known weight of liquid rather than (as later in Dumas' method) the weight of a known volume.

Gay-Lussac did not publish a detailed description of his apparatus but fortunately this omission was made good by Biot.[40] Gay-Lussac blew glass bulbs drawn out into a neck. The bulb was weighed first containing air and then filled with the liquid under investigation. The neck was then sealed with a flame and the bulb was re-weighed. Knowing in this way the weight of the liquid, it remained to find its volume in the state of vapour. A long graduated tube closed at one end was filled with mercury and inverted into a mercury trough as in a simple barometer. The bulb containing the liquid was introduced at the bottom of the tube and it immediately floated to the top. The barometer tube was surrounded by a heating jacket. The heat communicated to the glass bulb made it break, thus releasing the liquid which vaporised in the space above the mercury. The volume could be read off directly from the graduations on the barometer tube. It was also necessary to find the pressure and this was done by means of a copper scale reading from the surface of the mercury in the reservoir to a cursor level with the mercury in the barometer tube. The difference in height of mercury at the beginning and end of the experiment gave the vapour pressure, after making allowance for the temperature difference. Thus the experimenter knew the pressure of a known volume at a given temperature.

The possible objection that not all the liquid had vaporised was dealt with and in the calculation the expansion of the glass was allowed for. Biot reproduced some of Gay-Lussac's experimental results. He had found that 1 g of water occupied 1.6964 l at 100°C and 76 cm mercury. But if 1 g water occupies 1 cc, then 1 cc water becomes 1696.4 cc steam. His data gave the ratio of the density of water vapour to air as 10 : 16 (almost exactly the same as the modern figure), compared with Saussure's earlier figure of 10 : 14.

A particularly important determination of vapour density was that of 'sulphuric' (i.e. diethyl) ether, since Gay-Lussac used it to throw light on the constitution of this well-known substance, which was obviously related to (ethyl) alcohol and possibly to olefiant gas (ethylene), but what the relation was was not clear. Gay-Lussac found that when 1.993 g was vaporised at 100°C it occupied 194 divisions of the tube, each division being 0.00499316 l. The mercury level in the tube was 7.6 cm when the barometer read 76.62 cm and the temperature was 10°C. From these data[41] Gay-Lussac worked out the vapour density as 2.586.[42]

It had been suggested by Théodore de Saussure that ether was a compound of olefiant gas (ethylene) and water. He had given the composition in gravimetric terms as: olefiant gas 100, water 25, these numbers being rounded off from experimental values on the assumption of a simple gravimetric ratio. When Gay-Lussac[43] converted this to a volumetric ratio (102.49 : 40.00) he found that it could not be reconciled with his experimental value (2.581) for the vapour density of ether. His calculations were based on the premise that the density of the compound was the sum of the densities of its constituents, allowing for a possible contraction in the formation of the compound. He expressed the composition of ether as

Ether   =   Olefiant gas   +   Water vapour
1 volume       2 volumes          1 volume

By a further brilliant extension of this method he was able to express in volumetric terms the conversion of sugar by alcoholic fermentation into alcohol and carbonic acid, a reaction sometimes still known as 'Gay-Lussac's equation'.

### The influence of the volumetric approach

Gay-Lussac's volumetric approach is important not only in understanding his own ideas but also the thinking of other chemists. Avogadro's hypothesis was a direct outcome of the law of combining volumes, but as the value of Avogadro's work was not appreciated in his own lifetime, it was through Berzelius that Gay-Lussac's volumetric ideas had their greatest immediate influence. Lacking the concept of a diatomic molecule, Berzelius could not see how a single 'atom' of, say, hydrogen chloride could be formed except from half an atom of hydrogen and half an atom of chlorine. But to speak of 'half an atom' was to contradict Dalton's idea of the atom. There was no difficulty, however, in speaking of 'half a volume' and one could say as a direct result of an experiment that

1 volume hydrogen + 1 volume chlorine = 2 volumes hydrogen chloride
or
½ volume hydrogen + ½ volume chlorine = 1 volume hydrogen chloride

It is interesting that it was in connection with a theory derived from Gay-Lussac's work on combining volumes of gases that Berzelius first introduced his chemical symbols to the world. Berzelius wrote:

It is known that bodies in their gaseous state either unite in equal volumes, or one volume of one combines with 2, 3, etc., volumes of the other. Let us express by the initial letters of the name of each substance a determinate quantity of that substance; and let us determine that quantity from its relation in weight to oxygen, both taken in the gaseous state, and in equal volumes; that is to say, the specific gravity of the substance in their gaseous state, that of oxygen being considered as unity...It is obvious that this comes to the same thing as Mr Dalton's weights of atoms; but I have the advantage over him, of not founding my numbers on an hypothesis, but upon a fact well known and proved...[44]

Berzelius was, therefore, attracted by the direct experimental evidence on which Gay-Lussac's memoir on combining volumes was based as opposed to some arbitrary assumptions in Dalton's atomic theory. In his 'Essay on the cause of chemical proportions' Berzelius amplified the reason for his allegiance to Gay-Lussac. One of these was his difficulty in assigning a simple formula to oxalic acid on the basis of his own analyses, in which he had every confidence.

I have already, in preceding memoirs, made mention of another method of viewing chemical proportions – a method founded on a fact discovered by Gay-Lussac; namely, that bodies when in the state of gases unite either in equal volumes, or 1 volume of one combines with 2, 3, etc., volumes of the other. This fact has been already verified by several distinguished chemists. From what we know respecting definite proportions, it follows that it would hold with all bodies in the temperature and pressure at which they would assume the gaseous form. Hence there is no other difference between the theory of atoms and that of volumes, than that the one represents bodies in a solid form, the other in a gaseous form. It is clear, that what in the one theory is called an atom, is in the other theory a volume. In the present state of our knowledge the theory of volumes has the advantage of being founded upon a well constituted fact, while the other has only a supposition for its foundation. In the theory of volumes we can figure to ourselves a demi-volume, while in the theory of atoms a demi-atom is an absurdity. On the other hand, the theory of volumes has a disadvantage from which the atomic theory is free; namely, the existence of compound bodies, especially of an organic nature, which we cannot suppose ever to have existed in the form of gas.[45]

Berzelius went on to mention how few were the atomic weights of elements which could be determined directly from their vapour densities – 'We must therefore endeavour to discover the weight of their volumes by other means. Our results will be, doubtless, very uncertain but not altogether unsuccessful. . .'. Berzelius was, therefore, substituting one uncertainty for another. He must presumably have felt that it was at a less fundamental level. His subsequent argument, however, that lower oxides are composed of 'equal volumes of their elements', and higher oxides with small integral multiples seems to have a clear parallel in Dalton's arbitrary rule of simplicity. Altogether Berzelius developed his theory of volumes for five years but after 1818, although he did not abandon it completely, he laid more emphasis on the gravimetric approach in his corpuscular theory. He was able to do this the more readily as he had always insisted (as Gerhardt did later) that *volume* was just another word for atom.

The work of Dulong and Petit (1819) gave particular encouragement in France to the acceptance of Dalton's atomic theory and the young chemist J. B. Dumas began to advocate an atomic theory. He recommended the application of Gay-Lussac's law of volumes as the principal method of finding the atomic weights of gaseous elements.[46] He undertook to determine a wide range of atomic weights by means of vapour density determinations of elements and compounds, a method in which he then placed supreme confidence:

Indeed it is the only method by which we can know their true composition.[47]

Yet by 1837 he found that all his careful quantitative work produced only contradictory results. He could have attacked the method of vapour density determination but, on the argument that this represented a 'fact' as opposed to a 'hypothesis', it was the atomic theory which had to be sacrificed, thus giving occasion to his famous saying that the term 'atom' should be erased from science.[48] The reason for this change was the anomalous results presented by such elements as mercury, phosphorus and sulphur. In Germany Liebig made use of Gay-Lussac's volumetric approach. He wrote:

The laws of the combining proportions of gaseous bodies can be derived from the volume in which they unite with even as much accuracy as by means of the balance. In this point of view the knowledge of the specific weights of simple and compound bodies in the gaseous state is one of the most important means to submit to the sharpest proof, the constitution of compounds obtained by other methods.[49]

However, in view of the recent work of Dumas showing anomalies with sulphur, phosphorus and arsenic, Liebig was not willing to apply the volumetric approach universally and he said that when there was a conflict between the equivalent deduced from other methods and the theory of volumes, then the other evidence was to be preferred.[50]

Dumas' predicament might have been solved if he had been prepared to accept the association of atoms and the dissociation of compounds. However this solution had to wait for Henri Sainte-Claire Deville.

Writing in 1860[51] Sainte-Claire Deville was full of admiration for Dumas' method of determining vapour densities. He said that by it

[Gay-Lussac's] law of volumes receives. . .daily applications.

It provided the best method of analysis and the best method of determining equivalents. It was particularly valuable in organic chemistry since few organic compounds had a boiling point higher than the convenient maximum temperature for this method of 300°C. Sainte-Claire Deville's purpose in this memoir was to confirm that 'the law of Gay-Lussac' could be extended to about 1000°C provided the bodies studied did not decompose. In the case of sulphur he took particular care as this seemed to be a case 'where the law of Gay-Lussac has seemed at fault'. His experiments showed that the density at 860°C corresponded to the formula $S_2$. It is interesting as an example of Gay-Lussac's influence, at least in France, to see his name appearing half a dozen times in a memoir written ten years after his death (and half a century after the enunciation of the law) when this has traditionally been conceived in the context of Avogadro's hypothesis. Deville even stated[52] Gay-Lussac's law in the form that vapour densities are exactly proportional to equivalents if one introduces the factor $\frac{1}{2}$, 1 or 2 according to particular cases to allow for volume changes.

Which multiplying factor to use was, of course, a vitally important question. Persoz was one of those who tried to solve the problem in 1835[53] but it was Gerhardt who did most before the readoption of the Avogadro–Ampère hypothesis to introduce a complete theory of chemical composition based on volumes. He rebuked those who had taken refuge in equivalents and pointed out that, for organic chemistry in particular, vapour density was the only practical laboratory method available in approaching the problem of composition.[54] He therefore tried in 1842 to standardise molecular formulae as 'four volume' formulae, i.e. formulae corresponding to molecular weights occupying the same volume as four grams of hydrogen. This meant writing water as $H^4O^2$ instead of $H^2O$ on the 'two volume' formula. The problem was that, in order to introduce consistency between inorganic and organic formulae, it was necessary either to double the former (as with water) or to halve the latter. Gerhardt later used 'two volume' formulae, which was a decided improvement for many compounds, although giving only half the correct atomic weights for some metals. The respective merits of 'two volume and 'four volume' formulae continued to divide chemists in the mid-nineteenth century. Although the volumetric viewpoint was of great utility, it failed to introduce unity into chemistry.[55]

It was in France that Gay-Lussac's concern with vapour densities had greatest influence. We have seen that Berzelius, after incorporating a volumetric viewpoint into his work, reverted to his gravimetric approach. German chemists too had found the volumetric approach useful but in 1838, after Dumas' renunciation of the atomic theory, Liebig wrote to tell Pelouze that after discussing the matter with other prominent chemists: Wöhler, Magnus, H. Rose and L. Gmelin, they had decided to abandon a system of atomic weights based on volumes and revert to equivalent weights.[56] Liebig in taking this step was escaping from endless discussions about molecular constitution and, impatient to pursue his laboratory work, was content to speak only of combining weights. The evidence of Liebig, like that of Berzelius, suggests that Gay-Lussac's ideas were only fruitful to a limited extent in the early nineteenth century. It was only in the 1860s that, with Cannizzaro and Deville, the value of vapour density determinations was to be fully appreciated. Cannizzaro showed how the atomic weight of an element could be determined from the vapour densities of its volatile compounds.[57] Agreement about atomic weights gave chemistry a unity and a confidence which had previously been lacking.

Gay-Lussac's law of volumes was to influence atomic theory in another quite different way. The multiplicity of Lavoisier's elements was to produce a reaction in the early nineteenth century which found expression in the work of William Prout. Prout's Hypothesis was concerned with the unity of matter and found its empirical base in the fact that many atomic weights were integers or almost integers. Prout's first paper published in 1815 was prefaced with the remark that his observations were 'chiefly founded on the doctrine of volumes as first generalised by M. Gay-Lussac'.[58] His paper was entitled: 'On the relation between the specific gravities of bodies in their gaseous state and the weights of their atoms'. By comparing the weight of successive gases with the weight of an equal volume of hydrogen under the same conditions, he arrived at whole numbers and, following Gay-Lussac's example, he extrapolated his investigations to include solid elements. It is interesting that this theory ignored the work of Dalton in preference for that of Gay-Lussac 'which, as far as the author is aware at least, is now universally admitted by chemists'. Prout's Hypothesis exercised a peculiar fascination on many chemists, including Dumas, throughout the nineteenth century.

But although Gay-Lussac's law of combining volumes was very influential, one person who refused to accept it was John Dalton. In view of what we have said earlier about Dalton's low standards of accuracy, it is ironic that he chose this argument as a basis for rejection: 'In no case perhaps is there a nearer approach to mathematical exactness than in that of 1 measure of oxygen to 2 of hydrogen; but here the *most exact experiments I have ever made*, gave 1.97 hydrogen to 1 oxygen'.[59] Dalton's dismissal of Gay-Lussac's work was criticised by Andrew Ure who said of the theory of volumes 'we know nothing in chemical science to which we can venture to attach the anchor of our belief, for nothing is better demonstrated than that theory'.[60]

# 6

# Scientific research

'One should not be miserly either with one's time or care'
Gay-Lussac[1]

## Scientific productivity

An appropriate place for an assessment of Gay-Lussac's contributions
to science is at the end of a book rather than in the middle and this
chapter therefore is concerned rather to outline the range of Gay-
Lussac's scientific work, which is far wider than one might think, since
it extends not only from the most theoretical to the most practical but
also covers a spread of subject matter ranging from physics through
chemistry to physiology. We will begin with the physical end of the
spectrum, which is where Gay-Lussac himself began after his mathematical training at the Ecole Polytechnique and his association with
Laplace, culminating in institutional recognition as a member of the
physics section of the Institute. The label of physicist, however, seems
increasingly inappropriate as Gay-Lussac's work became almost exclusively chemical.

Before considering the details of various aspects of Gay-Lussac's
work it might be useful to consider his productivity over his whole
career. He published approximately 150 papers which appeared under
his own name alone and about thirty joint papers with a variety of
collaborators. It may be of interest to present the publication of these
papers against a scale of Gay-Lussac's life span, omitting the first
twenty-two years, when, understandably, he published nothing. Fig. 2,
therefore, begins in the year 1800.[2] One of the dangers of attempting
to quantify scientific genius is that any representation may be interpreted in absolute terms. We therefore hasten to say that it is intended
to do no more than provide an indication of Gay-Lussac's *relative*
output of scientific papers over a busy life. An apparent drop in output
in his forties is one inference which may be drawn from the graph but
it should not be assumed that he was any less busy. It meant simply
that he was contributing more to applied science than to pure science
and he was giving advice privately for a fee rather than giving it
publicly in exchange for recognition which he now had in plenty.

On fig. 2 it was decided to show the number of papers published every two years rather than every single year to avoid anomalies such as an accidental calendar year of zero publications (1806) when Gay-Lussac was really very busy. Taking two-year units evens out such irregularities but not too much notice should be taken of slight variations in adjacent periods, which would obviously be slightly different if three-, four- or five-year periods had been chosen.

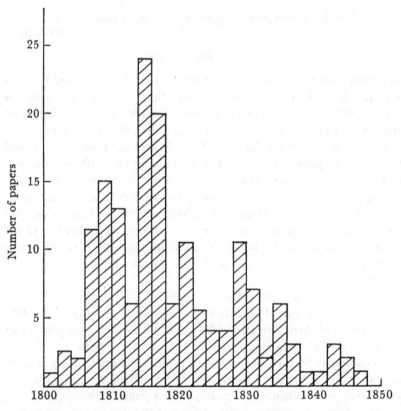

Fig. 2. Gay-Lussac's 'productivity' as measured by the number of papers published in scientific journals.

The graph shows that Gay-Lussac, after publishing his first paper in 1800 at the age of 22, reached a peak of publication in 1809–10 when he was 31–32 years old. He reached a second peak in 1815–16 when a total of twenty-four papers appeared under his name (eighteen papers in 1816). It should be noted, however, that 1816 was the year in which Gay-Lussac took up the editorship of the *Annales de chimie et de physique* so that many of these 'papers' are editorial notes rather than fundamental contributions to science. This raises the question of the validity of taking numbers of papers published as an index of productivity. One could meet the criticism that this measures quantity at the expense of quality by picking out two of Gay-Lussac's most brilliant and original contributions to science: his paper on combining volumes of gases (published 1809) and his memoir on cyanogen (published 1815). The publication of these papers coincides with the two peaks mentioned above and suggests that quantity in Gay-Lussac's output is not shown at the expense of quality.

As we would expect, a greater number of Gay-Lussac's later papers are concerned with applied chemistry rather than pure science but in so far as many of Gay-Lussac's important contributions to applied science were published separately as books or brochures they are not shown in this histogram.

However, Gay-Lussac was not someone who thought of a book as a standard form of publication. His only book on pure science (with Thenard as co-author), the *Recherches physico-chimiques*, was probably written up as an entry for the decennial prizes offered by Napoleon and was little more than a collection of papers. Gay-Lussac, like most of his professional colleagues in nineteenth-century science, saw the paper as the method of communicating his work to the scientific community and he never published a text-book.

## *Work in physics*

The modern demarcation of scientific subjects cannot be automatically transferred to the early nineteenth century and we should therefore not attempt to specify too precisely what in Gay-Lussac's work is to be regarded as 'physics'. However, in France by the late eighteenth century *la physique* had been recognised as a branch of science separate from mathematics on the one hand and chemistry on the other. In the institutional framework of post-Revolutionary France there were positions in the educational establishment described as 'professeur de physique' and the subject taught: properties of matter, heat, light, electricity, was basically elementary physics. Gay-Lussac's claim to be a physicist is much more strongly supported by his institutional affiliations than his scientific research. Whereas only a handful of his published memoirs is on physics, he became officially classified as a physicist in the sense that, in a period when there were no vacancies in the chemistry section of the Institute, he managed to be elected in the section of general physics and he stayed there all his life. In those early days, with the mathematical training of the Ecole Polytechnique and the stimulus of Laplace, his work lay as much in 'physics' as in chemistry. His first memoir, that on the thermal expansion of gases which has been described earlier, is certainly more physical than chemical.

Gay-Lussac's second claim to the title of physicist is provided by his teaching position at the newly-founded Paris Faculty of Science where he taught a part of the course of (elementary) physics from 1809 to 1832. When the Faculty was established there were more chemists available than physicists and the chemistry chair was given to his older friend Thenard. In the Institute Gay-Lussac was already labelled as a physicist and he was glad to have an additional paid teaching post when it was offered to him. He never wanted to lecture on the whole physics syllabus and he divided his course first with Hachette and then with Biot, so that he could teach those aspects which interested him, notably gases, heat and change of state, and properties of matter.[3] Turning to Gay-Lussac's actual research, it is his work on heat, and to a much lesser extent his electrical research, which figure prominently in the list of his publications. These may be considered in turn.

Gay-Lussac took a particular interest in heat phenomena at a time when heat was a study on the borderlines of physics and chemistry. Berthollet had devoted a section of more than one hundred pages of his *Statique chimique* to the matter of heat or 'caloric' and when Gay-Lussac came to give lectures for the physics course of the Paris Faculty of Science he too made use of caloric, explaining to his students that the term had been introduced in order to distinguish the effect (heat – what is felt) from the cause. Lavoisier and Laplace in their historic joint memoir on heat (1783) had introduced the term caloric without committing themselves to either of the alternative views: that it was a substance or merely something associated with motion.[4] Berthollet too had said that caloric might be either a substance or a force.[5] However it was usually treated as a substance or more precisely as a 'subtle fluid' which was material and yet did not have weight. Many experiments in which bodies were weighed first cold and then hot had failed to demonstrate conclusively that caloric had weight. Gay-Lussac proposed to his students an experiment in which a tightly stoppered flask containing a layer of concentrated sulphuric acid covered by a layer of water was weighed and then shaken; heat was evolved and the flask was reweighed.[6] This experiment also showed that the heat produced had no weight, or as Gay-Lussac remarked, if one wanted to be more precise and keep strictly to experimental evidence one could say that the caloric produced weighed less than was possible to detect by even the most sensitive balance.

Gay-Lussac had been able to write his first memoir, that on the thermal expansion of gases (1802), without involving himself in discussions about caloric. His contribution to debates on caloric arose through his interest in the specific heats of gases. In his research with Humboldt in the winter of 1804–5 on the proportions in which hydrogen and oxygen would react when sparked together, Gay-Lussac had noticed that the gases did not combine together completely when the proportion of oxygen to hydrogen exceeded 10 : 1 by volume. Moreover when excess nitrogen was substituted for excess oxygen the inhibiting effect was similar. As oxygen and nitrogen were so dissimilar in chemical properties, Gay-Lussac attributed this effect to the absorption of heat by the two gases so that the temperature fell below that necessary for combustion. It followed that the gases were thermally equivalent, i.e. they had the same specific heats. Gay-Lussac's first guess was that all gases might have the same specific heat. But, as gases expanded equally on heating, and at a given temperature their volumes were directly proportional to the pressure, a convenient way of testing this hypothesis was to allow different gases to expand equally and see if the temperature changes were the same.[7] He took two 12-litre flasks, which were thoroughly dried with anhydrous calcium chloride, and then evacuated. One flask was then filled with the gas under test, the flasks connected with a lead pipe and the taps controlling each flask opened. Doing the experiment with air successively at atmospheric pressure, half atmospheric pressure and one quarter atmospheric pressure, Gay-Lussac found that the fall in temperature of the first flask was equal to the rise in temperature in the second. In one case where they were not, he considered that these were equal within the limits of experimental error. His table of mean results was set out as follows:

| Density of the air measured by barometric pressure | Cold produced in first flask | Heat produced in second flask |
| --- | --- | --- |
| 0.76 m | 0.61 °C | 0.58 °C |
| 0.38 m | 0.34 °C | 0.34 °C |
| 0.19 m | 0.20 °C | 0.20 °C |

In view of its later significance for the law of conservation of energy,[8] the conclusion that the heat lost by expansion is equal to the heat gained by compression is particularly famous in the history of physics. Gay-Lussac also wished to conclude that the heating/cooling effect was proportional to the pressure of the gas allowed to expand into the vacuum. Although 0.34°C might seem a reasonable approximation to one half of 0.61°C, the final figure of 0.20°C did not fit into this pattern. Gay-Lussac, however, attributed this to the time taken (about two minutes) before the thermometer recorded the maximum difference. His provisional conclusion was therefore that the change in temperature was directly proportional to the change in density of a given gas.

Repeating the experiment with hydrogen produced an immediate effect on the thermometer when the connecting taps were opened, the speed with which a gas passed through an orifice being inversely proportional to the square root of the density of the gas. Gay-Lussac therefore had a special tap constructed by the instrument maker Fortin so that the orifice could be varied. In this way the time taken for the hydrogen to pass into the vacuum could be made equal to that taken by the air in the first experiment. Under these comparable conditions Gay-Lussac found the temperature difference produced by hydrogen greater than that produced by air, whereas in further experiments with carbon dioxide it was less; with oxygen too it was less and he felt justified in concluding that the specific heats of gases were inversely proportional to their densities. However, he put forward this conclusion only in a tentative way[9] and remarked in particular that the masses of the flasks and thermometer were very great in comparison with that of the gases, so that the thermometer readings were not a measure of all the heat absorbed or given out.[10]

Gay-Lussac's discovery that the heat lost by expansion of a gas was equal to the heat gained in the second flask is of historic importance. Perhaps if this memoir had been translated into English its implications would have been realised sooner and it would not have been left to Joule in 1845 independently to re-discover this fact. However, although much of the memoir is pioneering in spirit, the experimental work is rather crude, the quantification approximate and, as a historian of science has remarked, the memoir was 'hardly worthy of him'.[11] For anyone who sees Gay-Lussac essentially as a physicist it must be puzzling that for five years during his most creative period he had nothing further to contribute on the subject of heat. Indeed his return to the study of the specific heats of gases seems to be connected with his position as a member of the physics section of the First Class of the Institute. Being elected in December 1810 as a member of a commission to choose a subject for the physics prize for 1812, it may well have been he who suggested the subject of the specific heats of gases as an area requiring further study and also one which would throw more light on the caloric theory.[12] Entries for the prize were to be submitted by 1 October 1812 and it might be expected that Gay-Lussac as a potential judge would not himself make any contributions on the subject. It is therefore surprising to find him presenting a short memoir on the subject to his colleagues at the Institute on 20 January 1812.

In this[13] he applied to gases the method of mixtures used in eighteenth-century calorimetry for liquids. Equal volumes of two gases were taken, one at a high temperature and the other at a low temperature. After mixing, the final temperature was recorded. Gay-Lussac devised a piece of apparatus consisting of two gasholders of 8-litre capacity with a tube designed to allow the two gases to mix in equal volumes. Before mixing they passed through spiral tubes, one immersed in a freezing mixture at say 21 degrees below room temperature and the other immersed in hot water 21 degrees above room temperature. Equivalence was extended to having tubes of equal length, etc., so that heat gains of the one gas would be compensated for by heat gains of the other. Hydrogen, oxygen, nitrogen, carbonic acid and air were mixed with each other in successive experiments, each experiment being repeated at least four times. He concluded:

It seems to follow from these experiments that the above gases, and probably all elastic fluids have at the same volume and pressure, the same capacity for heat. From the point of view of weight this result agrees with that which I announced five years ago, namely that the less the density of the gas the greater is its capacity for heat. But I had not discovered then according to what law this capacity varied...

His conclusion was that the (volume) specific heats of all gases were the same. This took him back to his guess of 1805 and involved a rejection of his own experimental work of 1806. However, he said that he would have liked to have repeated the experiments on a larger scale but had been prevented by various circumstances. It was just as well he entered this caveat because he soon found that he had been mistaken.

In his efforts to make a permanent contribution to the laws of specific heats of gases he had been too hasty. He inserted a note in the number of *Annales de chimie* dated 31 July 1812[14] admitting that the specific heats of gases were not equal, although his experimental figures varied according to which of the gases were heated and which cooled. There was obviously room for further research. However, in view of the research then being carried on by his friends Bérard and Delaroche, he was happy to leave further exploration of the question of specific heats of gases to them.

Gay-Lussac seems to have abandoned different parts of his early work on heat as they were taken on by other younger colleagues associated with the Arcueil group. At least it would be true to say that they took further work he had started. His work on the specific heat of gases was continued by Delaroche and Bérard. In the same year, 1815, Dulong and Petit came forward with a further study comparing the mercury and air thermometers. Gay-Lussac at the beginning of his career had undertaken a correlation of these two thermometers over the temperature range 0–100°C.[15] It was only appropriate that in 1815 when Dulong and Petit extended this work to higher temperatures and presented their memoir to the First Class of the Institute, Gay-Lussac should have been one of the two commissioners appointed to examine and report on the research. The report signed jointly by Gay-Lussac and Biot[16] provides a further – though little known – contribution to the subject of heat.

Another contribution by Gay-Lussac to the study of heat was read at the Institute on 6 March 1815. He was obviously not satisfied with it as he deliberately withheld it from publication until 1822, when he realised that he was not likely to have the time and energy to carry out the further experiments he had planned. He thought, however, that it might be useful to place in the public domain the results of his experiments of 1815 on the cold produced by the evaporation of liquids.[17] He pointed out that under certain conditions the heat absorbed by the evaporation of a liquid would be equal to the heat passing from the walls of the vessel. However if the liquid were allowed to evaporate into a vacuum surrounded by a freezing mixture the degree of cold could be increased indefinitely as long as the liquid exerted an appreciable vapour pressure. John Leslie had reached a temperature almost as low as the melting point of mercury by evaporation of ether, but Gay-Lussac now succeeded with ease in freezing mercury to a solid block by the evaporation of water vapour in a vessel surrounded by a freezing mixture. Gay-Lussac had no doubt that with a volatile liquid it would be possible to obtain even lower temperatures.

If the liquid were allowed to evaporate in a gas instead of a vacuum the cooling would not be so great. He worked out a formula for the cooling produced by a liquid and drew up a table to compare the cooling produced by a dry current of air on a thermometer bulb covered with moistened cambric with that worked out from his formula.

Gay-Lussac returned to the subject of the artificial production of cold in 1818.[18] Once again one can see in it a development of his fruitful paper of 1806. His earlier hopes of almost unlimited cooling by vaporisation having not been realised, he looked again at his important result of the equivalence between cooling caused by expansion of a gas and heating caused by compression. It had been found that by compressing air to one fifth of its original volume it was possible to attain a temperature of 300°C and Gay-Lussac thought this might be increased to 1000 or even 2000°C. Now if the heat produced is carried off and this compressed air is allowed to expand, it should absorb as much heat as was given out in its compression and its temperature should be lowered by 300°C (assuming the specific heat of air is constant). Therefore, said Gay-Lussac,

If we take a mass of air compressed to 50, 100 atmospheres, the cold produced by its instantaneous expansion will have no limit.

This paper was quite outspoken in its optimism. Gay-Lussac concluded:

If it is beyond doubt that an unlimited cold may be produced by the expansion of gases, the determination of the absolute zero of heat must seem a completely imaginary question.

Gay-Lussac had therefore put his authority against the concept of an absolute zero. Desormes and Clément, however, argued that there was an absolute zero and that Gay-Lussac had himself shown what this temperature was.[19] According to his own work the coefficient of thermal expansion of gases was $1/266.66$ per deg C. Hence there was a limit of contraction at $-266.66°C$ which was the absolute zero.

The other question over which Clément and Desormes took issue with Gay-Lussac was whether a vacuum could contain caloric. Views about the effect of volume changes in a gas on its heat content divided supporters of caloric theory, Gay-Lussac taking one view and Clément and Desormes taking another, following the interpretation of the Scottish school of Irvine and Crawford.[20] Gay-Lussac gave his opinion in his 1806 paper[21] that there was no truth in the argument that a vacuum contained more caloric than when air was present. In their entry for the 1812 prize on specific heats of gases Clément and Desormes made much of the doctrine of vacuum caloric and their decision in 1819 to publish not only their (rejected) prize memoir but a further memoir on the subject brought a reply from Gay-Lussac.[22] He considered that the idea of vacuum caloric could be refuted by means of a crucial experiment. On the implicit assumption that caloric was an elastic fluid, there should be a heat change when its volume was changed.[23] As an increase or decrease in the volume of a vacuum had no effect on a thermometer, Gay-Lussac considered that this negative result confirmed his view.

Gay-Lussac had not quite finished with the study of heat. On 29 April 1822 he announced to his colleagues at the Academy that he was undertaking further research with his friend Welter on the thermal effects of allowing a gas to expand.[24] Although they had not completed their research, they had concluded that when air escapes from a container under pressure its temperature does not change, since the heat produced by escaping through the hole was equal to the cooling produced by expansion. This research was never fully written up but they evidently continued their work, since Laplace in a note on the velocity of sound, published later that summer, was able to use their latest figures for the ratio of the specific heat of a gas at constant pressure and at constant volume:

This important ratio may be deduced with great exactitude from the interesting experiments which MM. Gay-Lussac and Welter are carrying out at this moment on the compression of air. Four of these experiments, done at an atmospheric pressure of 757 mm and which these learned physical scientists ('physiciens') have kindly communicated to me, have given this ratio as 1.3748. The extreme results only differ from this mean result by one part in 136.[25]

Since Gay-Lussac had been using dry air, Laplace allowed a correction for the water vapour in the atmosphere; he deduced a theoretical value of 337.8 metres per second for the velocity of sound, which compared quite well with the experimental value of 340.9 metres per second. Although Laplace's regard for the high accuracy of the experimental value obtained by Gay-Lussac proved to be over-optimistic, he can be given credit for ending the century-old scandal due to blatant discrepancy between the experimental value of the velocity of sound and the theoretical result from Newton's formula.[26]

Gay-Lussac's later work on heat was marked by greater caution, particularly on the question of caloric. On the question of the absolute quantity of heat in a vacuum he remarked that this

in the present state of our knowledge cannot be resolved except with the aid of numerous hypotheses which are all the less plausible as *we do not know the nature of heat*; and if we cannot regard the question as completely fanciful ('chimérique') we should at least give it very little importance. Observations are still too few for us to be able to tackle such a question successfully...[27]

Thus although Gay-Lussac had always imagined caloric as a subtle fluid, he now admitted that the nature of heat was an open question and in his later physics lectures he told his students of two views of caloric. The first was the traditional one of a subtle fluid, but, he continued:

There is a second hypothesis which has also been adopted for a long time and which begins to be held in increasing favour. This second hypothesis is that there exists an extremely subtle fluid spread out indefinitely in space, a fluid which was denoted a very long time ago by the name of *ether* or *ethereal fluid*, and it conceives all the molecules of a heated body in a special state of vibration, of to and fro motion, undergoing a small

oscillatory movement, which motion produces the phenomena, the sen-
sations and the effects of heat...The latter hypothesis receives more
support from the discoveries which have been made on light, yet we shall
not use it at all in this course for the explanation of phenomena because
science is not yet sufficiently advanced in this respect for this hypothesis
to lend itself equally to the explanation of all phenomena. Consequently
we shall retain the old hypothesis, which considers caloric as a particular
fluid which can remain in a body.[28]

Considering the contributions to electricity and electrochemistry by
many European men of science including Ritter, Davy and Berzelius,
in the fifteen years following Volta's electric pile of 1800, it may seem
surprising that Gay-Lussac did not contribute more in this field. However, if one relates research to national background one finds the
leading contributions from German, British or Scandinavian sources
with comparatively little interest in France. Gay-Lussac was unusual in
doing anything in this area in the Napoleonic period and an article on
electricity in nineteenth-century France singles out the research of
Gay-Lussac and Thenard as practically the only work done on electricity in France for more than a decade after the initial enthusiasm in
1801–3 following Volta's presentation of his views to the Institute and
the subsequent report by Biot.[29]

Gay-Lussac's earliest interest in electrical effects is reflected in a
joint paper with Humboldt published in 1805 on the electric torpedo
or gymnotus.[30] In 1807 Gay-Lussac was interested in the chemical
effects of current electricity, particularly the question of whether acids
or alkalis were formed in the electrolysis of water. The Italian
Pacchiani had made such a claim but it had been challenged. Gay-
Lussac presented to the Philomatic Society a review of recent research
on this problem,[31] and mentioned contributions by Thenard, Biot, Pfaff
and Davy, whose work translated by Berthollet had appeared in a
recent issue of the *Annales de chimie*.

In this and succeeding issues of the journal of the Philomatic Society
Gay-Lussac was concerned merely to give an exposition of work done.
Meanwhile news of Davy's decomposition of the alkalis had reached
Paris and this called for action rather than passive commentary. The
action taken by Gay-Lussac in collaboration with Thenard was to
repeat the decomposition of the alkalis on a larger scale by *chemical*
means and their conclusion was that 'chemical agents are at least as
powerful as the electric fluid'.[32] This might suggest a repudiation of
the latest tool of research; yet that summer Gay-Lussac and Thenard
were entrusted with the large new voltaic pile constructed at the Ecole
Polytechnique. In the general essay prefacing their published research

they accepted Biot's electrostatic interpretation of galvanism but also described some new experiments which were difficult to reconcile with it. Collecting volumes of gases evolved in electrolysis, they argued that this was a measure of chemical energy which was different from the electrostatic force of the pile.[33] They thus raised the interesting question of whether a voltaic pile had chemical energy which was distinct from electrical energy. The term 'energy' at this time should not be given a post-1840 meaning yet it was, of course, questions such as this which were to lead to the concept of energy existing in different forms.

In 1816 we once again find Gay-Lussac writing on the subject of the voltaic pile.[34] However this was not a research memoir but a historical review prompted by a paper read to the Institute. Gay-Lussac was particularly concerned to claim priority for the French in constructing dry batteries, although he pointed out that some liquid must be present for the battery to work.

As 'physics' one cannot help being disappointed by Gay-Lussac's contribution to electricity. When he was making original contributions it was the chemical properties of the pile on which he, like Davy, concentrated. Electricity never seems to have caught his imagination. Whereas he felt that heat was a fundamental problem affecting chemical reactions, he could not think of electricity as having the same importance. His friend Thenard had a greater love of the spectacular but he too failed to make any really significant advance in this science. It was Berzelius who introduced a fundamental electrical viewpoint into chemistry but this came only after Gay-Lussac's most creative period.

Our conclusion on Gay-Lussac's physical work must be that much of what appears unambiguously as 'physics' was on closer examination more like simple physical chemistry. He did important work on heat but this study had been considered a part of chemistry in the eighteenth century and Gay-Lussac was still working largely within this tradition. The simple quantitative work of Gay-Lussac is in a quite different world from the sophisticated mathematics of Joseph Fourier or the equations which one finds as a regular feature of the memoirs of J. B. Biot.

Gay-Lussac's main interest in 'physics' came early but his interest in the middle of his life in heat and related phenomena continued. One does find an increasing interest under the Restoration with meteorology, which Gay-Lussac regarded as a part of 'physics'. As he continued in the Academy in the physics section, it was understandable that he should retain some links with physics.

The final judgement can come from Gay-Lussac's friends and contemporaries. They recognised a distinction between physics and chemistry and Gay-Lussac's respective contributions were nicely expressed by Arago when he wrote that Gay-Lussac was a clever physicist but an outstanding chemist.[35] In other words it was as a chemist that he was outstanding although the fact that he contributed to physics may be independently confirmed by turning to any detailed history of that subject. Biot considered that Gay-Lussac was only really involved in physics up to about 1807 – but long enough to have a life-long affiliation at the Institute. Biot described his contributions to chemistry as 'the most brilliant as well as the most permanent part of his work'.[36]

If Gay-Lussac was not unambiguously a physicist in the content or motivation of his work, he was perhaps more in his approach to science. He often reduced problems to the investigation of variables, either implicitly or explicitly. His first research on the thermal expansion of gases involved investigating the correlation between changes in temperature and changes in volume at constant pressure. In his memoir on the expansion of liquids he comments:

Bodies. . .obey opposing forces, the relation and the intensity of which are very variable; and yet it is only by placing them exactly in the same circumstances that one may hope to discover some relation between their properties.[37]

It is in considering Gay-Lussac's methodology that we discover a unity of his work which might not be so obvious if we stressed subject boundaries too much. Gay-Lussac's interest in correlation and general laws discovered by precise measurement provided a unifying framework for investigations which might now be classified as physics, chemistry and meteorology.

A few words must be said about Gay-Lussac's work in meteorology. Weather phenomena were an ever-present challenge to anyone who considered the physical world should be reduced to physics and chemistry. Meteorology also constituted a possible field for the discovery of new laws. The balloon ascents of 1804 provided some scope for this investigation but meteorology was an interest which Gay-Lussac kept up throughout his life. In this he was influenced by Humboldt and later by Arago, his co-editor of the *Annales de chimie et de physique*. Gay-Lussac's interest in thermometry and the gaseous state also had obvious implications for meteorology. It was in the early years of the Restoration, the year marking the beginning of his co-editorship with Arago, that he designed a new maximum and minimum thermometer and, more famous, a portable barometer. Indeed many portable barometers are referred to as 'Gay-Lussac type', even though they differ from his design. Gay-Lussac's siphon barometer[38] was a development of the simple 'J' type, consisting of two vertical tubes joined by a bent heavy narrow-bore tube with the two large tubes one nearly above the other. The lower tube which acted as the reservoir had a tiny hole above the mercury, thus allowing the free access of air. Unfortunately, once the glass was sealed, the mercury could not be cleaned and this presented a long-term problem. However, Gay-Lussac's barometer was popular for the rest of his lifetime and was manufactured in large numbers.[39] Although Gay-Lussac made valuable contributions to instrumentation he must have felt dissatisfied that in meteorology he had failed to make any breakthrough. In 1818 he was complaining that in meteorology (unlike chemistry and physics) 'the phenomena are so little accessible to our researches'.[40] In these circumstances he considered himself entitled to the luxury of indulging in conjectures within certain limits.

## The physical chemist

It may be permissible to describe Gay-Lussac as an early physical chemist. To call him this would be an attempt to characterise a whole approach to chemistry rather than to claim him exclusively for any one branch of the subject. Indeed Gay-Lussac was a distinguished organic chemist, inorganic chemist, analytical chemist and technical chemist. To characterise Gay-Lussac's chemistry as physical is to emphasise its generality. He felt that the business of the chemist was to do more than prepare and investigate the properties of individual chemical species. His ultimate goal may have been a 'chemistry without substances'. He was interested in the gaseous state rather than individual gases. His first research was to show that all gases expand equally on heating. Thus, regardless of chemical differences (whether a gas was acidic or basic) or physical differences (whether a gas was soluble in water or comparatively insoluble) a gas could be treated as an example of the general concept of gas rather than as an individual.

In the neutralisation of an acid by a base any base will serve provided the appropriate equivalent weight is used:

We can use any base indiscriminately provided that we take the appropriate quantity.[41]

In one of Gay-Lussac's late memoirs he took this further as a principle of 'equipollence' or 'indifference of permutation' saying that 'acids combine indifferently with bases and vice versa: the order of combination has little effect provided that acidity and alkalinity are satisfied'.[42] Ultimately the course of a chemical reaction depended on physical factors: insolubility, density, volatility, etc. Gay-Lussac had earlier made a study of the effect of light on gaseous reactions, thus helping to lay the foundations of photochemistry.[43]

## Prussic acid and cyanogen

Gay-Lussac's research on prussic acid, presented in a memoir to the Institute in September 1815, is characterised as 'outstanding' by Partington[44] and as 'beautiful' by another historian of chemistry.[45] Perhaps the most impressive commentary on the research came from Humphry Davy. He was happy in 1816 to 'offer...experimental confirmation of the very elaborate and ingenious researches of M. Gay-Lussac on the prussic acid and the prussic base' – 'a subject which is peculiarly M. Gay-Lussac's'.[46] It was not that Gay-Lussac had discovered the compound – it had been first isolated in 1783 by C. W. Scheele. Some forty years later Gay-Lussac could write that 'few bodies have been more studied and yet few are less known'.[47] Among those who examined the new acid and its salts was Berthollet, who had been unable to obtain oxygen from the acid.[48] However, before concluding definitely that this was an example of an acid containing no oxygen, Berthollet had said that it would be necessary to prepare pure prussic acid. This was one of Gay-Lussac's achievements, but we can see his work not merely as a simple extension of Berthollet's work but also as an immediate response to interesting but partly erroneous work of the British chemist Robert Porrett, published in Thomson's *Annals of Philosophy* in 1814 and 1815, to which Gay-Lussac himself refers in his introductory historical section.

If the memoir deserves to be described as 'beautiful' it might be partly because of the orderly presentation of the research but probably more because of the economy and simplicity of the research itself. Gay-Lussac showed conclusively that prussic acid was simply hydrocyanic acid, a compound of carbon, hydrogen and nitrogen, and he announced the isolation of the cyanogen radical.

Gay-Lussac had already prepared pure prussic acid in 1811 by the action of hydrochloric acid on mercuric cyanide.[49] His method of volumetric analysis of the vapour 'taking advantage of the hot days of the month of August' has already been described. He was able to compare his analysis with that of Porrett who had given the same elements but in quite different proportions, the hydrogen varying by a factor of ten.[50] He concluded that it was a hydracid for which he proposed the name hydrocyanic acid from the radical which he called cyanogen (from κυανος = blue).

One of Gay-Lussac's triumphs in this memoir was to report the isolation of cyanogen, which he presented as the radical of prussic acid. He prepared it simply by heating mercuric cyanide, which decomposed into mercury and cyanogen gas. Several other chemists must have heated mercuric cyanide before Gay-Lussac, but the latter observed two main precautions. He first took care to obtain it as a perfectly pure neutral crystalline salt by digesting red oxide of mercury with Prussian blue and removing all traces of iron. Proust had gone so far but had not dried his salt before testing its reactions. Once again (as with his investigation of the thermal expansion of gases) the care taken by Gay-Lussac to ensure dry conditions paid dividends.

Cyanogen was a gas with characteristic smell. It burned with a bluish flame and was soluble in water. Its specific gravity compared with air was 1.8064. Gay-Lussac found that sulphur, phosphorus and iodine could be volatilised in the gas without undergoing any reaction but if iron was heated to white heat it partly decomposed cyanogen. It combined directly with potassium to form potassium cyanide.

However, the most important single question about cyanogen was its composition. Gay-Lussac found by detonating it with excess of oxygen that cyanogen contained a sufficient quantity of carbon to produce twice its volume of carbonic acid gas, 'that is to say, 2 volumes of the vapour of carbon and 1 volume of nitrogen, condensed into a single volume'.

One volume of cyanogen combined with 1 volume of hydrogen to produce 2 volumes of hydrocyanic acid vapour and Gay-Lussac pointed to the analogy with chlorine and iodine, each of which reacted in 1 volume with 1 volume of hydrogen to form 2 volumes of the corresponding acid. Gay-Lussac also investigated the compound formed between chlorine and cyanogen, a reaction discovered by Berthollet. He called the compound chlorocyanic acid and remarked on the fact that aqueous solution produced no precipitate with silver nitrate. He took some trouble to determine the composition of chlorocyanic acid, which he expressed as:

1 volume of the vapour of carbon
$\frac{1}{2}$ volume of nitrogen
$\frac{1}{2}$ volume of chlorine

Two of the most interesting aspects of Gay-Lussac's work are his use of analogies and his contribution to the theory of radicals, a fundamental concept of Lavoisier's chemistry. When he burned potassium metal in hydrogen cyanide gas he found that 100 parts decomposed to form 50 of hydrogen. A solution of the product (potassium cyanide) was identical with that obtained by the action of potash on the liquid acid. Potassium reacted similarly with chlorine and iodine and he pointed out the

very great analogy between prussic acid and muriatic and hydriodic acids. Like them, it contains half its volume of hydrogen; and, like them, it contains a radical which combines with the potassium, and forms a compound quite analogous to the chloride and iodide of potassium. *The only difference is, that this radical is compound, while those of the chloride and iodide are simple.*[51]

Gay-Lussac had done more, however, than merely point out an analogy between radicals. He went on to isolate the radical as cyanogen (in modern symbols $(CN)_2$):

This gas, when it combines with hydrogen, shows us a remarkable example, and hitherto unique, of *a body which, though compound, acts the part of a simple substance* in its combinations with hydrogen and metals.[52]

He made a rare claim – that the discovery of cyanogen opened a new field of research of particular importance. Ampère described the study of compound radicals as 'a new branch of chemistry, of which M. Gay-Lussac had the first idea'.[53] Liebig, writing in 1840, said that the discovery of cyanogen and its compounds was one of the most fruitful of all in organic chemistry.[54]

Yet Gay-Lussac's work on cyanogen did not have the impact that it might have had on organic chemistry since most chemists thought of these compounds as belonging to *inorganic* chemistry. Gay-Lussac complains of lack of time in preparing his memoir. If he had had more leisure to speculate about the implications of his work he might have seen that, not only because cyanogen contained carbon, but *because it was a compound radical*, it was analogous to other compound radicals which were organic. This distinction had to wait more than twenty years for the famous joint manifesto of Dumas and Liebig; with its famous phrase:

In mineral chemistry the radicals are simple; in organic chemistry the radicals are compound; that is all the difference.[55]

### The problem of acidity

Until Gay-Lussac's research on iodine (1813–14) and prussic acid (1815), he had accepted the general principles of Lavoisier's theory of acidity based on oxygen as the unique acidifying principle. It is true that Berthollet had pointed out the curious case of hydrogen sulphide which formed an acid in solution apparently independently of oxygen. Yet Berthollet had not pressed the point.

Gay-Lussac's discovery and investigation of hydriodic acid reopened the whole question and his main discussion of acidity was published as an appendix to his monograph on iodine.[56] Gay-Lussac did not doubt that oxygen was an acidifying principle and felt in cases of compounds of oxygen which were not acidic that the acidic characteristic of the oxygen had been drowned by other constituents. It was not possible to work out a fully consistent theory in terms of property-bearing principles. There was no simple correlation even between the proportion of oxygen in a compound and its acidity. In gravimetric terms the composition of neutral water was (according to Lavoisier's analysis) an embarrassingly high 85 per cent oxygen although, when seen through Gay-Lussac's eyes in volumetric terms, it was only one part of oxygen to two of hydrogen. Gay-Lussac, drawing on Berthollet, introduced a second parameter – 'condensation'. A compound which contained more than 50 per cent by volume of oxygen *ought* to be acidic but it might not be if it was sufficiently condensed. When viewed historically, we can see Gay-Lussac as inheriting concepts from both Lavoisier and Berthollet and trying to fuse them into a synthesis to tackle an increasingly wide range of chemical phenomena; when viewed philosophically, his hypothesis seems very *ad hoc*. It is a struggle to save the oxygen theory of acidity and his theory, far from having any predictive value, was insufficient to account for all the existing phenomena.

However, Gay-Lussac should not be dismissed too easily as a reactionary. His second extension to the oxygen theory is more interesting and also has a greater empirical content. Although he did not abandon the idea that oxygen was an acidifying principle, he referred to this element significantly as 'the *principal* acidifying substance'.[57] In other words oxygen was no longer the *unique* principle of acidity. He was prepared to extend this property to chlorine, iodine, sulphur, phosphorus and carbon. Thus Gay-Lussac departed significantly from Lavoisier's oxygen-centred chemistry. At a time when Berzelius considered oxygen unique (and still thought of chlorine as a compound of oxygen), Gay-Lussac had moved to a modified position.

Gay-Lussac's single most important contribution to the theory of acidity was to introduce the concept and the name 'hydracid'. After his detailed work on the compound of hydrogen and iodine including a study of the salts there could be no doubt that this was a new acid: hydriodic acid.

Gay-Lussac recognised clearly that 'hydrogen was constantly necessary to convert iodine to the state of an acid'. He therefore concluded that hydrogen 'plays the same role in nature in respect of a certain class of bodies as oxygen does for another class'.[58] One may note here that it would have been possible for Gay-Lussac to have saved the oxygen theory of acids on the hypothesis that iodine was really a compound containing some oxygen. He rejected this possibility in favour of a new theory which constitutes an important landmark in the history of post-Lavoisierian chemistry. He divided acids into two classes.[59] The first we should call oxy-acids' but Gay-Lussac, unwilling to break with the past, calls them 'real acids' ('acides proprement dits'). The second category consists of acids containing hydrogen and another substance. This class includes hydrochloric acid, hydriodic acid and the acid formed by hydrogen sulphide. Gay-Lussac marked this important new classification by introducing the term 'hydracid' to describe these and it was used in chemistry for the next hundred years.

We must return to the difficult question of the cause of acidity. No one could unblushingly argue that in the first class of acid described above oxygen was the cause of acidity, while in the second class the acidity was explained by the presence of hydrogen. Gay-Lussac says that in the case of the three hydracids mentioned above it was probable that the chlorine, iodine and sulphur respectively were the acidifying principles. This was a very reasonable claim since he had argued elsewhere the close analogy of these elements with oxygen. If oxygen had strong acidifying properties, elements similar to oxygen would also be expected to produce acidic properties in combination. Electrochemistry would have enabled Gay-Lussac to express this more clearly as the properties of electro-negative elements. But he refused to do this because it was too absolute. Acidity and alkalinity were relative and Gay-Lussac took temporary refuge in a purely operational definition of an acid:

Considering the word 'acid' in the most general way, for a long time an acid has been for me no more than a body, whether containing oxygen or not, which neutralises alkalinity...[60]

However, Gay-Lussac also dallied with a new and original theory of acidity according to which the property of acidity was a consequence of the internal arrangement of the atoms in a compound. He writes:

The arrangement of atoms ('molécules') in a compound has the greatest influence on the neutral, acidic or alkaline character of the compound.[61]

Such a statement did not come easily from the pen of Gay-Lussac. As he said two years later in another context:

Bodies hide from us the number, the form and the disposition of their particles.[62]

His lack of confidence in this corpuscular hypothesis is suggested by the fact that he was simultaneously trying to construct a theory of acidifying principles.[63]

Gay-Lussac's consideration of the arrangement of particles is less important in the history of acidity than for its general implications in chemistry and particularly in the theory of isomerism.

## Isomerism

Gay-Lussac made some important contributions to the recognition and explanation of isomerism. He was involved with this newly recognised phenomenon in his respective roles as researcher, editor and teacher. As a researcher he was one of the first to call attention in 1814 to the existence of compounds with apparently identical chemical composition but quite different properties. Then years later, as editor of the *Annales de chimie et de physique*, he published a paper of Wöhler with an analysis of silver cyanate and pointed out that, if the analysis was correct, this compound had the same composition as Liebig's silver fulminate.[64] Finally, as a teacher he had to explain to his students this phenomenon, to which Berzelius in 1830 had given the name isomerism. Once the phenomenon was well recognised the next stop was to distinguish different kinds of isomerism. Gay-Lussac, however, belongs to the first generation who helped to establish the existence of the phenomenon.

Gay-Lussac's first contribution to ideas on isomerism came in 1814. He found that acetic acid and cellulose contained the same elements in the same proportion by weight.[65] This struck him forcibly because he had previously classified organic compounds into different groups according to their composition, acids having an excess of oxygen while neutral compounds contained hydrogen and oxygen in the same proportions as water. It was because acetic acid and cellulose were quite different types of compounds that he remarked on their similar composition. In 1811,[66] he and Thenard had produced the analyses:

| Sugar | Gum arabic |
|---|---|
| C = 42.47 | C = 42.23 |
| O = 50.63 | O = 50.84 |
| H = 6.90 | H = 6.93 |

This agreement to 1 part in 250 would often have been interpreted by Gay-Lussac as identity within the limits of experimental error but curiously he made no comment at that time, leaving the first claim of identity of composition until his note of three years later.

A second case involved two forms of stannic oxide and stannic chloride of identical composition. Gay-Lussac again attributed this to the different arrangement of the particles.[67] This apparent difference can now be explained in terms of polymorphism and different water content rather than isomerism. On the other hand, Gay-Lussac, who gave racemic acid that name, attributed to it a slightly different composition from tartaric acid although it was soon found to have identical composition.[68] Again in the case of urea and ammonium cyanate, which Wöhler (1828) claimed as chemically identical, Gay-Lussac's immediate reaction was to prefer a simple chemical explanation – that urea was ammonium cyanate with one molecule of water of crystallisation.[69] Such judgements are a reminder that pioneers often make mistakes. What is clear, however, is that Gay-Lussac recognised the phenomenon of isomerism and did his best to account for each apparent case.

Research on cyanogen in 1815 led Gay-Lussac to announce that cyanogen combined with oxygen to form two distinct series of salts: the stable cyanates and the unstable fulminates (mercury fulminate was used as a detonator).

This is how he explained the possibility of isomerism:

There cannot be different kinds of combination between two bodies but there can be differences between three and even more between four bodies. If you have two bodies A and B and combine them together, you always have the same result, AB. If you have three bodies, A, B and C, you can form several kinds of combinations; first you can combine A with B or A with C or even B with C and then combine the binary compound AB or AC or BC with the third body.

It may ('peut') be like this in the combination of cyanogen with oxygen. We have three elements: carbon, nitrogen and oxygen; they can form different binary compounds and consequently produce different ternary compounds.[70]

We might call this an algebraic rather than a geometrical approach. Gay-Lussac thinks that AB = BA but ABC $\neq$ ACB. He had no concept of different numbers of atoms within a molecule and did not consider that AB $\neq$ $A_2B_2$. Gay-Lussac was not thinking in spatial terms. The basic modern concept of structure came only a full generation later. Without wishing to claim that Gay-Lussac anticipated the work of Butlerov, Kekulé and van't Hoff, he was contributing cautiously to the interpretation of evidence and stands in the front rank with Berzelius as one of the fathers of isomerism.

## Atomic weights and equivalents

Nineteenth-century chemistry has sometimes been interpreted in terms of 'atomic debates'. Dalton's atomic theory, published in 1808, was to make possible, as we now appreciate, a major advance in chemistry. Dalton's book was in the hands of the Arcueil circle very soon after it was published and we must consider what Gay-Lussac made of it, remembering, however, that as has been pointed out, 'only with hindsight is it clear that Dalton's was the most fruitful path to tread'.[71]

Dalton saw weight as the fundamental parameter. He applied this to Lavoisier's elements which he supposed to consist of atoms. The existence of such atoms was encouraged in his mind by his being able to conceive a mechanical model of them, like lead shot. These first two factors are present in the French tradition but not the last two. Thus Lavoisier made extensive use of the balance and one of his most important contributions to chemistry was to draw up a list of elements. Gay-Lussac grew up to accept both of these ideas but he, like Lavoisier and Berthollet, could have little sympathy for such speculative ideas as those of invisible atoms and for a Frenchman the fact that one could describe a mechanical model was totally irrelevant.

Most of Dalton's contemporaries regarded his atomic theory with caution and among the more sceptical was Berthollet. Berthollet had already come across some of Dalton's work when he had been writing the *Statique chimique.* Dalton, using his vivid pictorial sense, had drawn up a scheme representing the relation of the gases of the atmosphere. This arbitrary scheme had been scornfully described by Berthollet as 'un tableau d'imagination'.[72] Berthollet was a little kinder to Dalton in 1808 but equally sceptical when he heard about the atomic theory. It was, he said, an ingenious hypothesis 'but the more seductive it is, the more necessary it becomes to submit it to a close examination'.[73] Berthollet concluded his brief analysis of Dalton's theory by saying that what was wanted was further precise experimentation 'rather than devoting oneself to hypothetical speculations about the number, arrangement and the shape of atoms ('molécules') which escapes all experience'.[74]

Gay-Lussac's first reference to Dalton's atomic theory came in a memoir presented on 31 December 1808; here he explicitly rejected the theory since

It would follow from this mode of looking at compounds that they are formed in constant proportions, the existence of intermediate bodies being excluded, and in this respect Dalton's theory would resemble that of M. Proust.[75]

Proust's ideas on definite proportions had been combated by Berthollet, and Gay-Lussac loyally supported his patron. Thus in addition to his distaste for atomic speculations Gay-Lussac had a second reason (or prejudice) which prevented him receiving the theory sympathetically. Since there was a certain similarity between the idea of ratios of combining volumes of gases and the constant proportions of the atomic theory Gay-Lussac thought it necessary to issue a disclaimer that 'his [Dalton's] researches have no connection with mine'.[76] Nevertheless he took the theory seriously enough to relate his results to it. He called Dalton's theory 'ingenious'[77] and admitted after mentioning the evidence of Wollaston and Thomson that it had 'a large number of facts in its favour'.[78] It is clear therefore that even in Berthollet's own 'house journal' Gay-Lussac did not dismiss Dalton's theory out of hand. For Gay-Lussac it was to remain as one of a number of possibilities.

In his lectures[79] Gay-Lussac first mentioned Dalton to his students as the discoverer of the law of multiple proportions. Having given due credit in that direction, he turned to the atomic theory, pointing out that it was only 'speculation' and a 'supposition' of Dalton. Nevertheless by the 1820s Gay-Lussac could say that 'this theory has been confirmed by a very large number of experiments'. In a later lecture he said that the atomic theory had been 'guessed at' ('devinée') by Dalton and then made the following antithesis between fact and theory:

Dr Wollaston has supported by evidence ('les faits') the principles of the theory of proportions which Mr Dalton had based only on speculations.[80]

It is interesting that this distinction still seemed an important one for Gay-Lussac in the 1820s when he had come to accept the main implications of Dalton's theory. By this time Berthollet's objections seemed less telling and Dulong and Petit had discovered a law relating atomic weights to specific heats.

Dalton fully appreciated Gay-Lussac's scruples. In a lecture of 1830 he remarked:

Gay-Lussac and the leading French chemists. . .[do not] deny the existence of atoms and their combination. . .They go no further than the expressions of facts.[81]

We must now consider how Gay-Lussac approached atoms. It was hardly likely that he would attempt some sort of historical reconstruction of the path which led Dalton to them – considerations of gases in the atmosphere and the relative solubility of gases. Gay-Lussac was, however, deeply interested in the gaseous state and one might think that, having arrived at his law of combining volume of gases, he might relate it to Dalton's theory as Berzelius did. The fact that he did not was probably because he saw his work as a vindication of the volumetric approach over the gravimetric ideas of many of his contemporaries and particularly Dalton. He therefore approached the atomic theory from quite a different direction yet one with a *logical* relation to atomism so that it has been argued that there might be a direct *historical* connection with the development of Dalton's ideas.[82] This alternative approach was that of reacting quantities of acids and bases and a crucial factor here is that the doctrine of chemical equivalence, although originating in the work of the German J. B. Richter, only became widely known when it was incorporated in Berthollet's *Essai de statique chimique*. This was the grammar of Gay-Lussac's chemical education and it is not surprising therefore that he should have taken a great interest in the problem. Even before the publication of Dalton's atomic theory Gay-Lussac had formed several hypotheses about 'the capacity of saturation' of acids and bases.

Berthollet had been interested in equivalent quantities as a measure of their affinity. But because he had at the same time shown that many other variables were involved, affinity studies went into a sharp decline in the early nineteenth century to be replaced by ideas about units or atoms. What has sometimes been considered as a sharp break in chemical studies is found to have some continuity in the study of equivalents. Gay-Lussac later wrote:

Berthollet's principle that the affinity of different acids for a given base is inversely proportional to the weight of each necessary for the neutralisation of equal quantities of that same base...has been abandoned as a measure of affinity. At the time when Berthollet wrote his *Statique chimique* the atomic theory was still little understood; and several years later Berthollet would certainly not have proposed as a measure of affinity a method which gives nothing else than atomic weights or equivalents which we know are independent of chemical attractions [*sic*] or at least only distantly connected with them.[83]

Gay-Lussac came to use the terms 'atom', 'proportion' or 'equivalent' as synonyms[84] when dealing with the reactions of acids and alkalis. The quantity of each required for neutralisation was a constant ratio, independent of units:

We feel ('on sent') therefore that, as bodies combine in large or small quantities atom to atom, and as these ratios are the same, the numbers which we have found to indicate the ratio of the weights could be transposed to the atoms themselves.[85]

Gay-Lussac therefore, by now convinced that 'what is called affinity is something which cannot be measured',[86] was using Berthollet's method to obtain data corresponding to Dalton's atomic weights. The main practical difference was that Gay-Lussac's equivalents were, in the Lavoisier tradition, on the oxygen scale (O=1 or O=10) whereas Dalton had compared the weights of his atoms with those of hydrogen. Another difference between the values of Gay-Lussac and Dalton was that in the case of such compounds as water and ammonia, the French chemist was able to draw on his law of combining volumes of gases to obtain what corresponds to our modern formulae of $H_2O$ and $NH_3$. Dalton, relying on a hypothesis of simplicity and refusing to accept Gay-Lussac's law, considered these as HO and NH.

In 1827 Gay-Lussac and Dulong were appointed by the Academy of Sciences to examine a memoir by Dumas on the atomic theory. The choice of commissioners was a good one. Not only were Gay-Lussac and Dulong eminent chemists but each had a particular interest in atomic theory. Moreover, they held different views on the subject, Gay-Lussac maintaining cautious reserve while Dulong had been in the vanguard of those young Frenchmen who felt that the atomic theory had not been given a fair hearing in France. Yet the report which they drew up[87] represents clearly the opinion of Gay-Lussac:

For some years under the name of 'atomic theory' there has been generally understood [first] a series of well established facts, and [second] more or less probable systematic views which have a necessary connection with the intimate constitution of bodies. These two orders of ideas, which do not afford the same degree of certainty, should be distinguished with care so that laws founded on observation should not be confused with philosophical speculations which are always susceptible to objections.

This preamble with its obsession with the problem of certainty and its distinctions is pure Gay-Lussac. It gives unqualified support to 'laws founded on observation' and treats 'philosophical speculations' with the utmost reserve. The report goes on to mention the law of equivalents. Combining proportions backed up by quantitative analysis were an important advance in chemistry but Gay-Lussac was at pains to distinguish this from the atomic theory which he preferred to call the atomic *system*. Gay-Lussac's very first published reference to Dalton's atomic theory had been to disclaim any connection with his own ideas and to say that Dalton's ideas were part of a 'system'.[88] There is here a reflection of eighteenth-century distrust of 'systems'. Even in the early nineteenth century some Frenchmen could not forget the over-ambitious Descartes, who had tried to explain the entire world of nature in a simple mechanical system.

After expressing all his doubts, Gay-Lussac was prepared to concede the principle that matter was probably composed of atoms typical of each element. Yet the chemist wanted to know how many atoms of each constituent element there were in a compound. This was of course a weak point in Dalton's theory and Gay-Lussac did not neglect to point out that his work on combining volumes of gases could provide invaluable evidence in many cases.

## Fermentation

Gay-Lussac's research extended to areas which would now be considered as biochemistry and bacteriology, although in the early nineteenth century it was not generally appreciated that fermentation involved living organisms. His research on fermentation was inspired by his examination of Appert's process for preserving vegetable and animal matter by putting the substance in a bottle, exposing to the heat of boiling water and then corking very tightly.[89] What struck Gay-Lussac particularly was that grape juice which had been preserved for a whole year began to ferment within a few days of being poured into a fresh vessel. He guessed that fermentation was connected with the presence of air and devised a number of experiments to test his hypothesis.

He transferred grape juice which had kept for a year from the original bottle to a second bottle which he corked tight and then warmed. Fermentation took place but it did not happen in a second bottle warmed but not exposed to the air. He broke off the neck of another bottle of grape juice under mercury so that it was opened in the absence of air. One portion of the juice was then passed through the mercury into a jar containing some oxygen while a second sample was introduced into a jar in the absence of air or oxygen. The first portion fermented in a few days while the second sample gave no sign of fermentation after 40 days. Gay-Lussac now took a further step of considerable value. He absorbed the carbon dioxide given off in fermentation and remarked on the little residue left. Thus he showed that most of the oxygen gas was absorbed. In a further experiment he noted:

I obtained a quantity of carbonic acid gas equal in bulk to a hundred and twenty times the oxygen gas I had added to the grape must; whence it is evident, that, if oxygen be necessary to the commencement of the fermentation, it is not to its continuance; and that the greater part of the carbonic acid produced is the result of the mutual action of the principles of the ferment and those of the saccharine matter.[90]

All this seemed to show a chemical cause for fermentation. Gay-Lussac however commented:

It is very remarkable, that, when a fermentable juice which has been kept a long time, is poured into another vessel, so that it would ferment from having been exposed to the contact of the air, it may readily be deprived of this property, by exposing it anew, in bottles closely corked, to the heat of boiling water.[91]

This experiment should surely have suggested that *heat* rather than the absence of air was the reason for the lack of fermentation. However, as he was looking for a chemical effect, he considered that the heat had such an effect – it caused absorption of the oxygen:

the oxygen absorbed produces a new combination which is no longer capable of exciting fermentation or putrefaction.[92]

This idea fitted in well with Seguin's idea that albumen was the true principle of fermentation since albumen was radically changed by heat. To confirm his original hypothesis he ingeniously contrived to go back to the source. He arranged to press grapes with an iron rod in an atmosphere of hydrogen. Although the vessel was kept warm (15–20°C), after twenty-five days there was no sign of fermentation, although when a little of the juice was exposed to oxygen, it began to ferment after the first day. Gay-Lussac admitted, however, that in a further experiment fermentation began after three weeks in the absence of oxygen; this he attributed to the grapes being in 'a very advanced stage of ripeness'. As oxygen had caused this sample to ferment within thirty-six hours he concluded that at the very least 'oxygen gas is singularly favourable to the development of fermentation'.

Gay-Lussac went on to argue that animal and vegetable substances could be preserved in the absence of oxygen, a conclusion which is strictly true. But he went further than this in assuming that it is the presence of oxygen which causes fermentation. As Pasteur showed two generations later, Gay-Lussac's explanation was incomplete.[93] Some oxygen was required in the initial stages of fermentation of grape juice but it was not an ordinary chemical substance but an organism which brought about the actual fermentation. It was not only Gay-Lussac who failed to consider this possibility. When a quarter of a century later it was suggested that yeast was a living organism, the suggestion was ridiculed by most leading chemists and in particular by Berzelius, Wöhler and Liebig. It was a major achievement when Pasteur was able to give a bacteriological explanation of fermentation. To achieve this he had to repeat more carefully some of the experiments of Gay-Lussac.

Gay-Lussac also made a contribution to the study of respiration, challenging in 1844[94] some of the work of the German H. G. Magnus.

Magnus claimed to have shown that in respiration carbon dioxide was formed in the blood rather than simply in the lungs as Lavoisier had assumed. Gay-Lussac saw that it was not sufficient to prove, as Magnus had done, that the blood contained carbon dioxide. If the products of oxidation only entered the lungs after the return of venous blood, there should be a much higher concentration of carbon dioxide in venous blood than in arterial blood. Gay-Lussac's data, using horses and calves, showed that this was not necessarily the case. Magendie agreed with Gay-Lussac and further research was obviously necessary before one could claim that the physiology of respiration was understood.[95]

# 7

Professor, Academician and editor

'I compare the *gaylussacites* (*gaylussaciens*) of the Institute of
France, or the leaders of this society, to those for whom
boasting takes the place of knowing. If they are *gaylussacites*
they would say: "You cannot attack the principles for which
we stand, because we are everything, and in so far as you
can see anything, it is only through us".'

H. Bodelio[1]

In the life of Gay-Lussac the Revolutionary period (or more precisely
the period after 1795) was the time when he received his basic scientific
education. During the succeeding period of the Consulate Gay-Lussac
was a research assistant to Berthollet, while the ten years of the Empire
(1804–14) were for him a period of intensive scientific research. By the
Restoration of 1815 Gay-Lussac had established an international
reputation for himself. A French 'Who's Who' published in 1817
described him as 'one of the most distinguished physicist-chemists of
the capital'.[2] He was now a part of the establishment. He came to
replace Berthollet all the more naturally as his master had taken the
eclipse of Napoleon as a time to withdraw from the public scene.
Although Gay-Lussac had benefited indirectly through Berthollet by
Napoleonic patronage he had never been formally associated with the
imperial regime. Thus he could serve the restored Bourbon monarchy
with a clear conscience, although perhaps not without a tinge of regret
for old times. It was under the constitutional monarchy of Louis
Philippe that Gay-Lussac reached his peak of power and influence. But
we must return in time to understand not only the basis of Gay-Lussac's
authority but also his livelihood.

In the post-revolutionary period the scientist was primarily a professor. In the new society education had a new importance. The various
institutions of higher education required teachers. In the Jardin des
Plantes, one of the few institutions to survive the Revolution, the
teaching function of the staff was emphasised by each being given the
title of professor. Teaching, once considered a menial employment (or
alternatively the task of the clergy), now enjoyed esteem. Indeed it was

not below the dignity of councillors of state and government ministers, notably Fourcroy and Chaptal, to combine this office with a professorial one. Even Berthollet, who had few talents as a teacher, was enrolled to lecture at the short-lived Ecole Normale of 1795 and later at the Ecole Polytechnique. Although elementary and popular courses of lectures were quite common under the old regime, what distinguished the new movement was the high level of teaching. It was now possible to lecture at the level of research. Even in the Faculty of Science, where instruction in science began at an elementary level, Gay-Lussac and Biot divided the physics course into two parts according to their research interests. All these lecture courses were given considerable publicity and it would have been natural for the public to think of scientists as lecturers rather than researchers. Only in a few outstanding cases, such as that of Gay-Lussac, would accounts of their research be published for all to read in the *Moniteur*.

It was in his multiple teaching posts that Gay-Lussac first claims our attention as a member of the French scientific establishment. He held posts at the Ecole Polytechnique and later at the Muséum d'Histoire Naturelle, but it was as a professor at the Paris Faculty of Science that he came to be most widely known. His nomination to the newly established Faculty in 1808 was a major step towards his recognition as one of the leading figures in French science.

## The Faculty of Science

When Napoleon created the so-called 'University of France' in 1806–8 three levels were clearly distinguished: elementary or primary education, the *lycées* and the Faculties. The Faculties, which we would describe as providing a university-type education, were thus an integral part of the whole educational structure. Moreover, whereas in Britain a gulf has existed between schools on the one hand and universities on the other, in France the gulf lay between the elementary system with its teachers (*instituteurs*) on the one hand and the elitist *lycées* and Faculties (each with their *professeurs*) on the other hand. The Faculties of Letters (corresponding to the traditional Faculty of Arts) were established primarily to certify *lycée* teachers and the historic parallel establishment of Faculties of Science marked an important stage in the recognition of the place of science in the French educational system. Three grades were recognised by the University: *bachelier*, *licencié* and *docteur*. In order to enter the Faculty of Science one had to have the *baccalauréat* in letters.

In the Faculties of Science the doctor's degree was introduced as the official qualification of the university professor. But as the first professors appointed to the Faculties in 1808–9 obviously did not have this qualification, it was decided that the diploma of doctor should be awarded immediately to those who had ten years of teaching experience. This took care of Biot, Thenard, Haüy, Desfontaines, Lacroix and Francoeur. Although Gay-Lussac (aged 30) and Poisson (aged 27) did not have long enough teaching experience, Cuvier, as vice-rector of the Faculty of Science, proposed that they should be treated like the others.[3]

Accordingly imperial approval of the nomination of Gay-Lussac and Poisson as professors at the Faculty of Science was given on 11 May 1809. The Paris Faculty of Science and the Faculty of Letters now had an administration, an impressive list of professors, and rooms in the College of Plessis; there remained the problem of whom they were to teach. Fontanes, the Grand Master of the University, thought that they should be used to teach in the re-constituted Ecole Normale which was then being planned. A 'normal school' for training teachers had been set up by the Convention in 1795. Napoleon wished to re-establish such an institution as an integral part of the educational system. When the Ecole Normale opened in November 1810 it was in the same building as the Faculties of Science and Letters and with largely the same professors. Only gradually was the Ecole Normale to win its independence from the Faculties. Meanwhile here was a marriage of convenience. The thirty-seven students of 1810 (rising to seventy-seven in two years) had twenty-seven professors of two faculties to lecture to them.[4] Gay-Lussac's original Faculty salary of 1500 francs was supplemented in 1812 by a salary of 3000 francs payable to those members of the Faculty of Science attached to the Ecole Normale.[5] Henceforth Gay-Lussac's salary for this one chair was 4500 francs. His basic duty was to deliver two lectures a week. In 1811 he was lecturing on Tuesdays and Thursdays from 8.30 to 10.00 a.m.

A large part of the work of the Faculties was concerned with examinations. In 1811 there were thirty-two candidates for the *baccalauréat* in science, of whom thirty passed. There were seven successful candidates for the *licence* and the first doctor's degrees were awarded. Although in the period 1825–30 there was a sharp increase in the number of *baccalauréats* because of the new requirement that medical students should have this initial qualification[6] the number of *licenciés* graduating remained only a trickle, only once in the first twenty-five years of the Faculty reaching the twenty mark (1832).[7]

As the lectures at the Faculty were attended by hundreds of students, those who actually took the examinations constituted a small minority.

What then was the function of the Paris Faculty of Science? This was the question considered by Dumas in a report of 1837:

It is obvious that one purpose is to prepare young men for the examinations of the *licence-ès-sciences,* the *agrégation*[8] and the doctorate in science. Further it spreads among the public a knowledge of the exact sciences by its courses, to which everyone may be admitted without any distinction. It is the latter aspect that the Paris Faculty of Science presents to an observer who has not made a special study of its organisation. Indeed this mixture of auditors belonging to the highest classes of society and of young men who have a positive goal, and a position to obtain, together with this multitude of young foreign students who come to complete their studies at the Faculty, all this forms an ensemble which gives the teaching of the Faculty a special character.[9]

Dumas' comments help us to see the student audience of the Faculty as comprised of three main groups. First there were those following the course for examination purposes. After the foundation of the Faculties under Napoleon they were firmly established in the Restoration. Science for the first time in a France at peace had become one of the recognised professions. Science lectures would therefore have a larger number of serious students than any courses given under the *ancien régime.* Raspail speaks of the student population of Paris in the early Restoration as 'largely composed of the young remnants of our armies who tried to rebuild a solid future for themselves in the honourable professions of the law and the sciences'.[10]

As the professors were the examiners, and as examinations were closely related to the lecture courses, these students would be expected to follow the lectures regularly and with close attention. Less assiduous would be the younger members of the aristocracy and the upper bourgeoisie who felt that in the nineteenth-century world some knowledge of science was desirable. In their search for a general education they might have been compared with some sections of the audience at the Royal Institution in London. Finally there were the foreign students, who would usually be older than the French students. Although they often attended elementary expositions of science they had in many cases studied science in their own countries and they went to these lectures, where, without payment, they could see distinguished French scientists. They might be already generally conversant with the subject matter of the lectures, but if they were deficient in spoken French, they could listen to an orderly exposition of a branch of science by an expert.

It is difficult for the historian to estimate the numbers of students attending the lectures, as it was not even possible at the time to obtain precise data. Lacroix, as Dean of the Faculty of Science, explained the difficulty in a letter of 15 September 1815:

As regards the number of students who have attended the lectures, I cannot give any precise information. The courses of the Faculty, intended specially for students of the Ecole Normale and for young men who aspire to the grade of *licencié-ès-sciences*, are open to any serious person who wishes to take advantage of them; in this class of auditors the greatest number divide between the different courses according to the attraction of the subject matter. The chemistry course is the most popular, followed by the physics course. They are given in a rather large amphitheatre which is nearly always full.[11]

In fact they were often overflowing, as is suggested by a letter of 12 August 1819 by the chemistry professor Thenard to Cuvier, asking for an additional appointment. He provides a graphic account of the difficulties of dealing with a large audience:

What efforts the professor must make to maintain attention especially of those who are obliged to stand for about $2\frac{1}{2}$ hours.[12] The large number of auditors who turn up from all over, the obligation to raise one's voice to be heard, the necessity of making continual efforts to capture attention, make the task difficult, but they become much more painful because of the heat from the furnaces with which the professor is surrounded. These discomforts are felt particularly in summertime. Then the amphitheatre becomes for the professor like a cauldron, which he only leaves covered in perspiration.[13]

Thenard goes on to say that although the amphitheatre can hardly seat 300 students, as many as 550 have attended his lectures.

One of Gay-Lussac's auditors, Bodelio, then past middle age, complained about the habit of the young men there who saved places for each other, so that even if an outsider like himself arrived early, he would not be certain of a good seat. According to him there were as many as 700 at Gay-Lussac's lectures in 1818 but some exaggeration must be allowed for in this testimony.[14]

Although one could specialise for the *licence* in either the 'mathematical' sciences or the 'physical' (= experimental) sciences, the physics course was common to both. Thus Gay-Lussac had a key course with large audiences and in December 1812 he petitioned successfully for an assistant (Hachette). In 1815 Hachette changed to descriptive geometry and Gay-Lussac now shared the physics course with his old friend Biot. They asked the Faculty of Science to be allowed to divide the syllabus according to their personal interests:

M. Gay-Lussac will deal this year with heat, gases, hydrometry, electricity and magnetism. M. Biot in the remainder of the time will then teach magnetism, acoustics and optics.[15]

If to modern eyes the level of science taught in the Faculty does not seem very advanced, it was partly a reflection on the state of the subject and even more a consequence of the education system. One of the educational decisions of the Napoleonic period had been that the proper place for science was not at the secondary level but the tertiary level. Thus when students left the *lycées* and entered the university, science courses had to be of a fairly elementary standard and, in order to make it accessible to a wide audience, the physics course could not afford to make more than the most rudimentary use of mathematics.

Gay-Lussac (assisted by Pouillet) in the physics course, and Thenard (assisted by Dulong) in the chemistry course, provided the focus of teaching in the physical sciences in the Paris Faculty in the 1820s. Several of their students placed on record their assessment of the teaching. The German student C. F. Schönbein (1799–1868), who was in Paris from 1827 to 1828, was most impressed by the science lectures:

The lectures themselves are the best that I have ever heard and what a contrast to German verbosity and lack of substance. I am speaking here particularly of the scientific lecturing of Gay-Lussac, perhaps the most outstanding scientist of our days.[16]

He praised the 'clarity and elegance' of the French lecturing style and was impressed by the use of demonstration experiments.

Schönbein's only reservation in his praise of the lecturers in the Paris Faculty of Science was that their liveliness sometimes bordered on affectation and he was thinking particularly of Thenard:

Mr Thenard's lectures are certainly excellent too, but this man has something of a mountebank about him, and his mimic gesticulations are disagreeably contrasted by the English simplicity and German modesty of Mr Gay-Lussac's manners.[17]

Another student, who attended the science lectures in Paris in 1820, was the Scotsman Sir Robert Christison, who wrote:

Gay-Lussac was perhaps the most persuasive lecturer I have ever heard. His figure was slender and handsome, his countenance comely, his expression winning, his voice gentle but firm and clear, his articulation perfect, his diction terse and choice, his manner most attractive, and his lecture was a superlative specimen of continuous unassailable experimental reasoning.[18]

He too compared the lectures of Gay-Lussac and Thenard and stated his preference for the former. Christison admired Thenard's energy,

But the incessant vigour, *sans relache sans repos*, makes one long for a little of his friend's [Gay-Lussac's] no less persuasive quiet occasionally.

The difference seems, therefore to have been little more than a question of temperament. Thenard was an extrovert with the mannerisms of Mediterranean peoples. Gay-Lussac was more reserved and his quiet authority tended to have a greater appeal for students from Britain and Germany.

However if the less demonstrative British and German visitors preferred Gay-Lussac's lecturing manner to that of Thenard, some of the French students preferred the latter, although it is sometimes difficult to say whether it was not their preference for the subject matter which led to their assessment. Charles de Remusat, who registered for the two courses of chemistry and physics in the academic year 1815–16, was immediately won over by Thenard's chemistry course (which began before the physics course). Coming to chemistry as a complete beginner, Remusat felt that with such inspiring lectures and a good text-book one could master the subject. He continued:

This is less true of physics. Gay-Lussac and Biot, who taught us, were not equal as lecturers to Thenard, but they had their good qualities. Gay-Lussac's attitude of attention and concentration when he described an experiment gave one the impression that he was making a discovery. There was in his manner the genius of the inventor. Biot was more relaxed and more animated. These two courses were as brilliant as was possible for a course of physics without mathematics.[19]

Pouillet later suggested that Gay-Lussac had something of a classical style in lecturing – it was always simple, logical and 'severe' with never a chance word.[20] And yet Pouillet spoke of his brilliant improvisations. Gay-Lussac, it seems, never got carried away in his lectures unless perhaps in describing some scientific discovery. If it were his own, however, he was noted for the modesty with which he described such work anonymously and in the third person. Not all of Gay-Lussac's audience was equally well-disposed towards him. His critic Bodelio, who took advantage of the right of the public to attend lectures of the Faculty of Science, complained that Gay-Lussac tended to copy calculations from his notes onto the blackboard instead of working everything out.[21] However Gay-Lussac's manuscript lecture notes show that he took every opportunity to make his lectures more intelligible with the help of diagrams. These pedagogical aids are unfortunately omitted from the published versions of his lectures.

One consequence of the freedom of the public to attend lectures of the Faculty was that a lecturer's course notes could be copied and sold.

Gay-Lussac was annoyed when stenographers attended his course in 1828 and his lectures were published. He was doubly vulnerable since in that year he had exchanged his physics course for Dulong's chemistry course which was shared with Thenard. This arrangement suited Dulong and it was also welcome to Gay-Lussac, whose interest in physics had dwindled. Gay-Lussac was thus presented to the public as a leading authority on both physics and chemistry. As a safeguard against error Gaultier de Claubry was engaged by the publisher to check the text but Gay-Lussac himself would have nothing to do with the venture.[22] He emphasised that his lectures were only a part of a course and that he was not sufficiently satisfied with them to have them published. Unlike Pouillet, who took legal action against the publishers, Gay-Lussac contented himself by refusing to take any responsibility for the contents and refusing the money he had been offered for his co-operation. However, as in the case of Boerhaave a century previously, the production of an unauthorised edition of his lecture notes did stimulate him to undertake to produce a text-book himself. Unfortunately this stimulus had come too late in his career and he never released the manuscript of the text. One of his last acts was to have this manuscript burned, a sad loss to the historian of science.

The formal establishment of the Faculty of Science in 1808 was more than a paper innovation. The professors were drawn from existing institutions; two from the Collège de France, two from the Muséum d'Histoire Naturelle, two from the Ecole Polytechnique and two from the lycées. Being a parallel creation to the Faculty of Letters it was not understood at first that science needed apparatus. Initially the expenses of all Faculties were included under either salaries or administration. However complaints from the science professors did produce small sums of money. In the period 1813 to 1819 the total expenditure on physics apparatus was 4565 francs and Biot was taken to task for exceeding his annual allowance of 700 francs.[23] Gay-Lussac complained that he had no microscope and indeed the only way to give a good lecture course was to bring along one's own apparatus or else borrow apparatus from a better endowed institution such as the Ecole Polytechnique or the Muséum, both of which had benefited from Revolutionary confiscations and were favoured by the government as having a research as well as a teaching function. The Faculty continued as a poor relation and in 1837, when Dumas in a state of exasperation asked for a library to be established, his shocked colleagues felt that he had gone beyond making a reasonable demand. The Faculty had no research function. It administered an examination system, granted diplomas and provided lectures but it never existed as an organic entity concerned with the advancement of knowledge. The confusion only arises because the Faculty from its foundation drew on the most famous names in French science for its professors. They were glad of the position and emoluments but they did their real work elsewhere.

## The Ecole Polytechnique

The death of Fourcroy in 1809 had provided the vacancy Gay-Lussac and his friends had been hoping for. His nomination on 17 February 1810 as professor of chemistry at the Ecole Polytechnique set the seal on his brilliant early career, confirmed his right to use the research facilities of the school, and provided him with an additional income.

Gay-Lussac now developed his lecture course. Although this was intended exclusively for the carefully selected students of the Ecole Polytechnique, there were several instances of people from outside the school attending lectures. Thus in 1812 Gay-Lussac asked the Minister of the Interior for permission to allow several teachers from other institutions to attend his course.[24] They were Marcel de Serres (professor at Montpellier), Sementini (professor at Naples) and Rotland (professor at Munich). Such an audience suggests a course at the highest level for as international an audience as might be possible at the height of the Napoleonic wars. The most famous visitor from Britain was Humphry Davy, who visited the Ecole Polytechnique during his stay in Paris in 1813 and who attended one of Gay-Lussac's ordinary lectures. We are fortunate to have an account of this from young Michael Faraday, who accompanied Davy and made the following entry in his diary for Wednesday 8 December 1813:

I went to-day with Sir H. Davy to L'Ecole Polytechnique, the national school of chemistry [sic!] to hear the leçon given to the scholars. It was delivered by M. Gay-Lussac to about two hundred pupils. The subject was vapour, and treated of its formation, elasticity, compressibility, etc. Distillation both by heat and cold was introduced. It was illustrated by rough diagrams and experiments and occupied about an hour. My knowledge of French is so little I could hardly make out the lecture, and without the experiments I should have been entirely at a loss.[25]

At the Restoration one might have expected that Gay-Lussac, as a prominent professor, might have been made a member of the governing body, the Council. But one of the changes introduced by the Bourbons was that no professors should sit on the Council. Instead three members of the Chambre des Pairs were nominated. This was indicative of a change of values under the Restoration when the nobility reasserted themselves above any products of the Napoleonic meritocracy. In the period 1816–30 the Ecole Polytechnique reflected in several ways the re-establishment of the monarchy. This creation of the Convention became the Ecole *Royale* Polytechnique and the royal ordinance of 4 September 1816 put the school under the 'protection' of the King's nephew, the Duke of Angoulême. The opening of the school on 17 January 1817 was celebrated with a mass and the students were enjoined to serve 'God, King and Country', a formula new to nineteenth-century France.

There were various tendencies reminiscent of the *ancien régime*. It would be an exaggeration to claim that the scientist from the former position of prophet or technician had descended to becoming a courtier, but nonetheless, in contacts with the royal family, the savants assumed the role of subject rather than citizen. Thus the *Moniteur* of 8 July 1819 reported that the Duke of Angoulême had been to the Ecole Polytechnique and had sat in on one of Gay-Lussac's lectures. According to one report he remarked publicly on the great discomfort Gay-Lussac must have felt in his balloon assent of 1804 because of the heat at such altitudes! Poor Gay-Lussac did not dare contradict him. In the course of his lecture, adapting his phraseology to the occasion, he made the unfortunate assertion: 'Thus, Your Highness, these two gases will have the honour of combining before you.'[26]

After the revolution of 1830 a commission was set up to examine possible improvements to the Ecole Polytechnique. Gay-Lussac was one of the members of the commission and Arago also was on it. The commission recommended a return to the system of a governing body with representatives from the Academy of Sciences. The Academy was accordingly approached in 1833 and Gay-Lussac, Arago and Mathieu were elected as its representatives. Gay-Lussac was re-elected in succeeding years. He had grave reservations about the majority decision to transfer the school from the Ministry of the Interior to the Ministry of War. The Ministry of War was able to take responsibility for the transfer of less-able students to branches of the army not requiring mathematical competence, but at the same time it detracted from the status of the Ecole as an institution for higher education.

The teaching situation at the Ecole Polytechnique was described by Arago as follows:

The professors of science at the Ecole Polytechnique have no direct personal contact these days with the students. Each professor arrives on the appropriate day of the week and at the appointed hour; he finds the benches of his lecture theatre filled, he gives his lecture and withdraws. This process is repeated forty, fifty or sixty times in the course of the year, according to the programme, without the professor ever having the occasion to speak individually to a single one of his audience. When the course is finished the students are divided by drawing lots between the professor and the *répétiteurs*, who examine them individually for several minutes and award them marks. The professors and the *répétiteurs* themselves would find it so difficult to check whether the student who presented himself was the one on the list, that this is verified by a signature.

One day I asked one of the professors of chemistry [Gay-Lussac?] to tell me how many students he had occasion to speak to in his last course. The reply was brief: 'To no-one'.[27]

One must allow for some exaggeration in this account, since the context shows that Arago was defending himself against the suspicion that the liberal tendencies of the teaching staff might corrupt the students. Yet this picture of impersonal teaching is probably an accurate one, with allowance for an occasional exception. It is, however, a reflection on the system rather than on the personality of an individual professor.

Some idea of the content of Gay-Lussac's mature course may be gained from the report of an American visitor.[28] We reproduce *verbatim* his report on the chemistry syllabus.

*First year*   General principles. Division of the course.
        Examination of the principal simple substances.
        Mixtures and binary compounds. Laws of definite
            proportions, etc. [*sic*]
        Hydracids. Oxacids and oxides.
        Bases. Neutral binary compounds. Salts.
        Principal metals.
*Second year*   Reciprocal nature of acids and oxides.
        Action of water upon salts.
        Laws of Berthollet discussed.
        General properties of the carbonates and special study of
            some of the more important borates and silicates.
        Glass and pottery. Nitrates. Gunpowder.
        Phosphates, etc. Sulphates. Chlorates.
        Chromates and other classes of salts with details as to the
            more important.
        Extraction of the metals from their ores, methods of
            refining, etc.
        Organic chemistry. Vegetable substances. Animal
            substances.

This course is accompanied by manipulations in the laboratory of the institution, in which the most useful preparations of the course are made by the pupils themselves. They are also taught the principles of analysis, both mineral and organic, practically.

Although this syllabus represents an attempt at comprehensive coverage, there are several indications of Gay-Lussac's own special interests including hydracids and glass. The inclusion of gunpowder was an example of the applications of chemistry, but the syllabus is marked by its refusal to teach chemistry purely in terms of its utility. However, since the majority of his students at the Ecole Polytechnique went on to become engineers or administrators, one can understand Gay-Lussac's feeling, later publicly expressed, that it was hardly essential for such students to acquire a detailed knowledge of chemistry.[29]

We are fortunate to have on record Gay-Lussac's considered opinion, delivered in the Chambre des Députés in 1835, of the state of the Ecole Polytechnique, forty years after its foundation. Gay-Lussac says that France no longer had a Polytechnic School in the exact meaning of the term. It no longer prepared men of science or men who might use their scientific training in industry. As the Medical School produced doctors and the Law School lawyers, the Polytechnic School should produce scientists. Instead it was limited to a training in those sciences which would be useful in the public service. Gay-Lussac went further and suggested that the Ecole had virtually *become* one of these *écoles d'application*:

It is necessary to point out that the Ecole Polytechnique, having become today a school specially directed towards the public service, is now very narrow; it has departed from its origin. It is distressing, after having called upon the whole of France and having taken the elite of young men, to see that the most able of them are lost for the cultivation of science.

The phrase 'lost for science' occurs several times in Gay-Lussac's speech. He felt that French science would be enormously advanced if the four or five best students each year were given posts in higher education. Instead they became engineers or army officers or employed their talents in the administrative bureaucracy.

Gay-Lussac's concern is of course partly autobiographical: as a graduate of the school he had become a professional scientist through his teaching appointments and he wanted others to follow in his footsteps. Gay-Lussac's testimony, however, is of even greater value as a diagnosis of one of the reasons for the failure of French science to live up to its earlier promise. There was no educational ladder to feed back into the system the ablest young graduates of the Ecole Polytechnique. From the beginning a very few graduates (like Poisson) had gone *directly* into teaching posts. The usual method was an indirect one, even in Gay-Lussac's own case. Meanwhile a safe salaried employment in government service was held out as the main goal and the majority of able young men were thus siphoned off and became administrators rather than scientists.

Gay-Lussac's resignation from the Ecole Polytechnique in 1840 attracted some public attention and a critic writing in the *Constitutionnel* suggested that there was something wrong at the school. Arago felt himself called upon to give some explanation and said that his friend had resigned 'because at his age one needs rest'.[30] It is true that Gay-Lussac was now 62 but a study of the archives of the Ecole Polytechnique shows that Gay-Lussac's resignation was not simply retirement. It is worth reproducing *verbatim* an extract of the minutes of the meeting of the Conseil de Perfectionnement on 19 October 1839:

One member urges the advantage of the adoption of chemical equivalents as representing factual information susceptible of being translated into theory. This method of expounding definite proportions has received general consent among scientists of the north.

The same member argues the advantages of certain groupings which facilitate, abbreviate and throw light on chemistry and its applications. He wants several compounds of little importance to be omitted and emphasis placed on several products such as gunpowder, concrete, dyestuffs, etc.

Another member emphasises the same point and mentions several chemical compounds which it is useless to study.

Several members give testimony of the high standard of instruction of the students in chemistry.[31]

Thus Gay-Lussac was facing penetrating criticism from his colleagues. The chemists present were Chevreul, examiner to the school since 1831, Thenard and Pelouze. The emphasis to be placed on dyestuffs suggests the intervention of Chevreul who obviously felt that Gay-Lussac should not be indulging in the luxury of pure chemistry but should, like Chevreul in his own chair at the Muséum, be concerned with applied chemistry. Chevreul seems to have been able to attack Gay-Lussac's science as distinct from the man, since that same year when he published his book *De la loi du contraste simultané des couleurs* (Paris, 1839) he presented a copy to his colleague with the inscription: 'A M. Gay-Lussac Hommage d'amitié et de profonde estime. Chevreul'.

We have already reported Gay-Lussac's unease about the Ecole Polytechnique being more a preparatory school for the *écoles d'application* than a broad training in pure science and there was obviously a fundamental conflict in goals. A further criticism was made at the meeting that Gay-Lussac sometimes exceeded the time set aside for his lectures. Perhaps the most fundamental criticism however was the first one given above, which implied that Gay-Lussac was not keeping abreast of current developments. One is reminded of the later case of Marcelin Berthelot (1827–1907) who refused to use atomic weights long after his contemporaries had agreed on them.

It seems that the critics won the day, since a report of the Conseil de Perfectionnement of July 1841 speaks of recent changes made in the chemistry syllabus which emphasise utility and avoid going into great detail.[32] It also refers to a distinguished professor who previously occupied the chemistry chair. Gay-Lussac's resignation had been accepted by the Ministry of War on 18 November 1840.[33]

### The Muséum d'Histoire Naturelle

Each of Gay-Lussac's teaching positions was different, modified not only by the subject of his course but by the institution in which the course was given. Only an understanding of the different levels of scientific instruction and more subtle factors such as freedom to teach one's own syllabus will enable us to understand his respective positions as Professor of the Ecole Polytechnique, the Faculty of Science and the Muséum, and what it meant to abandon the Faculty in favour of the Muséum.

Pouillet later contrasted the difficulties of Gay-Lussac's task at the Faculty of Science with the straightforward lecturing at the Ecole Polytechnique:

At the Ecole Polytechnique where he taught chemistry, his task was easier, he was addressing a prepared audience chosen for their intelligence. But at the Faculté des Sciences it was necessary to educate the public and make popular the great laws of physics by raising all the minds of his audience to these high levels.[34]

However the Muséum too had a fairly general audience of students.

One reason why Gay-Lussac sought the position at the Muséum was probably that there was greater academic freedom than at the Faculty. The professors at the Muséum tended to be their own masters whereas the Faculties were run as government departments of education. At the Muséum Gay-Lussac was able to draw up his own syllabus and there was no examination. There were also material advantages, notably a house. Gay-Lussac's Paris address was now to become 'Au Jardin des Plantes', not, it is true, at the centre of Paris, nor even in the main Latin quarter, but a comfortable house in the capital provided by the state was not something to be overlooked. Gay-Lussac also had a position in the Gunpowder Service which included the provision of living quarters at the Arsenal. But when he had been given notice in

1828 that his post was to be abolished he was also given notice of eviction from his home. Although his position was retrieved, it must have come as a shock to Gay-Lussac to think that the security of his home might depend on the whim of a future minister. When, therefore, a few years later a teaching post appeared on the horizon which carried with it a house with security of tenure, the possibility was worth serious consideration.

The story of how Gay-Lussac gave up his chair at the Faculty of Science as an exchange for a chair at the Museum of Natural History is a complex one, but it is worth describing not only as a part of his biography but also as an illustration of some of the hurdles in the French academic world of the nineteenth century. The Academy had the right of nomination to many posts in higher education including the Muséum. At the meeting of the Academy on 21 May 1832, a letter was read from the appropriate government minister reminding them of the recent death of Laugier, professor of chemistry at the Muséum, and asking the Academy to nominate without delay a candidate for the vacancy. The chemistry section got together during the meeting and at the end drew up for discussion by the Academy a list of three names: Serullas, Dumas and Robiquet. The election would normally have taken place at the next meeting but a few days later (25 May) Serullas died, a victim of cholera. The election was therefore postponed, but when the subject was mentioned again on 4 June, Chaptal on behalf of the chemistry section announced their new recommendations in order of preference: Gay-Lussac, Dumas and Robiquet.[35] Some explanation seemed to be called for as to why a man who held two chemistry chairs was being recommended by his friends for a third; it was said on his behalf, that if nominated he would give up one of his other chairs. Accordingly on 11 June Gay-Lussac was elected as the nominee of the Academy by thirty-five to Robiquet's two.[36]

Gay-Lussac now passed to the second stage of his candidature. The Muséum prided itself on its democratic organisation, dating back to the Revolution. The running of the Muséum with a rotating chairman or director was in the hands of the body of professors who also had the right to veto appointments. He therefore had to make himself acceptable to his future colleagues. At a meeting of the professors of the Muséum on 29 May 1832 a letter from Gay-Lussac was read out saying that if he received a majority vote for the vacancy of professor of general chemistry he would give up his chair at the Faculty of Science.[37] The professors were suitably impressed by his preference for association with them, and when a vote was taken, Gay-Lussac was unanimously elected. The final stage was a royal ordinance of appointment which came on 16 June. We find him attending his first meeting of the professors on 3 July.

He began his lecture course on 2 April 1833, giving three lectures a week on Tuesdays, Thursdays and Saturdays and he fell into the Muséum tradition of an early start by lecturing from 7.30 to 9.00 a.m. We know from a letter of Jules Gay-Lussac of 1835 that his father was already dissatisfied with his lectures at the Muséum since he felt obliged to cover a large amount of material rapidly and incompletely in the time available.[38] He was then thinking of substituting a course on the general principles of chemistry under the title *philosophie chimique*. His friend Thenard had decided to include a section with this title in the latest edition of his text-book,[39] but unfortunately the manuscript of Gay-Lussac's work was later burned under his orders and we can only guess at its contents.

His chemical colleague at the Muséum was Chevreul, who lectured on the same days of the week at 10.00 a.m. The bills and placards of the Muséum are much more explicit on the content of Chevreul's course than Gay-Lussac's. In 1837, however, Gay-Lussac advertised that he would be dealing with mineral or inorganic chemistry, whereas in 1838 he lectured on organic chemistry. In the first lecture of this course[40] he stressed that in organic chemistry only the ordinary physical and chemical forces should be considered. He explicitly excluded the action of any vital force and with this the mysterious catalytic force. He then introduced the principle of organic analysis, typically omitting to specify his own contribution. An exhaustive treatment of organic acids covered nearly twenty lectures, after which alkaloids and miscellaneous neutral substances were described. It thus seems that he took seriously his own fundamental classification of organic compounds into acidic, basic and neutral. In the final lectures gums, resins, dyes and saponification were discussed.

In the 1840s the subject matter of Gay-Lussac's lectures generally alternated between inorganic and organic chemistry. In the year 1839–40 he took his turn as director of the Muséum and, with the burden of administrative duties, does not seem to have given his lecture course that year. He occasionally allowed someone to substitute for him in his lectures. Thus some of the final lectures in the summer of 1838 were given by Pelouze. In the late 1840s Gay-Lussac introduced Fremy as his *suppléant* and when the British chemist Edward Frankland visited the Muséum in July 1848 'in the expectation of hearing Gay-Lussac lecture in chemistry' he was reported to be unwell and Frankland listened to Fremy instead.[41]

A view of chemistry in the Muséum as essentially a descriptive science is suggested by a speech made by Gay-Lussac in the Chamber of Deputies in 1834.[42] He asked for funds for a collection of the elements which he described as 'conquests of the human mind'. Although Gay-Lussac was himself associated with certain elements, notably boron, iodine and chlorine, this view of chemistry was not typical of him. He said that each professor at the Muséum had a collection for which he was responsible and among his duties was that of describing the specimens within the collection. There were now some fifty-four elements known and he urged that specimens should be bought and placed alongside those of mineralogy and geology. An additional estimate of 4000 francs to the budget for that year was agreed by the Chamber for that purpose.

## The Academician

Only three months after his election to the First Class of the Institute[43] Gay-Lussac found himself involved in commissions passing judgement on the work of others. On 16 March 1807 he was appointed together with Fourcroy and Haüy to judge the effectiveness of a pendulum for searching out water and metals in the ground. At the same meeting he was appointed in conjunction with Berthollet and Vauquelin to present a report on a memoir on the decomposition of salts by an electric current. At the next meeting Laplace and he were appointed to examine several memoirs by Erman of Berlin. Thus by association with senior members of the Institute, Gay-Lussac was soon initiated in the role of member of the Academy, which combined some features akin to literary criticism with a judicial function. By the meeting of 28 September 1807 he was considered worthy of being a member of a commission to judge awards for the annual prize of 'galvanism'. Although he had as colleagues for his task Haüy, Laplace, Hallé and Rumford, the report which was first presented to the First Class on 7 December 1807 was the work of Gay-Lussac. This report[44] marks not only the coming of age of Gay-Lussac as an Academician but also constitutes an informed and deep study by him of the work of Davy, to whom the commission unanimously recommended the award of the prize. The report, because of its importance, was chosen to be read again at the next public meeting of the Institute on 4 January 1808. In several other instances in this intensely active and fruitful part of Gay-Lussac's life we find him not only sitting on commissions but taking the leading part by drawing up the final report. Over the next few years he took his full share of membership of commissions, having this task on average nearly once a month.

Although a member of the physics section, Gay-Lussac was recognised from the beginning as at least an equal authority on chemical matters. In 1810 he sat on a commission to report on oxalates, phosphorescence, on inflammable powder and the waterproofing of leather. During the next few years other chemical topics he helped consider included affinity, gold compounds, indigo and iodine. In the area of physics he was concerned with electricity, the light obtained by sudden compression of gases and also many problems of heat ranging from freezing to pyrometers. In April 1813 he helped to examine memoirs on congelation, respiration and animal heat.

After the Restoration Gay-Lussac was more prominent in applied science but he continued to give his advice as a member of the Academy on many questions of pure science. In 1817 it was alkaloids, in 1818 crystallography and in 1820 Prussian blue which were among the more important subjects on which he reported. Eighteen-twenty was the year of Oersted's experiments on the magnetic effect of a wire carrying an electric current. Gay-Lussac was one of those selected (with Arago) to report on this work to the Academy.[45] Another important assignment was to verify Balard's claim to have discovered another substance similar to iodine and chlorine. Balard announced his discovery at a meeting of the Academy on 3 July 1826. Five weeks later the Academy's commission (Vauquelin and Thenard with Gay-Lussac as spokesman) reported favourably.[46] They suggested, however, that the new element should be called 'brome' (bromine) from the Greek word *bromos* (bad smell) rather than 'muride' which was Balard's name for it.

A major problem in analysing the joint report of a commission is to distinguish the contributions of each member. In most cases this is hardly possible. One member of the commission, however, was sometimes designated *rapporteur* and charged with the job of drawing up the report. We should obviously pay more attention to reports presented to the Academy by Gay-Lussac than reports in which he had merely been a member. Thus although Gay-Lussac was only one of four scientists appointed in September 1807 to report on lightning conductors, the fact that he was the *rapporteur* when the commission reported back on 2 November[47] suggests his special involvement in this issue. In fact the maiden speech by Gay-Lussac as *rapporteur* was to be remembered by his colleagues so that he came to be considered as the expert on lightning conductors.

In many of the commissions on which Gay-Lussac served one may see the emergence of applied science in areas which suggest a beginning of the modern world. We find Gay-Lussac a member of commissions concerned with the problem of pollution from chemical factories, on the determination of the alcohol content of wines, on an early internal combustion engine, on gas lighting and on steam engines. In the last three cases Gay-Lussac took a particularly prominent part and it is therefore worth examining briefly each in turn.

Gay-Lussac collaborated with Lazare Carnot in 1810 on a gas combustion engine, a description of which had been submitted to the Academy by M. de Rivaz.[48] The idea was to make use of the explosive force of an ignited mixture of hydrogen and oxygen in place of steam to drive a piston. Gay-Lussac was the *rapporteur* on this early combustion engine and pointed out several basic defects: (1) there was no regular supply of hydrogen; (2) manual force was necessary to expel excess gases after combustion; (3) the electric spark for ignition could not be relied upon for continuous operation; (4) the presence of the water formed in condensation was neglected. Carnot and Gay-Lussac concluded by saying that this was not the first time someone had had this idea; what was wanted was a working machine based on it. Until the difficulties they had mentioned could be overcome the steam engine provided a better source of power. In this as in other reports the Academy might seem to be performing a negative function but such reports were usually written in as constructive a way as possible. When the scientists were themselves practical men they realised, like Gay-Lussac, the enormous gulf between an idea and its execution in practice.

In France Philippe Le Bon (d. 1804) had taken out a patent in 1799 for gas lighting and two years later he had lit up the whole interior of a large town house in Paris by this means. It was, however, in England that the major progress in this field was made. Already in the 1790s William Murdoch had used gas generated from coal for lighting purposes and from 1806 F. A. Winsor, who had come to England from Germany, tried to launch a joint stock company for gas lighting. He secured the help of a chemist, F. C. Accum, whose *Practical Treatise on Gas Lighting* was published in 1815. A copy of the book was sent to the First Class of the Institute in Paris and it was Gay-Lussac who was called on to present a report on the book.[49] Five years later Gay-Lussac was a member of a commission considering illuminating gas from oil and in 1823 he was asked to consider the problem of using hydrogen for illumination.[50] The Minister of the Interior turned to the Academy for advice on the special safety problems involved in the production, distribution and use of gas and Gay-Lussac was one of a commission of five appointed by the Academy which presented its report on 2 February 1824.[51] The general attitude of the commission was that the dangers of fire and explosions in gas works had been exaggerated. They were safe and presented no special health problems provided certain precautions were taken. They wished to support the industry not only because it produced superior lighting in towns but because it provided a new use for coal, it stimulated the iron industry (through the construction of pipes) and in its by-product, coke, provided an alternative domestic fuel which might prevent further deforestation. One aspect not made public in this enthusiastic endorsement of the new gas industry was the business interests of the authors of the report. Gay-Lussac, employed as a part-time consultant (*chimiste-conseil*) at the Charenton ironworks, was hardly an impartial judge. The Charenton company, which manufactured steam engines and which amalgamated in 1828 with the famous Creusot ironworks, was owned by the expatriate British ironmaster, Aaron Manby. In 1821 Manby and Daniel Wilson had obtained the concession for gas lighting in Paris.[52]

There was some general consciousness in France under the Bourbon Restoration of the supremacy of British industry. Part of this could be attributed to the development of the steam engine as a source of power. Occasionally an improved steam engine would be submitted to the Academy for judgement. Gay-Lussac was one of the members of a commission reporting on this subject in 1821. The commission were enthusiastic about the design but added with a caution characteristic of such savants as Berthollet and Gay-Lussac:

on such matters the more one is disposed to allow oneself to be won over by the novelty of a machine, the more one should be on one's guard against this kind of seduction and the more it is necessary to base one's judgement on comparative experiments.[53]

They accordingly compared the steam engine with others and thus satisfied themselves of its superiority.

As attempts were made to improve the efficiency of the steam engine, higher pressure steam was used and accidents became more frequent. At a meeting of the Academy on 31 March 1823 a report was read on the safe use of steam engines. Gay-Lussac, who was a member of this commission, intervened to say that he did not agree with the report.[54] After further discussion in which Gay-Lussac took part it was agreed to introduce a wall to protect neighbouring property as an additional safety measure.[55] Accordingly a few weeks later a revised version of the report was read to the Academy,[56] a most unusual step and apparently a direct outcome of Gay-Lussac's intervention.

As an Academician Gay-Lussac helped scotch some of the more extravagant scientific claims made. Thus in 1813 he told the First Class of the experiments he had carried out to try to confirm the claim of

Morichini that ultraviolet rays could magnetise needles. He had been unable to confirm this.[57] In 1818 he was appointed together with Cuvier and Berthollet to examine the claim that living bodies had been produced in a flask containing distilled water, nitrogen and hydrogen. At the next meeting the claim was speedily disposed of by Gay-Lussac's pointing out that the flask had not been properly sealed.[58] Some chemists were fascinated in the early nineteenth century by the diverse colours which manganese compounds could assume and potassium permanganate was fancifully termed the 'mineral chameleon'. This compound was known to contain manganese, potassium and oxygen, but it was not clear how these combined together and the authors had not made any suppositions about this. The official comment on this clearly indicates the voice of Gay-Lussac:

One cannot but praise this wise reserve: it proves that their intention is to advance nothing which is not perfectly demonstrated. The new facts which they will gather will constitute the subject of a second memoir.[59]

In their report in 1832 on Gaudin's imaginative paper on the arrangement of atoms in molecules which revived Avogadro's ideas and linked them to Gay-Lussac's own work, Gay-Lussac and Bequerel showed characteristic caution.[60] They neither condemned nor bestowed inordinate praise. Instead of suggesting publication, Gaudin was rewarded with a pat on the back and told to carry on. What was needed was more facts rather than conjectures.

If membership of a commission reporting on a memoir submitted to the Academy implied the role of a judge, the judicial function of the Academician is even clearer in the awarding of prizes. Gay-Lussac was frequently elected as a member of commissions to judge prizes in physics and chemistry but his competence in neighbouring fields was also recognised as, for example, when he was chosen as a member of a commission to award a prize for work on animal heat or the ripening of fruit. The prize system of the Academy involved two stages: the choice of a particular subject and (one or two years later) the judgement of memoirs. Thus in December 1810 Gay-Lussac helped choose the subject of the specific heat of gases and then helped judge submissions in 1812. Indeed, as the only member of the commission particularly interested in this subject, he probably chose it.[61]

One of the most famous examples of an Academy competition in which Gay-Lussac was involved was the 1817 prize for diffraction of light proposed at a critical time in the history of the wave theory. The commission elected to judge the prize consisted of Biot, Arago, Laplace, Gay-Lussac and Poisson. All five were members of the former Society of Arcueil but within the commission one could mark out Laplace, Biot and Poisson as committed to a corpuscular theory of light, Arago as sympathetic to a wave theory and Gay-Lussac the least committed. The fact that the prize was awarded to Fresnel in 1819 for work which contradicted the Laplacian view may be seen as a decline in Laplace's influence, the triumph of a superior theory, or a vindication of the standards of impartiality of the Academy (or all of these) according to one's interpretation.

Gay-Lussac continued to serve on prize commissions in the 1820s and 1830s. In 1845 he was a member of a commission to examine entries for a prize for the best study of heat given out in chemical reactions, a subject close to his own interests. Entries did not then justify the award of a prize, so the subject was re-advertised. One of Gay-Lussac's last duties as an Academician was to act as a member of a commission appointed in May 1849 to examine entries for this prize. The first prize was awarded to Favre and Silbermann and a second prize to Thomas Andrews of Belfast. In this way, through the Academy, Gay-Lussac helped encourage some pioneers in the history of thermochemistry.

In 1822 and again in 1834 Gay-Lussac was president of the Academy of Sciences. Since the Revolution there had been an annual president and the useful system was adopted by which the vice-president of one year became president in the following year. Thus when Gay-Lussac was elected on 2 January 1821 as vice-president for that year, he had due notice of his special responsibilities in 1822. Twelve years later he was again successively vice-president and president.[62] We may compare Gay-Lussac's double presidency with the case of Thenard who was only elected president once or that of Biot, who never became president. Gay-Lussac had considerable influence in the decisions of the Academy and on Fourier's death in 1830 was one of those appointed to a committee to recommend a successor as secretary.[63] Their choice of Arago as secretary was approved by the Academy. The Academy also exercised influence outside, for example in the Ecole Polytechnique in electing representatives to the Conseil de Perfectionnement and in nominating a Director of Studies, and it was through Gay-Lussac that Dulong was recommended for the post.[64] We have an account of a visit to the Academy under Gay-Lussac's presidency but it does little more than confirm the picture we might have of competent chairmanship and adherence to the customs and rules of that august body.[65] One could, however, say that by 1834 the Academy not only honoured Gay-Lussac but his fame added to the renown of that institution.

In 1835 the Academy was changed by the publication of weekly bulletins, the *Comptes rendus*, and Gay-Lussac, having just completed his second presidential term, came to contribute less and less to the official body of French science. His public utterances on scientific and technological matters were now in the political assemblies. He no longer needed the platform of the Academy although such contributions as his memoir on affinity (June 1839) were no doubt intended to show that he had no wish to retire from the field of pure science.

In 1820 a Royal Academy of Medicine was established with three sections: medicine, surgery and pharmacy. Gay-Lussac was nominated one of ten *associés libres* in the latter group. He did not pretend to any medical knowledge but was able to bring his chemical training and common sense to medical problems. The greatest of such problems confronting France in the second quarter of the nineteenth century was cholera.

The cholera epidemic which struck terror into western Europe in 1831–2 affected the life of Gay-Lussac in various ways. First, as a scientist, he was concerned to apply his own expertise to this terrible problem. Secondly, as a member of the Chamber of Deputies, he was able publicly to urge the government to take action before it was too late. Finally, as a holder of a chair, he, like his colleagues, was drawn into an academic system which depended on 'dead men's shoes'. There were to be an unprecedented number of deaths in Paris in the spring and summer of 1832.[66]

When the cholera broke out in Russia in 1830 the desire of the Academy to send representatives to investigate the progress of the disease was frustrated by the Tsar Nicolas I, who refused to recognise the legitimacy of Louis Philippe's recent accession to the throne and would not grant passports to his subjects.[67] On 28 February 1831 the Academy of Sciences appointed a committee to report on the disease. France already had a Conseil Supérieur de Santé, which was concerned to advise the government on general matters of health, but it was now felt that this should be strengthened by bringing in additional members. From 26 July 1831 its membership was increased from eleven to twenty-two, including Gay-Lussac. The committee was asked, in view of the threat of cholera, to meet at least once a week. In October cholera reached England and Magendie went as an official observer who reported back to the Academy of Sciences. However, although he and other observers were able to describe the symptoms of cholera and the progress of the disease they were not able to agree on its cause. Some thought it contagious but many said it was not.

It was in this atmosphere of uncertainty and fear that Gay-Lussac made his maiden speech in the Chamber of Deputies. On 24 February he urged the government to take further action to combat the threat.[68] He criticised sanitary regulations as antiquated and imperfect. The Minister of Public Works suggested that there was little cause for alarm. Some measures had already been taken to improve sanitation and in any case the disease had lost most of its intensity! Government optimism was shattered when on 29 March cholera broke out in the crowded centre of Paris. What steps could science and medicine take? One of the first was a revival of the idea of disease caused by bad air and on 2 April Magendie asked the Academy to organise the analysis of samples of air from different parts of Paris. A commission consisting of Magendie, Gay-Lussac, Thenard and Serrulas was appointed. Meanwhile the cholera spread and the Minister of Commerce, exploring another avenue, asked the Academy to appoint a commission to examine any possible relationship between the outbreak of cholera and meteorological conditions. Again Gay-Lussac was on the commission appointed (2 July), this time with Dulong, Arago, Bouvard and Magendie. Of course neither commission was able to find the immediate cause of cholera. Such a step forward had to wait for the development of bacteriology and the discovery of the spirillum of Asiatic cholera by Koch in 1883. However the earlier work is interesting to the historian as an example of the failure of chemical reductionism. Magendie tried to introduce more chemistry into medicine with the help of Gay-Lussac.

## The Annales de chimie et de physique

By his editorship of the *Annales de chimie et de physique*, Gay-Lussac exerted a powerful influence over the development of science in France in the period after 1816. The *Annales* was edited and owned jointly by Gay-Lussac and Arago, the former being responsible for the chemistry contributions and the latter generally for the physics. These two ambitious young men had a financial interest in the success of their journal, and it provides one more example of how in the life of Gay-Lussac science was not an activity divorced from commerce and industry. We see Gay-Lussac as the entrepreneur responsible to no one except in the most general way to the international scientific community. The *Annales* was a French publication but its circulation became international, that is as far as science was cultivated throughout the world at that time. Britain, Sweden, Germany, Italy, Switzerland, the United States all received copies in research centres as well as private homes.

The *Annales de chimie et de physique* held a commanding position in France in the first half of the nineteenth century for the publication of papers in the physical sciences, and it continues in existence in a modified form to the present day. The *Journal de physique*, edited by the ageing Lamétherie was not a serious competitor after about 1815 and the later scientific journals of Ferrussac, Quesneville, Gerhardt and Laurent and Reiset were by comparison ephemeral affairs, which never attained an international status. During the Napoleonic wars the *Bibliothèque britannique*, published in Geneva, had often managed to obtain scientific news from Britain before it was available to the Paris savants. Once the war had ended, however, Paris had a geographical advantage over Geneva for contact with Britain. The *Bibliothèque universelle*, which succeeded the *Bibliothèque britannique* after 1815, did its best to compete with the new *Annales de chimie et de physique* but was openly derided by Arago.

The original *Annales de chimie* had been founded as the journal of the new chemistry in 1789 with an interesting precedent in the publication of chemical journals in Germany edited by Lorenz Crell. Crell's *Chemische Annalen* stopped publication in 1804 in face of competition from Scherer's rival chemical journal. In early nineteenth-century Germany there were several small competing pharmaceutical journals which tended to replace each other or amalgamate. The *Annalen der Pharmacie*, which Liebig was to make so famous, was formed in 1832 by the amalgamation of the *Archiv des Apotheker-vereins im nördlichen Deutschland* and the *Magazin für Pharmacie und Experimentalkritik*.

In France Gay-Lussac had none of these troubles. The active participation of Lavoisier had helped to confer an immediate importance on the *Annales de chimie*. Lavoisier's associates Guyton de Morveau, Fourcroy and Berthollet were also on the editorial board and, after Lavoisier's death and the temporary cessation of publication, they continued to make the journal one of central importance for chemists. When a break was made in 1816, Gay-Lussac was able to inherit the prestige of the former *Annales de chimie*.[69] The start of a new series was the outcome of a number of factors. Much of the organisation of the *Annales de chimie* had been in the hands of Descotils. When he died in December 1815 a new appointment had to be made. The political changes of 1815 and the virtual retirement of Berthollet under the Restoration came at a time when the number of volumes in the old series had reached nearly a hundred. It was therefore decided to start a new series and to give editorial responsibility jointly to Gay-Lussac and to Arago.

The decision to make the *Annales* a joint journal of physics and chemistry was a brilliant one. It completed the tendency of Lavoisier to associate chemistry with the techniques of physics rather than natural history. Not only was Gay-Lussac's chemistry biased towards physical chemistry but in the dynamic personality of his colleague in the Arcueil group, François Arago, there was a physicist-astronomer who could give his full attention and expertise to this side of the journal. Without ceasing to be a journal of central importance to chemists it also drew its readers from those interested in physics.

Gay-Lussac brought to the *Annales* the experience he had gained as one of the co-editors of the *Annales de chimie* in 1814 and 1815. He had the advantage of being able to ride on the high reputation of the first series of the *Annales*, rather than building up the journal from scratch. The publisher, Crochard, was the same as that responsible for the last volumes of the *Annales de chimie*. The new *Annales*, like the old, was to be a monthly publication. Even more than the old *Annales* it came to enjoy a virtual monopoly of French physical science.

The old *Annales de chimie* had had a circulation of more than 600. For the new *Annales de chimie et de physique* we are fortunate to have found a notebook giving the accounts for the first two or three years of its existence. This shows that the new series began inauspiciously in 1816 with only ninety-one copies sold.[70] It was with the *Annales* for 1817 that the business really got off the ground with Gay-Lussac and Arago taking over completely from the old editorial committee. Gay-Lussac and Arago decided to make their venture known by giving away copies of the *Annales* to selected individuals and institutions; this accounted for fifty-five copies. By 1 January 1818 they had sold 525 copies and a further 109 went as back numbers during the following year. In their second year, therefore, they had a circulation of the order of 600 and this increased over the years. Optimistically the editors had had 1250 copies printed. Their accounts for 1817 are set out opposite.

Thus in purely business terms the *Annales* was soon a success. After a poor start they soon regained the circulation figures of the old *Annales*. Now that the scope had been increased from chemistry[71] to chemistry *and* physics, there was a prospect of a larger market. This expansion was achieved without increasing the number of pages of each issue simply by using smaller type. The editors calculated that in this way they had gained the equivalent of thirty-six pages per issue.[72]

The main part of the journal consisted of original memoirs, usually research papers, as in the previous series. The *Annales* included short book reviews and announcements of recently published or forthcoming books. The high quality of these reviews may be inferred from the reviewers, who included Berthollet and Biot in the early issues.[73] Meteorological data were published each month. A summary of the activities of the Académie des Sciences became a regular feature of the journal and, potentially even more important, the journal gave publicity to the subjects of prizes to be awarded by the Academy.

| | | |
|---|---:|---:|
| Printer | 3 370 f. | 55 |
| Engraver | 237 | 92 |
| Paper  200 | 1 600 | 00 |
| Carriage | 23 | 50 |
| Stamp [duty] | 154 | 00 |
| Art work | 35 | 00 |
| Making up at Crochard's | 349 | 00 |
| | 5 770 | 00[74] |
| Net receipts from Crochard | 11 620 | 00[75] |
| Hence profit | 5 850 | 00 |
| For each of us | 2 925 | 00 |

Many important communications in the *Annales* were published as letters, usually to either one of the two editors. These were not contained in a separate 'Letters to the editors' section but are generally to be found in the main body of the journal. The atmosphere is more personal and intimate than in a modern scientific journal. The scale of the scientific community would have much to do with this. Sometimes letters were virtually solicited, as when Gay-Lussac showed Liebig's memoir on organic analysis to Dumas for comment before publication.[76] Often letters would be protests from scientists whose work had been criticised in the *Annales*.[77] The effect of the inclusion of correspondence was to make the *Annales* a live publication and to make readers feel that they were potential participants in scientific debates. This provided a contrast with the *Mémoires* of the Academy which was more formal in presentation and where there was more the impression of knowledge being handed down by authorities.

A common theme among letters was a claim from their authors for priority. Often this would be over some obscure issue, but we find no less a figure than Michael Faraday writing a long letter to Gay-Lussac, claiming priority for his discovery of electro-magnetic induction. The first news of this had come to France in a letter of 18 December 1831 from Faraday to Charles Hachette, which had been read at a meeting of the Académie des Sciences on 26 December. The *Annales* had understandably taken the opportunity of publishing immediately an extract from this letter.[78] The way this had been followed up by two

Italian scientists Nobili and Antinori gave rise to some misunderstanding in priorities and the *Annales* consequently published a thirty-page letter from Faraday setting out his just claim to be the discoverer of electro-magnetic induction.[79] Considering the widespread interest, the *Annales* was rather slow in publishing a full translation of Faraday's classic memoir[80] but Faraday blamed himself for his hasty letter to Hachette, which had been the source of misunderstanding.[81]

The editors took advantage of their position to append critical comments to papers they published. These were usually signed 'R' (*Rédacteur*) but occasionally Gay-Lussac's comments are signed 'G.L.'. Gay-Lussac added a footnote to a memoir by Beudant, rebuking him for forming a conclusion on the subject of mixtures and compounds which conflicted with Berthollet's ideas.[82] On the other hand Gay-Lussac did not hesitate to devote half a dozen pages of the *Annales* to a table of Berzelius' values for atomic weights[83] although he did not accept such a system himself.

There is no evidence that Gay-Lussac exerted censorship on material submitted. In the case of the gunpowder commissioner Longchamp[84] Gay-Lussac opened the pages of the *Annales* to a contributor who had been snubbed by the Academy of Sciences. If Gay-Lussac thought a memoir had something to say but was badly mistaken, he would publish it, but might add a note stating what errors he considered had been made by the author. This happened with Théodore de Saussure,[85] whose reply to Gay-Lussac's criticisms was published in the following volume.[86] Of course the editor was always able to have the last word.[87]

In the interests of freedom of publication the editors were even willing to risk antagonising the government. In 1818 they published a letter addressed to Berthollet by a group of former students of the Ecole Polytechnique, asking that he should give his support to a subscription for a monument to his friend Gaspard Monge, who had just died. Monge was one of Napoleon's most unequivocal and ardent admirers and had become persona non grata under the Restoration and his name had been struck off the list of membership of the Academy by royal decree in 1816. Although the decision to publish the letter hints more of Arago's Republicanism than Gay-Lussac's conservatism, Gay-Lussac must have given his agreement in order for this letter to be published.

As an illustration of how perceptive Gay-Lussac's comments could be we may cite the classic case of his editorial observation that, if certain chemical analyses reported were correct, this was evidence of the importance of the arrangement of atoms within the molecule.

Although Gay-Lussac strikes a characteristic cautionary note, some historians consider that he 'discovered' isomerism:

M. Wöhler...concludes that cyanic acid is composed of two atoms of carbon, two atoms of nitrogen and one atom of oxygen, or of two atoms of cyanogen and one atom of oxygen. But this conclusion is certainly not exact; for cyanogen is formed of two atoms of carbon and one of nitrogen, and the two quantities of carbon and nitrogen given by Wöhler are precisely in this proportion. Thus cyanic acid, according to this result, would be composed of one atom of cyanogen and one atom of oxygen; that is to say that it would contain the same principles and in the same proportions as the acid which MM. Liebig and Gay-Lussac have designated by the name of fulminic or cyanic acid. But as these two acids are very different, since the compounds of the one detonate violently on a slight shock, whereas the compounds of the other do not possess this property at all, it would be necessary to explain their difference to admit a type of different combination between their elements. This is something which calls for further examination.[88]

When he published a French translation of Sertürner's paper on morphine taken from the latest number of Gilbert's *Annalen*, he remarked under the heading: 'Observations of the editor' on the lack of attention which had been given to a subject of such potential importance:

We are astonished that the first memoir of M. Sertürner has not attracted more attention from chemists, not only in France where it does not seem to have been known, but in the rest of the continent. The discovery of an alkaline base, composed of carbon, hydrogen, oxygen and nitrogen, in which neutralising properties are very pronounced, seems to us of the greatest importance; it is for this reason that we have hastened to inform our readers of it.[89]

Gay-Lussac went on to say that he had himself confirmed experimentally some of Sertürner's work on morphine. Other parts of the memoir he felt were not beyond criticism and he had asked Robiquet to undertake some further research. Gay-Lussac concluded boldly:

We are not afraid to predict that the discovery of morphine will open a new field and that soon we will have precise notions on the poisons extracted from vegetables or animals.

Gay-Lussac's calling attention to the importance of the alkaloids was probably decisive in ensuring that the subsequent discovery of other such compounds was made by French chemists, although the credit for the successive isolation of strychnine, brucine, veratrine and quinine must go to Pelletier and Caventou rather than Robiquet, who had been given a start in this research by Gay-Lussac.[90]

Among the characteristics of a good scientific journal one might list promptness of publication and the encouragement of new lines of research. Both of these hallmarks are to be found in the *Annales*. As an example of rapid publication we may cite a 'Note on the property possessed by certain metals of facilitating the combination of gases' read by Dulong and Thenard to the Academy on 15 September 1823. This memoir was squeezed in small type in the last four pages of the September issue of the *Annales*.[91] The reading of the memoir had been prompted by news published in the Journal des Débats of 24 August and also in a private letter from Kastner to young Liebig (then studying in Paris) about Döbereiner's discovery of the role of spongy platinum in bringing about the combination of hydrogen and oxygen at room temperatures. Although Dulong and Thenard gave full credit to Döbereiner, they were able to publish their extension of his work before his own memoir was published in France.

There remains a suspicion that Gay-Lussac made special efforts for his friends to help secure rapid publication. Thenard's first paper on hydrogen peroxide (then considered by him as an oxygenated acid), read to the Academy on 27 July 1818, was published in the August issue of the *Annales*.[92] However, another chemical paper, which had been read to the Academy in June, was held over to the September issue.[93] When Thenard found that he was really dealing with a neutral compound (i.e. a second compound of hydrogen and oxygen), his note to this effect read to the Academy on 18 January 1819 was published in the issue of the *Annales* published that same month.[94] Thenard's research on hydrogen peroxide illustrates very nicely the relative roles of the *Annales* and the *Mémoires* of the Academy in the Restoration period. The *Annales* provided rapid publication and communication and secured priority whereas the *Mémoires* became a repository for the well thought out account when the research was completed.[95]

Gay-Lussac was occasionally in a position to take advantage of his editorial position in reporting his own work. For example, his memoir on prussic acid read to the Institute on 18 September 1815 was immediately published in the *Annales de chimie*, but while it was being printed Gay-Lussac was carrying out further experiments which he reported in the same memoir.[96]

One of the major contributors to the *Annales* was Berzelius. Berzelius' first publications had naturally been in Swedish journals, but he soon began to publish in the *Annales de chimie* and in German journals. Having a strong admiration for Berthollet his relation with the *Annales* was more cordial than that of a mere contributor and the bond with the *Annales* continued under the editorship of Gay-Lussac and Arago.

Gay-Lussac's relations with Berzelius date back to 1811 when the Swedish chemist wrote a long letter to the young French contemporary in which he said very flattering things about the *Recherches physico-chimiques*.[97] Gay-Lussac in his reply told Berzelius how much he valued correspondence with him. Yet in an article in the first volume of the *Annales de chimie et de physique* entitled 'Observations on the oxidation of several metals', the editor cast doubt on Berzelius' analyses of the oxides of iron, manganese, tin and antimony.[98] Gay-Lussac's discussion of the oxidation status of iron is particularly important. He (correctly) insisted on the existence of three distinct oxides as opposed to the two admitted by Berzelius and he discussed the reversibility of the action of steam on iron. Berzelius, however, was justly proud of his analytical skill and therefore took offence over what he regarded as Gay-Lussac's unjustified intervention:

There is no means of defending oneself against the abuse of [such] an authority...since the majority of readers only study science by reading and must always be convinced by the opinion of those whom they are accustomed to believe.[99]

Although the draft of his reply to Gay-Lussac hardly conceals his indignation at the criticism, the final letter (or at least that part which was published, taking up eleven pages) put the dispute on the level of experimental evidence and personalities were left out.[100] Gay-Lussac took advantage of his position as editor to include a seven-page reply to Berzelius immediately afterward. He concluded:

If, according to such considerations, it was not permissible to suggest the examination of results not sufficiently demonstrated...it would be necessary to ban all kinds of criticism of chemistry and accept without examination all scientific news. M. Berzelius is mistaken about my motives. I love science as much as he does and I have always applauded his numerous and useful discoveries. However, if I thought it permissible for me to raise doubts on some of his results which in my opinion did not possess a sufficient degree of certainty, I have never pretended to share these doubts.[101]

Gay-Lussac also wrote privately to Berzelius apologising for any offence and assuring him that he never intended to impugn the accuracy of his work which he had always admired.[102]

Gay-Lussac was here being very tactful with a correspondent whose work he genuinely admired and whose contributions to the *Annales* he valued. The following year Berzelius was able to make his long-awaited visit to France and during his long stay in Paris he frequently visited Gay-Lussac in his laboratory at the Arsenal. There was therefore no breach between the two men, who might have been described as the two leading chemists in the world.[103] His important work on selenium which was done during his stay in Paris was published immediately in the ninth volume of the *Annales*, from which it was later translated into German and English.

Up to the mid 1830s more than 60 per cent of the volumes of the *Annales* contained some contribution by Berzelius. One of the last papers by him that was published in the *Annales* was in 1838, some ten years before his death. In this he attacked the new theory of substitution in organic chemistry and defended his electro-chemical theory.[104] Often Berzelius would have a paper published first in Sweden and it would then be translated in other journals, but occasionally the *Annales* was given priority and Gay-Lussac would add a note to say that the article came direct from the author's manuscript.[105] The journal certainly benefited from having such an important contributor as Berzelius. Although the *Annales* helped him by providing a French audience, he also helped the *Annales* and provided strong evidence against any accusation that it was a purely French journal.[106]

In conclusion we may report a criticism made by the British chemist Thomas Graham, who felt that Gay-Lussac was too much under the influence of the Paris scientific community. In a letter of 1836 Graham wrote:

The Institute is a coterie, all the members are familiar with each other and unwilling to give each other cause for offence. Gay-Lussac likewise, who is chemical editor of the *Annales de chimie* is an exceedingly mild man, and is certainly carried too far by the fear of offending his brother academicians. There is one of my papers which he has promised to publish (on phosphuretted hydrogen) but which I would not be surprised to find him restrained from doing when he finds that I give more credit to Henry Rose than will be agreeable to M. Dumas.[107]

Graham's memoir 'On phosphuretted hydrogen' had been published two years earlier in the *Philosophical Magazine*.[108] It had also been translated into German but it never appeared in the *Annales*. Whether this was for the reason suggested by Graham cannot be decided positively. If there was deep French antagonism to the work of Heinrich Rose of the kind suggested by Graham, it is curious that we should find a paper by Rose on 'phosphuretted hydrogen' published in the *Annales* in the following year.[109] Perhaps Gay-Lussac valued the work of Rose above that of Graham? Any journal like the *Annales* is faced with the problem of how many translations to publish of articles which had previously appeared in foreign journals, and one has to be selective.

Gay-Lussac and Arago continued as the sole joint editors of the *Annales* until 1840, when an editorial board was set up in which the original editors were joined by Chevreul, Savary, Dumas, Pelouze, Boussingault and Regnault. They had thus ensured the continued excellence of the *Annales* by associating with its production several of the leading French chemists and physicists of the next generation.

## Gay-Lussac under attack

It is indicative of Gay-Lussac's position in the scientific establishment that he was singled out for public attack by critics. Two particular instances of this will be described. If he was not attacked as often, as vigorously or as effectively as his friend Arago, it was because he was not so involved in academic politics.

An example of some of the difficulties Gay-Lussac encountered as editor of the *Annales* is provided by the case of Longchamp, a former commissioner of gunpowder, who on 24 November 1823 had presented to the Académie des Sciences a memoir entitled 'A new theory of nitrification'. In the paper he argued that saltpetre was formed in the soil not because of any constituent there but uniquely from the elements in the atmosphere. Half a century earlier such a theory would have been plausible, but the work of Lavoisier for the Régie des Poudres and subsequent work gave no support to such a view. To assess the value of Longchamp's memoir the Academy had dutifully appointed a commission consisting of Vauquelin as chairman with Chaptal, Gay-Lussac, Dulong and D'Arcet. Two years later they had done nothing about the paper, Vauquelin giving as his excuse that he was too busy. Longchamp therefore approached Gay-Lussac asking that the paper should be published in the *Annales* and Gay-Lussac agreed despite the fact that it contained a personal criticism of himself.[110]

Gay-Lussac little realised that in opening the pages of the *Annales* to Longchamp, he was doing more than providing some twenty-five pages of space for the presentation of a theory of very doubtful value. In fact, not content with his publicity in the *Annales*, Longchamp published a separate brochure setting out his ideas. It seems to have been this separate publication as much as his personal attack on him which prompted Gay-Lussac to reply to Longchamp, which he did point by point for ten pages in the next volume of the *Annales*.[111] Gay-Lussac only allowed his feelings to become evident in the last page of this open letter and here he accused Longchamp of insulting the work of the gunpowder administration, of which several members were no longer alive. This prompted the other man to reply in a letter, which Gay-Lussac published at the end of the next issue.[112] Gay-Lussac now withdrew a phrase accusing Longchamp of libel. Longchamp was also angry because the so-called 'letter' published had in fact never been sent to him. In this case one might say that Gay-Lussac allowed Longchamp to have both the first and the last word in the argument. This was all the more the case as Longchamp was not satisfied with this, since at least one paragraph had apparently been omitted. He therefore had privately printed a further brochure of some thirty pages in which he set out in full and tedious detail the substance of what had now become his dispute with Gay-Lussac.[113] The famous chemist was not only the editor of the journal who therefore had a certain control over publication but he had also defended[114] the gunpowder administration from which Longchamp was alienated. The latter complained that he had hardly met Gay-Lussac yet 'how is it that he suddenly leaves the pleasant research of his laboratory' in order to publish insults to him?'[115] He criticised the 'scientific impartiality' of Gay-Lussac.[116] His most forceful criticism of the scientist in this brochure is that, as editor, he has fallen below the standards set by the founders of the *Annales de chimie*, who had stated that they would never discriminate against an author because they did not agree with what he said.

In the end, of course, Gay-Lussac won and Longchamp with his curious theory was forgotten. Nevertheless the integrity of the scientist had been publicly called into question and some readers of the brochures – if not of the more strictly argued scientific discussion published in the *Annales* – would come away with some doubts about the scientific establishment. In so far as Longchamp was critical of the value of the work of the gunpowder administration there was some justification for his criticism. On the scientific plane Gay-Lussac had emerged not merely as representing one side of a case but as representing authority.

Gay-Lussac had earlier been the subject of attack in 1818 in a curious book of over 400 pages of small type by a certain Hyacinthe Bodelio, a former naval surgeon who had been dismissed for alleged incompetence. When living in India, his scientific ideas had been received locally with interest and he had decided to return to France after a long absence and to go to Paris to debate with the leading scientific authorities.

Bodelio's book, which was largely an attack on the concept of atmospheric pressure, is something one might have expected in the seventeenth rather than in the nineteenth century.[117] However, it concerns us because the author used his book to attack what he called the *gaylussacites* (*gaylussaciens*). He explained:

I call *gaylussacites* all those who think, see and reason like M. Gay-Lussac, one of the leading members of the Institute.

He had already published an attack on the physics of his day in a brochure but, by what he interpreted as a conspiracy, no publicity had been given to this in the Paris papers. He was now launching a larger scale offensive. His quarrel with Gay-Lussac was over a memoir which he had taken to the Institute with certain criticisms of the physics and medicine of the time.[118] The Institute had accordingly appointed a commission of three: Gay-Lussac, Charles and Hallé. Bodelio explains how he gave the commission a week to look at his work and, finding that no report had been given, he decided to try to hurry things up by going to see Gay-Lussac, who had the responsibility of drawing up the report.

Gay-Lussac happened to be working in his laboratory but he agreed to see Bodelio without an appointment and told him that he had begun to read the memoir but that he would not be able to make a favourable report; under the circumstances he suggested that Bodelio might like to withdraw his memoir. Gay-Lussac said frankly that he would have preferred the memoir to be an elementary compilation of rational physics ('la saine physique') than something he found unintelligible. After half an hour of conversation Bodelio found himself defending his idea that air did not have weight against the arguments and 'mocking smile' of his adversary. After further discussions Bodelio withdrew, taking his memoir with him.[119]

Having failed in his discussions – he attributed this to Gay-Lussac's 'rhetoric' – Bodelio fell back on the printed word. He explained in the preface to his book how he had tried to have an audience with Louis XVIII but without success. Pathetically he had hoped to persuade the King to accept the dedication of this curious publication, now incidentally a book of great rarity. His book is full of references to the *gaylussacites* whom he alternatively called the 'Gaylussac sect'.[120] Although he also referred to that 'deceiver' Torricelli and others, it was Gay-Lussac who, in his eyes, embodied the Paris scientific establishment. Indeed he referred to Paris as 'the capital of the gaylussacites'.[121] He used the figure of Gay-Lussac not only because of his personal encounter but because of his eminence. With some exaggeration he claimed that Gay-Lussac's students regarded him as the greatest scientist of the century.[122] For the purpose of this biography, Bodelio is an excellent example of how by the beginning of the Restoration Gay-Lussac had come to represent the establishment to the man in the street.

# 8

~~~~~~~~~~~~~~~~~~~~~~~~~~~~~~~~~~~~~~~~~~~~~~~~~~~~~~~

# A scientist in the service of government and industry

'The British understand material interests much better than we do and this is the secret of their great prosperity.

We, Gentlemen, are always concerned with abstractions. Of course I do not deny the realm of ideas but I insist that applications should not be too much despised'

Gay-Lussac[1]

During the Revolutionary and Napoleonic wars science had been applied by the French in many ways: in munitions, communications and the search for alternative raw materials. In the period of the Restoration the basic utility of science was something which was generally accepted but the problem was to catch up with the Industrial Revolution in Britain which was giving Britain a leading position in world markets. Although in chemical industries France was probably at least the equal of her neighbour, in most other areas France looked to Britain as the leading industrial nation. The contrast was nowhere more marked than the development and use of the steam engine. In the conditions of peace a new stimulus was given to industrial development. The industrial exhibitions begun in the Revolutionary and Napoleonic period continued on a larger scale and Gay-Lussac, who had already been a junior member of the panel of judges for the 1806 exhibition, was asked to take part in the judging for the 1823 and 1827 exhibitions. The Conservatoire des Arts et Métiers was expanded in the Restoration period and in 1819 a governing body (Conseil de Perfectionnement) was set up with Gay-Lussac as one of its members.

Not only government agencies but also the expanding sector of private industry supported by the new banking houses turned to scientists for advice. For example, the supply of raw materials to manufacturers and the production of a finished product often demanded the adoption of certain standards, e.g. of purity. Scientists, and particularly those trained in chemistry and physics, could make a valuable contribution as consultants to industry. For more routine tasks in industry a little knowledge of science might be useful among the ordinary employees of a chemical works but for innovation an expanding industry turned to the senior men qualified in the relevant science. Berthollet was no longer active and Gay-Lussac very naturally took his place. It is true that Chaptal (1756–1832), an elder statesman of chemical industry, had not disappeared from the scene but his main efforts were directed towards supporting the sugar beet industry threatened with collapse when cane sugar was again imported. It was in the nineteenth century that the scientist came to the fore as consultant and Gay-Lussac was to give a lead by using his talents in this way. Indeed the many cases where he was able to introduce a new or improved method in industry or commerce not only impressed industrialists but helped to improve the status of industry in the eyes of scientists.

Under the old regime scientists had occasionally been called upon by the government to give advice on new technical schemes. Also a few scientists had been employed in the *manufactures du Roi*; the state porcelain works at Sèvres and the Gobelins dyeworks had benefited from the appointment of chemists, notably Macquer and Berthollet. Tillet worked on problems of assaying for the Mint and after the Revolution Vauquelin and D'Arcet fils provided further technical advice. Gay-Lussac was fully aware of this technical tradition and made several references to the work of Tillet in his own research for the Mint. But only in the nineteenth century do we find scientific consultants beginning to be used in private industry. Gay-Lussac by his well-defined and permanent appointment at the plate glass factory of Saint-Gobain was among the first of this new breed of consultant.

There were some of Gay-Lussac's contemporaries who regarded his work in the Restoration as a decline if not a betrayal of his talents. Boussingault, who visited him in 1820, thought that he had become too concerned with aspects of science which brought financial reward.[2] Mitscherlich, who came to Paris in 1824, was shocked to find the great scientist involved in a firm making scientific instruments.[3] The view that Gay-Lussac had turned from the cause of pure science to serve mammon is a distortion of the truth, not only because it makes Gay-Lussac's motives too base in his later life, but because it implies an unrealistic purity in the earlier period. The view also has other faults. It misunderstands the concern with applied science. Although idealists could regard any abandonment of pure research as a betrayal, Gay-Lussac inherited from Berthollet a view of science in which there was no rigid boundary between the pure and the applied. Gay-Lussac had been interested in the applications of science from the very beginning of his career and had become a paid member of the Bureau Consultatif des Arts et Manufactures in 1806. Anyone, therefore, who wishes to divide Gay-Lussac's life into two distinct periods, one of pure science and the other of applied science, must bend the evidence to make it fit.

We should not overlook a certain continuity in Gay-Lussac's life. One might think that in discussing Gay-Lussac's applied science we enter a different world, very far from the discussion of his methodology in chapter 3. Yet, although different circumstances called for different emphases, we can detect some overall unity. Gay-Lussac the 'pure' scientist and Gay-Lussac the applied scientist are not like Dr Jekyll and Mr Hyde. Not only would it be very difficult in the work of many scientists to state where 'pure' science stopped and 'applied' science began, but it may be possible to detect some common elements. One can detect in Gay-Lussac's applied science as well as his more theoretical work a concern with correlation. Gay-Lussac in classical mould wanted to tidy up the world. His work on alcoholometry was a detailed study of the precise correlation between density and alcohol content at different temperatures. In form it is not dissimilar to his study of the solubility of salts at different temperatures.

An internalist view of Gay-Lussac in 'decline' has recently been offered:

Faced with the general decline of the influence of Berthollet's ideas in the years following 1815, Gay-Lussac went into a kind of funk which he did not recover from until he had delivered himself of his full criticism of Berthollet's system [1839].[4]

It is argued that after 1839 it was too late for Gay-Lussac (now in his sixties) to make any fresh contributions to chemistry. This is too negative an approach. Gay-Lussac's life between 1815 and 1839 was not a desert. It was a very active period in his life but one when applied science and administration came increasingly to displace pure science.

Gay-Lussac was now middle-aged and perhaps the enthusiasm of his earlier research had weakened. The eclipse of those two powerful influences on his earlier career, Berthollet and Laplace, could have had a dampening effect on his research. With this brake on one side and the incentive of financial gain with his increasing family responsibilities it is not difficult to understand a change in direction in Gay-Lussac's research. As a man of wide knowledge of physics and chemistry he was an obvious choice for membership of commissions and for consultancies on behalf of government departments. When one starts judging the feasibility of applying scientific principles to practical problems one becomes drawn into the peculiar fascination and satisfaction which this can afford. Gay-Lussac found all this attractive, all the more so as it did not mean that he had to abandon his teaching or pure research.

The situation in France may be compared with that in Britain where there were few jobs for scientists, where most industry was situated far from the capital and the ancient universities, and where industrial enterprise might provide an *alternative* to a career in pure science. This last point was made by the professor of chemistry in the University of Glasgow, Thomas Thomson, who remarked in 1830 of the British analytical chemist Richard Phillips (1778–1851), an exact contemporary of Gay-Lussac :

Unfortunately of late years he has done little, having been withdrawn from science by the necessity of providing for a large family, *which can hardly be done in this country, except by turning one's attention to trade or manufactures.* The same remark applies to Dr [William] Henry, who has contributed so much to our knowledge of gaseous bodies, and whose analytical skill, had it been wholly devoted to scientific investigations, would have raised his reputation as a discoverer much higher than it has attained; although the celebrity of Dr Henry, even under *the disadvantage of being a manufacturing chemist,* is deservedly very high.[5]

A notable aspect of Gay-Lussac's career is that he was able to act as a consultant to industry and commerce without abandoning his professorial positions. He never had to choose between pure and applied science.

## The Gunpowder Service

A considerable amount of Gay-Lussac's time in the Restoration was to be spent in a military context. There may appear to be some irony in the fact that it was only *after* Waterloo that Gay-Lussac became involved as a consultant on saltpetre, gunpowder and cannon but there is a simple explanation of this, namely that it is only after a man has made his name that he may expect to be singled out for a consultancy post. What is more, there were basic changes in the administration in 1816 and 1818 and Gay-Lussac, who would have had no place in the old system, was offered a position in the new one which suited him well and which enabled him to give his services without abandoning all his other interests.

The history of the French saltpetre and gunpowder service is a long and complex one and only the barest outline is necessary here. It had begun in the fourteenth century as a state concern. After attempts at decentralisation in the sixteenth century the right to make gunpowder had been conferred on a private company and when this system had failed the business was nationalised in 1775 as the Régie des Poudres with Lavoisier as a leading member. This system lasted with various modifications until after the Revolution. In 1800 Bonaparte found it convenient to place the manufacture of gunpowder under the Ministry of War. The end of the Napoleonic wars brought about complete demobilisation and as part of this peace-time rationalisation the production of gunpowder was reduced to a small section of the War Ministry. It was also felt appropriate that the gunpowder service should be attached to the artillery, with an artillery officer, Lieutenant-General Ruty in charge. In the previous forty years one or two scientists had been appointed to the Régie in an administrative capacity, notably Lavoisier and Chaptal, and their chemical knowledge had been an invaluable complement to their administrative duties. There was, therefore, nothing revolutionary about employing scientists, but the important innovation was to employ someone exclusively in a scientific capacity to give advice and to carry out research at the highest level. Once again Gay-Lussac's appointment announces the beginning of the age of the scientist as consultant.

Although control of the manufacture of gunpowder was placed in the hands of artillery officers in 1816, the implications of this reorganisation were only fully worked out in a royal ordinance of 15 July 1818 which, perhaps in imitation of the Bureau Consultatif des Arts et Manufactures, established a consultative committee to superintend quality control and development in the manufacture of gunpowder. This committee of three was to include a senior scientist, in fact a member of the Academy of Sciences. The Academy nominated Gay-Lussac, who had already shown interest in problems of applied science and whose expertise in chemistry and physics made him a particularly suitable nominee. The Minister of War decided that, whereas the refinery inspector and any artillery officers who attended the committee should receive fairly nominal payments, the salary of the member of the Academy should be 4000 francs per annum.

One of Gay-Lussac's first jobs was to help in the production of several manuals. When the Ministry of War had bought gunpowder from the Régie, the main scientific problem was to test the quality. Now that the Ministry, or rather a section of the artillery, was responsible for the manufacture, it was thought advisable to issue a manual or *Instruction*. Gay-Lussac's work on the manuals is described later.[6]

A new departure in 1818 was the investigation of the possibility of manufacture of saltpetre (i.e. potassium nitrate) by a double decomposition reaction involving sodium nitrate and potassium salts. The sodium nitrate was imported in increasing quantities from Chile ('Chile saltpetre') and spelled the ruin of the traditional saltpetre manufacturers. It was the consultative committee which approved these trials[7] and may thus have incurred the enmity of the established saltpetre manufacturers.

Another aspect was the purity of the saltpetre, the main impurities being potassium and sodium chlorides.[8] Gay-Lussac did not content himself with estimating the total chloride content but thought out a method of distinguishing the two depending on the different effect on solution in water. Both dissolved with absorption of heat but potassium chloride caused a much greater lowering of temperature than the same weight of sodium chloride. In an age which had not established the concepts of molecular weight and ions, and was a century away from sophisticated ideas of solvation, the method used was strictly empirical. Gay-Lussac provided a formula from which he claimed that the percentage of potassium chloride could be calculated.[9]

Gay-Lussac also developed a method of analysing gunpowder. This consisted of first extracting the saltpetre with water and then converting the sulphur to potassium sulphate by fusing with a mixture of potassium carbonate and potassium nitrate. The charcoal content was found by difference. There was no question of such research in a military context being considered as secret. Gay-Lussac published it in his own journal.[10]

Gay-Lussac was equally happy to show everything to visiting scientists. Berzelius was allowed to spend a year in Paris and was provided with funds by the Swedish government on the understanding that he would study the production of gunpowder in France. Berzelius wrote as follows about this part of his visit in 1819:

After my return to Paris, in company with colonel of artillery Aubert and Gay-Lussac, I made a trip to the French governmental powder factory at Essone. Here I learned the recently improved methods for manufacturing powder and refining saltpetre. To this end, however, I had previously under Gay-Lussac's direction gone through the processes at the arsenal. I prepared a detailed report on this to be presented to the government on my return home.[11]

The manufacture of gunpowder has obvious dangers and in 1822 there was a major explosion at one of the provincial powder factories. Gay-Lussac, having just returned to Paris from the country, told his wife the news:

When I arrived here I learned of the terrible accident which happened at Colmar. The gunpowder store, which contained 66 *milliers*[12] of powder blew up. Of 13 workmen 11 have died; the other two are badly injured. The commissioner has lost his eldest daughter, who was crushed under the wreckage of the house. His youngest daughter was badly wounded and her right arm has been amputated. He and his wife received several injuries. Everything has been destroyed by the explosion or by the fire which accompanied it and not a single stone is standing.[13]

Gay-Lussac was also concerned with the economics of saltpetre refining, giving attention to the efficiency of different furnaces used to concentrate solutions. Among his papers is a note entitled 'Expériences faites à l'arsenal sur l'évaporation des eaux (1824)' and another: 'Expériences sur les fourneaux faites à la raffinerie des salpêtres de Paris depuis le mois de nov. 1823 jusqu'au mois d'avril 1825'.[14] Most of this work would have been fairly routine and it is doubtful whether Gay-Lussac played more than a remote supervisory role.

In 1825 he was also involved in collaboration with Aubert and Pélissier in the study of fulminating powders as detonators. Fulminating mercury had been studied in 1800 by Howard and then by Berthollet. Gay-Lussac took this work further in 1824 and made a particular study of fulminating silver which could be prepared in a very pure state.[15] Reporting on research undertaken jointly with the young Liebig, he concluded that these compounds were cyanates, containing carbon, nitrogen and oxygen. His study of these compounds in the context of both pure and applied science reminds one of Lavoisier who studied nitrates both for their composition and their explosive properties.

The ballistic pendulum had been introduced in England in the mid-eighteenth century to test the quality of various samples of gun-powder but the French had uncharacteristically preferred to rely on the less scientific mortar for trials. However the period after Waterloo was a period of reappraisal and Gay-Lussac was asked to collaborate with Aubert on trials of the ballistic pendulum.[16] He privately consulted Poisson to get the best advice of an applied mathematician on the problem.[17] Gay-Lussac and Aubert carried out detailed trials of the ballistic pendulum and drew up a report of their findings. They recognised that with fine powders the old method gave misleading results and they recommended the adoption of the ballistic pendulum. This was accepted in 1824, the first pendulum for cannon being constructed in 1826.

Gay-Lussac was also involved in the problem of the composition of cannon. Cannon in France were traditionally made of bronze but news came from Russia that it was possible to make stronger alloys if iron was added to the copper and tin to make a triple alloy. The artillery consequently authorised experiments in which iron in a wide range of proportions was alloyed to copper and tin and the corresponding alloys submitted to a series of physical tests to determine their 'hardness, tenacity and fragility'.[18] Gay-Lussac measured their hardness by determining the height from which a weight with a steel spike had to be dropped to produce a standard size indentation. 'Tenacity' was measured by the weight in kilograms required to break a cylinder of 5 millimetres diameter. 'Fragility' was estimated approximately by counting how many times a weight had to be dropped at the middle of a suspended bar to break it.[19] At the end of November 1825 Gay-Lussac was sent with D'Arcet and an artillery officer to the foundry at Douai to repeat on a large scale the trials they had carried out in Paris on the manufacture of the triple alloy.[20]

Gay-Lussac remarked on the difficulty of incorporating iron in such alloys without oxidation. His notebooks reveal ample evidence of chemical analyses of different alloys. But although he was always ready to carry out chemical analyses, he distinguished himself not by his willingness to do routine work but by his inventiveness. He was always thinking of new ways of estimating useful data. Thus in his analyses of alloys it was important that samples from different parts should be tested to confirm that the alloy was of uniform composition. By a chemical method this would have taken a long time but Gay-Lussac conceived a method of estimating iron based on the premise that the magnetic effect of an iron–copper–tin alloy would be directly proportional to its iron content. This was found by the oscillations of a magnetised needle suspended over the alloy. The results were not fully satisfactory but they provide an example of Gay-Lussac's inventiveness and his understanding that in practical affairs a rapid test of moderate accuracy may be preferable to a time-consuming test of great accuracy.

By 1828 Gay-Lussac had spent ten years advising the Ministry of War but the death on 24 April of that year of his chief, General Ruty, gave the new Minister of War, De Caux, an opportunity of reconsidering the system of technical advice on arms and munitions. Some reorganisation may have seemed all the more necessary after a public attack on the Gunpowder Service by General Sébastiani. Corsican-born Sébastiani was an opposition deputy who had caused the downfall of at least one minister in his many criticisms of successive governments. It was in the Chamber of Deputies on Monday 23 June 1828 that Sébastiani took advantage of a debate on the budget of the Ministry of War to launch an attack on the gunpowder administration.[21] He attacked it as a state monopoly which had succeeded in doubling the price of gunpowder compared with what it had been in private hands. He pointed out that, although the Gunpowder Service might have rashly improved the explosive force of their product, this had had the effect of reducing the number of times any cannon could be fired without splitting. This was a very real problem. Gay-Lussac was taken aback by the criticism, all the more so as it followed so closely on the death of General Ruty. He rushed to the defence of his government department, not simply by writing to Sébastiani, but having a pamphlet published in the form of an open letter (25 June).[22] Gay-Lussac had been told that Sébastiani had asserted that if the gunpowder produced was any better than in former times it was because some potassium chlorate had been added. Gay-Lussac seized this as an obvious false-hood – potassium chlorate was far too dangerous. Sébastiani replied in another pamphlet, pointing out that he had said nothing about potassium chlorate.[23] In his open letter to Gay-Lussac of 30 June, Sébastiani wrote:

In the attack I made against monopolies I did not expect to meet an adversary such as you, a scientist whose name will be written in history side by side with those who have done most honour to our century and whose discoveries provide evidence of a powerful genius and an extensive knowledge of science. The rather bitter language of your refutation diminishes in no way my respect for your talents and your character.

Gay-Lussac, looking for support had sent a copy of his own pamphlet to the Dauphin, the Duke of Angoulême,[24] who was the patron of the Ecole Polytechnique. Gay-Lussac seems almost to have felt that it was science itself which was under attack. As a technical adviser to the administration he considered that part of the criticism fell on him and that he should have been immune from criticism. Later, as a member of the Chamber of Deputies, he was to feel that comments on scientific and technical matters were not out of place in a parliamentary assembly. Understandably, however, he never criticised personnel nor government departments.

There was already a general consultative committee for the artillery and it was decided to attach the gunpowder consultative committee to this. Gay-Lussac's position therefore disappeared and the Minister of War wrote to the scientist on 16 June regretting the decision which he justified on grounds of economy.[25] However, Gay-Lussac was told that he could continue to draw his salary until 1 October when his present work would have come to an end. The artillery intervened at this point, emphasising the value of Gay-Lussac to them, particularly in connection with the study of alloys which he was then undertaking. The representation had the desired effect and Gay-Lussac was officially confirmed as a member of the joint consultative committee.

The Revolution of July 1830 brought about a further shuffling in most government departments, but it was not until 1834 that time was found to draft a decree setting out Gay-Lussac's precise position.[26]

Since it is unusually explicit and not too long it is worth citing in full:

Art. 1. The member of the Academy of Sciences who, according to the ordinance of 18 September 1830 is a member of the Conseil de Perfectionnement established at the headquarters of the Gunpowder and Saltpetre Service, will terminate this appointment and henceforth be attached to the central depot of the artillery.
Art. 2. He will be under the orders of the president of the artillery committee and will carry out the following duties:
He will take part in commissions concerned with research or experiments on problems of physics or chemistry which are of interest to the artillery.
He will carry out or supervise analyses or experiments which are given him either on the orders of the Minister of War or by the committee acting in the interests of the artillery service.
He will edit *Instructions* for analyses or processes to be carried out in the different establishments.
He will examine memoirs and proposals relating to the applications of physics and chemistry to the artillery and to related technology.
He will submit his observations to the committee and will make such proposals to it as he deems are in the interest of the progress of the army.
He will attend the meetings of the committee when the president considers it appropriate to call him as a consultant ('conseil').

One must not be too alarmed by the official style used, although it is useful to remember that any civilian in a military establishment was in an anomalous situation. In practice Gay-Lussac's position was that of a senior consultant and as a member of the Academy his status was secure.

Gay-Lussac continued to work with the artillery until the Revolution of 1848. Such research would not ordinarily have come to public attention. An incident in 1836, however, is worthy of special mention since, once more, it brought Gay-Lussac's scientific work for the War Ministry into the public arena.

The incident arose from a modest attempt at improving the laboratory facilities available for Gay-Lussac's military research. This meant asking for extra money from public funds which had to be itemised in the annual budget. When this was submitted for approval to the Chamber of Deputies it was felt to be an extravagance. It was Arago who stood up to defend his friend and at the same time advocate the utility of science.[27] The opposition claimed that all the artillery needed was routine analysis of the raw materials and that money could be saved if, whenever any difficult case of metallurgical analysis arose, the problem was sent to one of the other Paris laboratories such as the Mint or the School of Mines. Arago argued that a modest sum only was being asked for – 1750 francs annually for laboratory supplies and the payment of junior laboratory assistants. It was not a question of giving Gay-Lussac a new post or increasing his salary but merely of making more effective use of him. Arago advocated the role of science not for routine control but for *research*. One important task was analysis of alloys, since some bronze cannon could no longer be used after a hundred or even fifty firings. The storing of shells without deterioration was another piece of research which could save the government money. Finally there was evidence that the use of certain charcoal in gunpowder could increase the range, although there were some harmful side effects. This too was a project worthy of further research.

Arago's speech provided an effective justification of Gay-Lussac's work but it was never an easy matter to obtain additional facilities for scientific research from public funds.

## The Mint

At the Paris Mint Gay-Lussac was able to introduce a reform in assaying which may be regarded as a dividing line between medieval technology and modern science. The post of Director of the Assay Bureau (Directeur du Bureau de Garantie) at the Mint was Gay-Lussac's most lucrative appointment. The post had previously been held by the veteran analytical chemist Vauquelin, the author of the standard French work on assaying, *Manuel de l'essayeur* (1799), who had become an assayer at the Paris Mint in 1802 and Director of the Assay Bureau in 1812. In 1829 in his sixty-sixth year his health was poor and he was absent from nearly all the meetings of the Academy of Sciences in February and March. His last attendance was on 11 May; he then left the capital for his native Normandy. At the Academy meeting of 19 October, Gay-Lussac and Deyeux informed their colleagues that his health was improved,[28] but on 14 November he died. Perhaps his long illness had warned his colleagues to be prepared, since already on 17 November the president of the Mint Commission wrote to the prefect of Paris pointing out that the nomination of a successor lay in his hands and asking in the interests of commerce for an immediate decision.[29] Gay-Lussac told the prefect, Count Chabrol de Volvic, of his desire for the position and on 20 November he was appointed.[30] Chabrol, a graduate of the Ecole Polytechnique, had been prefect of the Seine department since 1812 and held the post (with the exception of the Hundred Days) until the Revolution of July 1830.[31] He was a discreet man not noted for unpopular decisions and his selection of Gay-Lussac was probably what was expected. Gay-Lussac was already known to Chabrol for his willingness to give his technical advice.[32] Moreover on 18 November, Gay-Lussac had been nominated a member of a commission set up by the Minister of Finance to examine the problem of assaying.

In the assaying of silver discrepancies had been found between the expected and the reported assay and different assayers obtained different results. The method universally employed was that of cupellation, in which the silver or gold sample was strongly heated in a bone cupel with lead. Impurities were oxidised and carried off with the lead oxide. In France Tillet had pointed out the inaccuracies in cupellation in 1760 but the method continued to be used. The Revolutionary legislation was completely lax but the law of 7 germinal year XI (28 March 1803) confirmed the franc as a coin of 5 grams of silver with a 90 per cent silver content and a tolerance of 3 parts per thousand.

Unfortunately the standard method of cupellation continued to give results which were 4 or 5 parts per thousand of silver less than the known composition of test alloys. The Paris Mint sent an assayer to London to see if the English had been able to solve the problem and specimens of known content were sent to various European Mints for assay. By 1829 it was quite clear that cupellation would need to be critically examined and the Minister of Finance appointed a high-powered commission to examine the traditional method of assaying, to suggest improvements and to recommend any action to be taken. The commission of eight, with Chaptal as chairman, included three other scientists: Gay-Lussac, Thenard and Vauquelin.[33]

The commission confirmed the lack of constancy in assay by cupellation and were able to attribute this to a portion of silver which was carried away with the lead oxide. This in turn depended on the proportion of lead and the temperature. Each assayer used the cupel of his choice and the preparation of lead he thought fit. But having explained the inaccuracies in cupellation, the commission turned to possible alternatives and Gay-Lussac was ready to describe to his colleagues a method based on precipitation of silver chloride which he claimed he had already tried out successfully.

The commission was divided on the action to be taken but the majority, including Gay-Lussac, who drew up the report, recommended a reform. Analysis of forty samples of five-franc coins by two assayers had given a mean assay of 4 parts per thousand less than that obtained by the silver chloride method. It is noteworthy that in its original gravimetric form Gay-Lussac did not recommend his method for its speed but rather for its certainty and its accuracy, which he claimed was to within 2 parts in 1000. Also it required less skill than cupellation and hence a shorter apprenticeship.

The report of the committee was examined on 8 April by the Mint Commission under the chairmanship of Count Sussy. The commission, faced with indisputable evidence of the falsity of current assay methods and yet having the responsibility for accurate assaying, was forced to take action. Among the advantages they pointed out was that the government would stand to gain the value (88 centimes) of 4 parts per thousand in every five-franc piece, which, in remelting the entire currency in circulation, would give over $2\frac{1}{2}$ million francs. (This estimate was later given as 6 million for all coins.) Moreover in new coins the government would save 4 parts per thousand and keep within the law which fixed the silver content at 90.00 per cent. D'Arcet (Director of Assays) and Gay-Lussac were asked to collaborate and draw up as quickly as possible an *Instruction* which would include both the assay method and the 'wet method'. This *Instruction*,[34] which was ready on 7 June 1830, said that Gay-Lussac's method:

which has been applied with the greatest success in the laboratory of the Mint, is new, and gives to the art of the assayer the precision which was lacking. If it is not always adopted in preference to cupellation, it should at least be used in difficult cases and as a means of verification. Accordingly it should be described in this *Instruction* with all the necessary details to make it easy for assayers to put it into practice.

A new ordinance was required and was signed by King Charles X on 6 June 1830, the day before the *Instruction* had been completed. The law left the individual assayer to decide which method he would use for ordinary assays but for counter-assays Gay-Lussac's method was henceforth to be the only one legally acceptable.

The details of this hurriedly drawn up *Instruction*, together with Gay-Lussac's definitive *Instruction* of 1832, are discussed in the next chapter. More relevant here is the fact that the new method, later to be copied in other countries,[35] gave Gay-Lussac unassailable authority at the Mint. He worked there until 1848, helped by his son Jules, who would sometimes stay behind in Paris to keep an eye on the assay bureau while his father took his summer break in the Limousin. Family letters show that, although physically absent, Gay-Lussac constantly had his Paris responsibilities in mind and he would often give detailed directions on action to be taken.

## Alcoholometry

Even before man thought of taxing alcoholic beverages, it was a matter of understandable commercial interest to find an easy means of estimating their alcoholic content. Early methods of assessing the strength of alcoholic solutions had included soaking a cloth in the alcohol or pouring the solution on gunpowder and seeing if it would catch fire. In the seventeenth century Robert Boyle proposed determining the specific gravity of the mixture and in the eighteenth century hydrometers of increasing sophistication were used to estimate the alcoholic content of wine, beer and spirits.

In 1768 Lavoisier carried out a study of the use of hydrometers (*aréomètres*) to determine quickly and accurately the specific gravities of liquids and recommended this method of finding the specific gravity of alcoholic liquors.[36] Several hydrometers were in use for excise purposes under the *ancien régime* in the different provinces of France, one of the best known being that of Baumé. Finally in England in 1790 George Gilpin under the auspices of the Royal Society undertook a systematic study of the relationship between the specific gravities of alcoholic liquors and their alcohol content.[37] Gilpin's tables were used as the basis for determining excise duties.

By the early nineteenth century the hydrometer was generally accepted in both France and England as the best method of determining alcoholic content of dutiable liquids, but legislation required that a specific type of hydrometer or alcoholometer be employed. In 1816 the French specified Cartier's instrument, which was substantially the same as that of Baumé. By the law of 14 April 1816 alcoholic liquors were divided into three categories for the purposes of taxation: those below 22°, those between 22° and 28°, and those above 28°. This rather arbitrary classification was criticised by Chaptal in the Chambre des Pairs in 1824. Speaking in favour of an alternative proposal of Gay-Lussac, he pointed out that under the existing law the strongest brandy paid no more duty than an inferior quality. He advocated taxation on a continuous scale in proportion to alcoholic content. Whereas previously it was only near the borderlines that accuracy was important, there was now a good reason for a thoroughgoing and reliable system.

Gay-Lussac had been approached by the French government to advise on the revision of the method of determining alcoholic content and he undertook a long and painstaking study in 1821–2 of the relation between specific gravity and alcohol content. He was able to refer to Gilpin's earlier work and to improve on it, since what Gilpin had considered as absolute alcohol had in fact contained an appreciable amount of water. Gay-Lussac took the precaution of drying his alcohol over quicklime. He gave the specific gravity of absolute alcohol as 0.7947 at 15°C compared with Gilpin's 0.82547 and the modern value 0.7936. He carefully made a succession of mixtures of alcohol and water, twenty-one in all, and determined the weight of a particular volume. The flask containing the mixture was immersed in a water bath at different temperatures and determinations were made at five-degree intervals between 0 and 30°C. His concern with accurate data over this temperature range was another significant improvement on the work of Gilpin.[38]

However, Gay-Lussac was not only concerned with improving the accuracy of the data; he considered it important to simplify the whole system, by graduating alcoholometers on a scale of 100 so that at a temperature of 15°C (an average temperature for wine cellars in the south of France) the reading of the instrument would give the alcoholic content directly. Thus a reading of 10 would mean that at 15°C the liquid contained 10 per cent by volume of alcohol. He drew up tables for conversion at other temperatures. Arago reckoned that the pains-taking experimental work had taken his friend more than six months, although Gay-Lussac acknowledged that in the actual calculations involved in drawing up the tables he had been able to rely heavily on F. Collardeau, an instrument maker and ex-student of the Ecole Polytechnique.[39] He had to give not only the relation between specific gravity and alcohol content but also tables showing how the new system corresponded to readings of Cartier's alcoholometer.

Gay-Lussac's work was the subject of a report presented to the Académie des Sciences on 3 June 1822 by Arago as spokesman for the commission appointed by the Academy at the express wish of the Minister of the Interior.[40] The Academy's favourable report convinced the Ministry that new legislation based on Gay-Lussac's data was necessary. This was supported by Chaptal on behalf of a committee of the Chambre des Pairs[41] and a new law governing duty on spirits was signed by Louis XVIII on 24 June 1824, making Gay-Lussac's centesimal scale the official one.

In a letter to his wife of 1822[42] Gay-Lussac referred to his being busy in his little factory ('ma petite fabrique'). He said that he had been in to work there on a Sunday morning and again the following day. A few days later he wrote to his wife that his presence in Paris in the month of August was still necessary in order to get his business going.[43] The business was one concerned with the manufacture of scientific instruments and it had been set up in 1821 at 3, rue de la Cerisaie, in partnership with Collardeau.[44] The business was to flourish on the manufacture of alcoholometers. Gay-Lussac was already hoping to get approval of his own model in 1822[45] but it was not until two years later that he managed to corner the market. According to the law of 24 June 1824 relating to the duty payable on brandy and spirits, Cartier's alcoholometer was replaced by the centesimal alcoholometer of Gay-Lussac as the official standard. From that year the firm of Gay-Lussac–Collardeau supplied alcoholometers to the government at 5 francs each. Later in face of pressure from the administration this was reduced to 3 francs. The manufacturers guaranteed accuracy to 1/8 degree. Many other hydrometers on the market using the name of Gay-Lussac were sold for as little as 1 franc or even 50 centimes. However the accuracy of these cheap instruments was doubtful and eventually in 1855 the Ministry of Agriculture and Public Works asked the Academy of Sciences for advice on the incorporation of the specification of a particular hydrometer into the law. During Gay-Lussac's lifetime, therefore, no one was *obliged* to buy one of his instruments. The law merely gave official recognition to his scale. Nevertheless Gay-Lussac was able to trade on his scientific reputation and his instruments carried an effective guarantee of accuracy. He no doubt provided some capital for the firm but his main contribution was to confer prestige; the actual manufacture of the instruments was the day to day work of Collardeau.

Collardeau also proved a useful collaborator in Gay-Lussac's work on volumetric analysis. Such a firm could supply the new glassware and it had experience in accurate graduation. When in one of his scientific memoirs he described the difficulties of making a standard solution of sulphuric acid, Gay-Lussac added a note to the effect that acid of the required strength could be bought from Collardeau at 3, rue de la Cerisaie, Paris.[46] When Gay-Lussac later published his *Instruction* on the use of silver nitrate in assaying in 1832 he ended his introduction with a note that all the apparatus described in the book – and most was original and some complex – could be bought from Collardeau, now of 56, rue du Faubourg Saint Martin. The chlorometer too was sold by Collardeau.

This private business seems very much the exception in the pattern of Gay-Lussac's life. That he was reluctant to engage in private consultancy work is suggested by several surviving letters.[47] Certainly in his later career Gay-Lussac was not able to accede to the many requests which were undoubtedly made on his time, but throughout his professional life he liked to work within the framework of an official appointment.

### Stearic candles

Wax candles and tallow candles had long been standard sources of illumination, although the former were expensive and the latter, though cheap, were hardly satisfactory and gave out an objectionable smell. Michel Eugène Chevreul (1786–1889) discovered in 1813–15 that animal fats (such as tallow) are compounds of glycerin with fatty acids, notably stearic, palmitic and oleic acids. By boiling the fats with alkali (a process known as saponification) he was able to separate the latter constituents and use them to make new and improved candles. Chevreul's original research was done on his own but in its application to the candle industry he collaborated with Gay-Lussac[48] and in January 1825 they took out a joint patent to protect their new process:

No one having yet applied to illumination fatty bodies saponified by means of alkalis or acids, we wish to introduce our patent for this application, that is to say to reserve for ourselves the exclusive right of preparing for illumination acids from fatty bodies, whether solid or liquid, which are obtained by saponifying with potash, soda or other bases, by acids or by any other means, fats, tallow, butters and oils.[49]

They also took out a patent for the same process in England, where it was registered under the name of Moses Poole, a patent agent who testified that the invention had been 'communicated. . .by a certain foreigner, resident abroad'. The patent describes saponification and separation of the products:

I do declare that the Invention for the preparation of certain substances for making candles, including a wick peculiarly constructed for that purpose, consists in the mode or the modes of refining tallow or any kind of animal fat. In order to effect this, it is necessary first to convert the component parts of the tallow or fat into acids, and afterwards to separate one acid, which is in a liquid state, from the other, which is solid. The first acid is fit for most of the purposes to which ordinary oil is applied, and the second is the substance intended to be made into candles; there are several processes by which the two acids may be obtained, one by saponification and another by distillation of the tallow or fat. The first (viz.), the saponification, may be done by soda, potash or any alkali, as lime, for instance. The soap thus obtained is to be decomposed by any acid according to the alkaline base employed. This decomposition should be made in a large quantity of water, kept well stirred during the operation, and warmed by steam introduced into it; when allowed to stand, the component acids of the tallow or fat rise to the surface, while the water which is drawn off carries with it the saline matters, fresh water being thrown on if necessary to wash more completely any sorts which may remain; when sufficiently washed, if allowed to cool, the whole forms a solid mass which, on being submitted to the action of a press like oil seed, the liquid runs off in the form of a substance resembling oil, and there remains a solid matter similar in every respect to spermaceti, and which is fit for making candles. . .[50]

Thus after saponification the product separated into two layers, the lower layer containing glycerine, while the upper layer consisted of the salts of oleic, stearic and palmitic acids. The liquid oleic acid was separated from the solid acids (which Chevreul called 'margaric acid') by pressure.

The separation of the products was not so easy in practice and the first candles produced by this method were greasy and generally unsatisfactory. The process was conceived in laboratory rather than commercial terms. Thus for saponification three alkalis are mentioned: soda, potash, and lime, regardless of the considerable advantage on the large scale of the cheapness of the latter. It was partly by using lime that de Milly was able a few years later to make a commercial success of the production of stearic candles. Again, Gay-Lussac's patent referred to the use of alcohol for purification, an excellent laboratory method but not usually economic on a large scale. The patent does not seem to have had any immediate success and brought no financial gain to its authors.[51] The scientific principles on which the new stearic candles were based (the separation of glycerine from fats) was sound but it was left to others to overcome the technical problems involved and to make their manufacture profitable. Cambacérès, for example, quickly followed on the heels of Gay-Lussac and Chevreul with several patents specifying wicks suitable for the new candles. However credit is apportioned, there is no doubt that the introduction of stearic candles in the civilised world in the 1830s provided a vast improvement in illumination.

## The Saint-Gobain Company

At one time the manufacture of high-quality glass was the virtual monopoly of the Venetian Republic and draconian measures were enforced to prevent the technical secrets involved from passing to other countries. In 1665 Louis XIV at the instigation of Colbert granted special privileges to establish the manufacture of glass in France and break the Venetian monopoly. After some initial difficulty, a glass factory was established in 1693 at Saint-Gobain, near Le Fère, about 130 kilometres to the north east of Paris. The factory from the beginning enjoyed certain privileges as a *manufacture royale*. A remote location had been deliberately chosen in the old tradition of trade secrets. Regulations of 1716 strictly controlled the movement of workers within the confines of the factory grounds and even a visit to the nearest village required special permission from the director.[52] Saint-Gobain was the site of a ruined castle belonging to the crown and, most important, was surrounded by forests providing a vital source of fuel.

There was an expanding market for plate glass and mirrors in the eighteenth century, not only at the French court, among the nobility and haute bourgeoisie in France, but also in other countries. Large quantities were exported to Britain, the Netherlands, the German states, Russia and India and in the peak year of 1775 the company did business in excess of two million livres. Having accumulated a large reserve in gold, it was able to withstand the cessation of manufacture and sale in the 1790s in the wake of the Revolution. Under Napoleon the company resumed production, although it no longer held a national monopoly. In the early nineteenth century its most serious rival was the Saint Quirin Company, although such rivalry had the effect of spurring on the company to greater effort.

The expansion was to come by the use of 'artificial' soda made by the new Leblanc process,[53] and the setting up of soda factories by the Saint-Gobain Company. Previously soda had been imported from Spain, where it was obtained from marine plants like *barilla*. This costly raw material was no longer freely available in the Revolutionary and Napoleonic wars. By employing the Leblanc method using salt and sulphuric acid French manufacturers became independent of imports and the Saint-Gobain Company took the important step of manufacturing its own major raw material. The establishment of the Leblanc process may be regarded as a point of departure of modern chemical industry[54] and the Saint-Gobain Company by establishing itself as a soda manufacturer laid the foundation for its twentieth-century position as one of France's major chemical concerns.[55] Already in 1835 the sale of chemicals came to 1 694 658 francs, compared with the 1834 figure for the sale of glass of 2 420 000 francs. Chemicals manufactured then included sulphuric acid, bleaching powder, hydrochloric acid, sodium carbonate, sodium sulphate and stannous chloride.

By 1805 Saint-Gobain was using artificial soda from D'Arcet's Paris factory but in 1806 it decided to establish its own soda factory at Charlesfontaine. Although it took several years to begin production, by 1816–18 this works was producing about 500 000 kilograms of crude soda a year, almost sufficient for its entire needs.[56] However, Charlesfontaine was rather far from Saint-Gobain and in 1822 on the advice of Clément and Gay-Lussac a new soda works was established at Chauny, only a few kilometres from the parent factory, and the Charlesfontaine factory was given up.

In 1830 Saint-Gobain became a joint stock company with a board of directors of seven, including famous bankers and other public figures. The capital of the old company was transformed into 1152 shares divided between 157 shareholders. Company matters were always considered highly confidential and accounts were divulged only to a small group of shareholders. The guarded management fitted in well with the tendency of the company to assign profits to reserves rather than to dividends.[57]

Gay-Lussac was elected[58] on 26 June 1832 as one of the two *censeurs* of the company. The statutes of the company stipulated that there should be an administrative council of seven, of whom five were *administrateurs* and two were called *censeurs*.[59] All were to be nominated at a general meeting of shareholders. The *censeurs* had an advisory role but, far from being simply external auditors or consultants, they were allowed and indeed encouraged to attend meetings of the administrative council or board of directors. The encouragement consisted of an entitlement to 50 francs each time they attended a council meeting and signed the minutes. Meetings were held every few days so that, even if one missed a few meetings in the course of the year, one could reach an attendance of one hundred and hence an effective annual salary of 5000 francs. Thus from his first attendance as *censeur* at a council meeting on 29 June 1832 Gay-Lussac was involved in the running of a large chemical company.

Looking through the minutes for successive years, one finds that the council was concerned with a wide variety of problems. On the financial side accounts were analysed, shares were transferred (although not sold on the open market) and the remuneration of employees was discussed, including pensions. The management of the factories at Saint-Gobain and Chauny was a matter of constant concern and technical improvements were called for from time to time. The council had to purchase additional land and wood as fuel and was also concerned with buying salt and other raw materials. There was also the problem of selling the glass, mirrors and chemicals. Transport was a major problem from the isolated Saint-Gobain but Chauny, being on the river Oise, was better placed and a road was built to join the two factories, thirteen kilometres apart. From time to time there were accidents and disputes which the council had to settle.

The great problem of the administrators of Saint-Gobain was that they were in Paris while their factories were in the provinces over a hundred kilometres away. The directors of the factories were sometimes summoned to attend meetings in Paris but members of the council themselves were prepared to make regular tours of inspection of the factories.[60] On 11 August 1832 the entire council, including Gay-Lussac, visited the factories at Chauny and Saint-Gobain. The subsequent minutes read as follows:

M. Gay-Lussac, who was visiting our works and soda factory for the first time, gave them all the attention which his chemical knowledge enabled him to devote to them. He suggested that he should return to Chauny to follow the process of manufacture in detail.[61]

We have here the first written evidence of Gay-Lussac's value to Saint-Gobain. As a *censeur* he was officially concerned with financial matters but his colleagues were not silly enough to expect him to serve as an accountant or a lawyer. His value as a consultant was not with financial but with technical matters.

It was not until November 1833 that Gay-Lussac found the time for an extended visit to Chauny.[62] He went again in August 1834 and on his return gave a full account of his work there.[63] His first concern was to check the purity of raw materials and products, particularly the latter. Gay-Lussac insisted that the sulphuric acid used to titrate against the soda should be pure acid of the highest concentration possible by boiling. Among impurities in the soda there were likely to be hydroxide and sulphide. To estimate the sulphide he titrated the solution with a carefully made solution of sodium plumbite.

Gay-Lussac continued as *censeur* from 1832 to 1840. At a meeting of the council on 21 May 1840 the death was announced of Brochant de Villiers, one of the administrators, and Gay-Lussac was nominated in his place. This change of title was a promotion but it made little difference to his work and he continued to be of most use to the company as a technical consultant. A further change in title came in 1843. At a meeting of the council on 15 June 1843 Gay-Lussac, in his absence, was unanimously elected president of the company in succession to Baron Mounier, who had died the previous month. Mounier, a former private secretary to Napoleon, a councillor of state and a member of the Chambre des Pairs was obviously in the tradition of French higher administration. Gay-Lussac in his first speech to his colleagues as president was somewhat diffident in being called upon to hold such a post.[64] He spoke of its difficulties but gave an assurance of his own zeal and devotion and looked forward with the help of his colleagues to consolidating and increasing the prosperity of the company. We find Gay-Lussac given the task of presenting a report at the 'General Assembly' or Annual General Meeting held each year in April. However, he did not attend the majority of the meetings of council held every few days. After attending three meetings in June 1843, Gay-Lussac's presence is not recorded again in the minutes until 26 December 1843, the meetings usually being chaired in his absence by his colleague Gérard. In succeeding years his long absences in the summer and autumn months became increasingly obvious and he considered handing in his resignation. The reply by the council, drawn up in April 1847, is a useful indication of his colleagues' appreciation of him and seems to justify quotation *in extenso* in the Appendix.[65] Perhaps the most revealing sentence is one in which they said that it was his name as much as his presence which helped the company! They therefore wished him to continue as president, albeit in a nominal capacity. However, if Gay-Lussac had been lacking as an administrator, he had more than compensated for this as a consultant. It was therefore proposed that a new title should be created to describe Gay-Lussac's position. He was to be formally designated 'senior consultant' (*conseil supérieur*). The normal method of payment of administrators which took into account their attendance at council meetings was not appropriate for a part-time consultant. A special arrangement was therefore made for paying Gay-Lussac a regular salary of 6000 francs, which could be supplemented by being present occasionally at council meetings.

After his death the council agreed that it was important to continue the post of senior consultant, and they described the qualifications required for the post. [66] It required a man who was both discreet and technically qualified, a man of proved talent who could work freely, independently and with authority.

### The Gay-Lussac tower

The introduction by Gay-Lussac of a method of preventing the escape into the atmosphere of the oxides of nitrogen essential for the manufacture of sulphuric acid may be seen not only as a measure of economy but as an expression of rationality. It was not only wasteful but irrational and messy to allow the valuable but noxious oxides of nitrogen to be released into the atmosphere.

The manufacture of sulphuric acid by the 'lead chamber process'[67] can be traced back to the seventeenth century when it was prepared by burning sulphur in air and dissolving the fumes in water. It was found that the addition of saltpetre to the sulphur helped the process, although an explanation of the function of the saltpetre had to wait until after the establishment of Lavoisier's oxygen theory. In 1806 Clément and Desormes showed that the nitre formed oxides of nitrogen which oxidised the sulphur dioxide in the presence of water to sulphuric acid. As early as 1807 Gay-Lussac discussed the manufacture of sulphuric acid and he also made a special study of the oxides of nitrogen (1809, 1816). It was known that oxides of nitrogen could combine under certain conditions with sulphuric acid but Gay-Lussac and Clément disagreed on the nature of the compound formed.[68] Clément gave some thought to the practical problem of absorbing the valuable oxides of nitrogen given off in the manufacture of sulphuric acid in lead chambers but could think of no better solution than absorbing the gases in alkali.[69] Gay-Lussac's knowledge of sulphuric acid was more theoretical until he became associated with the Saint-Gobain Company. He then had the brilliant idea of using the very product, i.e. sulphuric acid itself, to absorb the oxides of nitrogen.

In Gay-Lussac's very first written report for the administration of Saint-Gobain of 9 September 1834, he had commented on the waste of oxides of nitrogen in the sulphuric acid plant. It is obvious that this offended his tidy mind. The minutes read:

Nitric acid which escapes from the lead chambers: There would be a very great advantage if one could make use of this acid. As the air which escapes from the chambers still contains much oxygen we have proposed several trials to make this air serve again for a new combustion of the sulphur.[70]

Here, then, is a record not only of Gay-Lussac's immediate interest in the problem at the beginning of his career with Saint-Gobain, but also an indication of the direction of his thought in first dealing with it.

Gay-Lussac wanted to absorb the oxides of nitrogen in sulphuric acid. It was not enough that such a reaction could take place in the laboratory. It had to be tried out on a large scale using not glass vessels but metal pipes which would not corrode under working conditions. Lead and cast iron were such substances. The next problem of ensuring maximum contact between the sulphuric acid and the gases to be absorbed was solved by pumping the concentrated acid to the top of a tall tower, placed at the end of the lead chambers, and allowing it to flow down a tower packed with balls of inert material; this broke the acid up into small drops where it met the ascending stream of oxides of nitrogen. By suitably adjusting the rate of flow of descending acid and ascending gas, almost complete absorption could be obtained so that the only gas emerging into the atmosphere at the top of the tower was atmospheric nitrogen.

Having absorbed the oxides of nitrogen, the next problem was to decompose the nitrated sulphuric acid so that the valuable gases could be used again. Before the invention of the Glover tower[71] Gay-Lussac had devised his own system for recovery of the oxides of nitrogen. The sulphuric acid was pumped (no easy task) to the top of a cylinder of Volvic larva 2 metres high filled with broken bottle ends. A descending stream of acid met a rising current of steam. Great attention had to be paid to the contact between the acid and the steam so that all the oxides of nitrogen would be set free without excessive dilution of the acid. In order to ensure maximum dispersion of the nitrated sulphuric acid in the form of drops at the top of the column, Gay-Lussac introduced at Chauny two small connected cones each of which, when full, would alternately deliver its load below on to one side and then the other of a lead plate which would break the acid into drops. This was a rather crude system and an alternative device which delivered a continuous flow from two rotating heads (rather like a modern lawn sprinkler) was later tried in Bohemia and found preferable.

Gay-Lussac had his first tower installed in the late 1830s, but it was only after it had been put in practice at Chauny and its economy proved that he felt confident enough to register it for a patent (1842). He could then claim that his invention resulted in a saving of two thirds of the potassium or sodium nitrate used as a raw material in the manufacture of sulphuric acid. Gay-Lussac was hoping to increase this saving even further, since in principle the same oxides of nitrogen could be used over and over again. His patent also mentions the more obvious method of absorbing the acidic oxides of nitrogen by using an alkaline solution; the resulting nitrate would then have to be decomposed with sulphuric acid to release the oxides of nitrogen.

Gay-Lussac's patent was not only concerned with preventing the escape of the poisonous nitrous gases into the atmosphere, but with re-using or re-cycling them. The patent was therefore presented both as a contribution to the economics of sulphuric acid manufacture and as a social contribution intended 'to remove from this process all that is dangerous and incommodious for property and for people'. The principles were set out in a patent of 12 October 1842 and a week later further details were set out in a subsidiary patent.[72]

Although this discovery was to be of the greatest value it was not widely used for some time, not even in France. There were several reasons for this. Some manufacturers even denied that nitrous fumes escaped into the air, but for the majority the Gay-Lussac tower seemed merely to introduce a further complication. It absorbed the oxides of nitrogen only to present the problem of decomposing the nitro-sulphuric acid. This could be done with hot water, steam or even sulphur dioxide but the wear and tear on the plant was considerable and, most important, the resulting sulphuric acid was considerably diluted. They therefore incurred the additional expense of concentrating the sulphuric acid again to supply the Gay-Lussac tower.

Gay-Lussac's process for capturing the oxides of nitrogen by absorption in sulphuric acid required for its commercial success a complementary process for the release of the precious gases under controlled conditions. This was the achievement of the British manufacturing chemist, John Glover, who in 1859 invented a process for reclaiming the valuable oxides of nitrogen. He suggested using a second tower ('Glover tower') in which the nitro-sulphuric acid from the Gay-Lussac tower would react with sulphuric acid from the lead chambers at such a temperature that the oxides of nitrogen would be released and could thus be re-used. This was a much more efficient method of recovery than Gay-Lussac's method of treatment with steam, which diluted the acid. The supplementation of the Gay-Lussac tower with a Glover tower made the former worthwhile although it took a shortage and consequent rise in price of nitrates ten years after Glover's invention to force the British manufacturers to consider once more the economics of the wasted oxides of nitrogen. They then appreciated the advantage of having excess oxides of nitrogen in the lead chambers, and the value of an absorption tower became much more evident. Thus it was not until about 1870 that the Gay-Lussac–Glover system was widely adopted in Britain. By this means an acid was produced of 78 per cent concentration which could be used directly in the expanding dyestuffs industry without further concentration.

To take the story thus far, however, is to ignore the original problem of Gay-Lussac's patent. Any discovery of industrial significance naturally raised the question of taking out a patent and, if so, where and in whose name. The Gay-Lussac tower raised all of these problems. A major difficulty was to reconcile the interests of the inventors and the Saint-Gobain Company, of which they might have been considered the employees. Saint-Gobain liked to have complete control over everything related to its works. Gay-Lussac, however, did not consider that the company owned him, and even less his ideas. He would do his best to raise the productivity of the chemical works but any scientific ideas he applied were not in his eyes the property of the company. Any patent rights would therefore be his.

On 15 June and 5 July 1842 Gay-Lussac had raised with the company the question of taking out patents for the process in his own name and that of his collaborator, the manager of the Chauny factory, Lacroix. The matter was put to one side but eventually on 1 September the council agreed to explore the taking out of a patent in Britain. As regards the French rights, Gay-Lussac and Lacroix registered a patent in their own names on 11 October followed by further patents on 19 October and 11 November 1842.[73] For British patents Maurice Sautter, the son of the secretary general of Saint-Gobain, was charged with this task. On 23 October he left for London to explore the British patent rights and to try to interest British manufacturers in the process. The

English patent was taken out on 15 December 1842.[74] As it was customary to register in the name of an individual rather than a company, Saint-Gobain authorised him to use his own name. The minutes of the administration record protracted negotiations with Tennants of Glasgow, one of the largest chemical companies in Scotland. Tennants were seriously interested and nearly bought the British patent rights. However, when presented with the detailed plans, they revoked the contract, saying that the process was too complicated to be profitably exploited. Sautter had better luck with the Newcastle company of John Lee. On 11 April 1843 it was reported that he had sold a licence to them for the use of the Gay-Lussac tower. One quarter of the saving made by the application of the invention was to go to the Saint-Gobain Company. Sautter sold a further licence in Birmingham.[75] It was precisely at the time of registering the patent in London that the president of the council died and on 15 June Gay-Lussac was unanimously elected to succeed him. Gay-Lussac was therefore left to preside over administrative decisions on the company's exploitation of the advantages of his own process in France. The minutes of 1 August 1843 read:

M. Lacroix speaks of the great advantage which he recognises daily in the new apparatus which M. Gay-Lussac has had constructed at the extremity of one of the lead chambers to collect the nitrous gases which were lost before this discovery. As a result of the new apparatus the No. 1 chamber which works in this way has for the last fortnight used only 2 parts of nitrate per 100 of sulphur. M. Lacroix asks for permission to set up similar apparatus for the other chambers to replace the first ones which had been built and which require more expenditure of manual labour and do not give as much economy of nitrate. The Administration authorises the replacement of the old apparatus by new ones for the four lead chambers which are not yet fitted with them.

It is clear that Saint-Gobain was now benefiting significantly from the Gay-Lussac tower and it was at this time that Gay-Lussac decided that the patent should be transferred from his name (or rather that of Jules to whom he had given it[76]) and that of Lacroix to Saint-Gobain. This was done at the end of July 1843 and the question remained of the compensation to be given to the inventors. After delicate negotiations an extraordinary council on 30 January 1844 decided that Gay-Lussac should be rewarded with a sum of 150 000 francs payable in three annual instalments from that date.[77] This was a very large sum, very much more than would be strictly justified by the savings to the company over a few years. However the treatment of Gay-Lussac as a director of the company may be contrasted with that of Lacroix, regarded as an employee. Arthur Lacroix, the son of a foreman at Chauny, had entered the works as a boy and had risen by ability, hard work and self-instruction to the post of manager of the soda works at Chauny. He was appointed in 1832 with a salary of 3000 francs. His deep practical knowledge had been of great use to Gay-Lussac on his visits to Chauny and they had worked together in the laboratory there. As late as 18 January 1844 the council had paid tribute to him and recognised explicitly that the patent was partly in his name. Yet only two weeks later we read his abjuration. He agrees that all his work on the patent was as an agent of the company and that he had no other rights. The treatment of Lacroix at a meeting under the chairmanship of Gay-Lussac seems remarkably callous. The best defence would seem to be that a company reward to Lacroix would have created a dangerous precedent.

# 9

# A new technique and the dissemi-
# nation of technical information

'Let us therefore march forward and not hesitate to be the
first to give the example of a useful reform'

Gay-Lussac[1]

We have considered Gay-Lussac's various links with commerce and
industry but it is now time to look at general principles. Although
Gay-Lussac was capable of applying most aspects of physics and
chemistry to practical problems, there was one technique which he
made peculiarly his own – that of volumetric analysis. The estimation
of purity of solids and liquids was a general requirement of commerce
and industry and Gay-Lussac developed a primitive idea of Berthollet's
generation along the path towards a sophisticated modern method of
general application.

In science new ideas are canvassed in scientific periodicals and
sometimes in books but in the field of technology it is not so obvious how
communication is to be established between the scientist who devises
new methods and the technician or artisan who is to apply them. In
the second half of this chapter we shall consider the dissemination of
technical information by means of technical booklets or *Instructions*,
a medium to which Gay-Lussac made several major contributions.

### Volumetric analysis

Gay-Lussac's contribution to volumetric analysis may be regarded as
typical of the chemist, for this new approach to chemistry, although
firmly based on theory, was important for its practical applications. It
was above all an embodiment of his volumetric approach to nature. It
could be recommended for its high degree of accuracy, yet its chief
virtue was the pragmatic one of rapidity of execution. Gay-Lussac
claimed, for example, that it took only 2–3 minutes to carry out a
titration with bleaching powder.[2] Other analyses by the volumetric
method would seldom take longer. In gravimetric analysis on the other

hand the successive stages of precipitation, filtering, washing, drying and perhaps combustion might take a full day for a skilled operator. In volumetric analysis the operation could be entrusted to the semi-skilled and was therefore ideal for application to industry and agriculture. One did not need to be a Gay-Lussac to titrate an acid with an alkali, although a genius of a high order was required to initiate the procedure.

Although the more remote origins of volumetric analysis may be traced back[3] as early as the seventeenth century, when there was considerable interest in the action of acids on alkalis, volumetric analysis really rose out of the practical context of chemical arts in the late eighteenth century. The question of who was the first person to indulge in something approaching volumetric analysis is not an easy one to answer nor does it seem of great importance. Probably the most important early figure was a Rouen pharmacist F. A. H. Descroizilles (1751–1825), who had followed up Berthollet's method of bleaching using 'oxymuriatic acid' or chlorine. In 1789[4] Berthollet described the use by Descroizilles of a solution of indigo as a test of the concentration ('la force') of the solution of chlorine water. Descroizilles called the bleaching solution 'lessive de Berthollet' and the graduated cylinder he used to measure its concentration a 'berthollimètre'. Berthollet seems to have reciprocated the high regard in which he was held by Descroizilles and described his work on several occasions. Berthollet also referred to work of Welter, formerly his assistant at the Ecole Polytechnique. Welter had developed a method of estimating sodium carbonate by pouring a solution of it from a graduated cylinder into a beaker and adding dilute sulphuric acid from another cylinder.[5] A glass rod was used to stir the mixture and a drop of the solution extracted periodically to spot on litmus paper to test for neutrality. If the litmus did not turn red more acid was added until it did.

We can see from the above that by the opening years of the nineteenth century the basic steps of primitive volumetric analysis had been taken and that Berthollet was a central figure in this new branch of chemistry. It was not inappropriate that his most famous pupil should develop the next stages: the introduction of rigorous measurement into the approximate and empirical procedures and the extension of the method to other reactions. Gay-Lussac is a key figure in the development of the science of volumetric analysis. Before his work trials or rough 'assays' had been introduced into chemical industry as a crude control over quality of potash and bleaching liquor. Gay-Lussac introduced scientific rigour and established volumetric analysis as a branch of chemistry.

The next important steps in this branch of analytical chemistry had to wait until the period of the Restoration – partly because Gay-Lussac in the earlier period was too busy with pure research. Also it was not until Gay-Lussac was in his forties that he was involved in so much consultancy work. It was above all the problems of industrial chemistry which prompted and justified the development of volumetric analysis, whether in the testing of raw materials or in the evaluation of the product.

The historical evolution of titrimetric analysis contrasts with a modern logical exposition. One starts today with atomic weights and with pure substances of definite composition which can be easily weighed on a balance to make a standard solution. In the late eighteenth century the starting substances were chlorine gas (or bleaching powder) and sulphuric acid. The context was practical rather than theoretical, and volumetric analysis tended throughout the first half of the nineteenth century to be looked upon with suspicion if not outright hostility by academic chemists. Gay-Lussac's sponsorship of volumetric analysis is a reminder that he was not simply an academic chemist.

If Gay-Lussac's problem had been to compare the concentrations of different samples of acid, his starting point would have been different and probably also his method. His paper on volumetric analysis of 1820 reminds us that the main problem in France was to determine the quality of commercial soda. The Leblanc process had been successfully developed in France under Napoleon and was now a major industry. Some method of control and evaluation was necessary. The first product of the Leblanc process, 'black ash', contained only some 40 per cent of sodium carbonate, the remainder being mainly insoluble calcium sulphide. After dissolving out the sodium carbonate, the main impurities remaining were soluble sodium sulphide and sodium sulphite.

In 1820 Gay-Lussac published a joint paper with Welter on the estimation of commercial soda.[6] By this time no commercial dealer would have been content to know that a substance was crude soda – he would want to know the percentage purity of the sample. The method used was to find what weight of the sample was required to saturate a given amount of sulphuric acid. This method did not, however, allow for the presence of such impurities as sulphides and sulphites, which reacted with the acid to give an erroneous result.

In 1819 a Marseilles pharmacist, L. Laurens, had published a brochure describing a method of solving this problem.[7] Laurens' method depended on the detection of sulphur dioxide by the smell. Gay-Lussac and Welter pointed out that smell was not a reliable test and suggested an alternative. The new method proposed had its antecedents in Gay-Lussac's method of organic analysis. It consisted of heating the caustic soda sample with potassium chlorate in a platinum vessel. This oxidised any sulphite (or sulphide) to the sulphate, which was neutral. To obtain a reliable estimate of the alkali content of the soda, the treated sample was acted upon by sulphuric acid. If an estimate of the sulphite content was required, two titrations could be carried out, one with a treated sample of alkali and one with an untreated sample.

Detailed instructions for the accurate determination of the soda were given. The indicator used should be litmus and its concentrated solution should be added to a solution of the soda. The sulphuric acid is then added in very small amounts with constant stirring. Gay-Lussac and Welter recommended a fairly large vessel such as a wide-mouthed jar so that the solution occupied only the bottom part of the vessel. The solution should be reasonably dilute, e.g. the soda might be dissolved in thirty times its weight of water. When sulphuric acid was added they appreciated that the carbonate was first converted to bicarbonate and the colour of the litmus was not changed. In the second stage of the titration, however, the bicarbonate was decomposed, there was effervescence with the carbon dioxide evolved and the blue colour of the solution slowly turned to red. (They recommended placing a sheet of white paper under the vessel in order to observe the colour change more clearly.) Here the sulphuric acid should be added in fifths or tenths of a measure and a glass rod dipped in the solution should touch a piece of litmus paper after each addition. This is continued until a definite red colour is obtained. This means that the point of neutrality has been exceeded but they pointed out that it was not difficult to subtract the excess if the sulphuric acid had been added in known small increments. Essentially the same method could be used to estimate sodium carbonate, hydroxide or bicarbonate.

In Gay-Lussac's early published work he was prepared to accept the standard introduced by Descroizilles of 100 grams of concentrated acid in a litre of solution. He could not help remarking, however, that it would have been preferable to have taken a given weight of pure acid ('acide réel').[8] Although unwilling to go against a system which had already been introduced into commerce, Gay-Lussac did specify more exactly what was to be understood by 'concentrated sulphuric acid'. He specified acid of maximum concentration, having a specific gravity of 1.8427 at 15°C.[9] One hundred grams of this was carefully made up to a litre. Descroizilles, thinking of the convenience of the industrial chemist, had been satisfied with *commercial* sulphuric acid for his standard, pointing out that it was cheaper and of approximately constant specific gravity.[10] Gay-Lussac also differed from Descroizilles in taking as his standard for potash not 100 grams per litre but 96.14 grams of pure potassium carbonate, since this would neutralise exactly the same volume of standard sulphuric acid. Gay-Lussac therefore went much further than Descroizilles, who had been concerned exclusively with the practical problem of comparing one sample of an alkali with another to see if it was up to standard.

Gay-Lussac was concerned with simplifying calculations as far as possible, so he specified such concentrations of solutions that the strength required could be read off directly from the burette. Thus he found that 50 millilitres of his standard solution neutralised 4.807 grams of potassium carbonate. He therefore took 4.807 grams of the sample of potash to be estimated and found the volume of standard acid which had to be added for neutralisation. On his burette 50 millilitres corresponded to 100 divisions and hence the number of divisions gave a direct reading of the percentage purity of the sample. The direct reading undoubtedly helped to make volumetric analysis popular in practical circles, although it would not have been a point of recommendation in pure science.

In his work on the estimation of bleaching powder Gay-Lussac, following Welter, took as his 'unit of decolourising power' the volume of chlorine at 76 centimetres pressure and 0°C. This was used to make the standard solution of indigo of such concentration that exactly 10 volumes of it was decolourised by 1 volume of chlorine. The indigo solution was placed in the burette and titrated against a sample of bleaching powder. The volume of standard indigo solution decolourised by 1 volume of the solution of bleaching powder would indicate the number of tenths of a litre of chlorine it contained. Thus once again Gay-Lussac had devised a scheme by which the data required by the industrialist (in this case the number of litres of available chlorine per kilogram of bleaching powder) could be obtained immediately from the reading of the burette.

Gay-Lussac considered that volumetric analysis should refer to weights (in grams or kilograms) where possible. For example, if one were evaluating the quantity of potash in a commercial sample, he preferred to express the result as so many kilograms of pure potash per quintal rather than on Descroizilles' scale, which referred to the volume of an arbitrary standard solution, 'because it is more in harmony with general usage to express the mass of bodies by their weight'.[11] Even in the case of available chlorine in bleaching powder he felt that he was stepping out of line in proposing a volume of gas rather than a weight:

I would have preferred to adopt another scale which would have indicated weights rather than volumes; but it would have been too far removed from what is accepted. For fear of resistance I have respected the established usage.[12]

This conclusion may be a reflection of Gay-Lussac's conservatism but it is even more a reflection of his refusal to incorporate any system of atomic weights or equivalents. In his 1824 paper he almost went as far as to give a formula to bleaching powder, saying that it was a sub-chloride composed of:

$$
\begin{array}{llll}
\text{2 proportions of quicklime} & = & 2 \times 35.603 & = & 71.206 \\
\text{2 proportions of water} & = & 2 \times 11.1435 & = & 22.487 \\
\text{1 proportion of chlorine} & = & \text{.................} & = & 44.2653 \\
\end{array}
$$

$$137.9583^{[13]}$$

Although in the light of subsequent knowledge of its complexity the example of bleaching powder was an unfortunate choice, it is a pity that, having worked out a plausible equivalent weight, he made no use of it.

His conservatism with regard to the atomic theory, however, meant that he could not make the vital step of going from 'artificial' units, such as grams, to the 'natural' unit of the gram molecule or mole. Gay-Lussac's volumetric work has, therefore, for the modern reader a very odd and homespun appearance. His standard solutions had no chemical basis and could only be used for specific analyses and for a particular weight of sample.

It is in the work of Gay-Lussac that we first find the use of the verb 'to titrate'[14] and the terms 'pipette' and 'burette'. Gay-Lussac also referred to 'normal' solutions, but not, of course, in the modern sense. His 'normal' solutions were what we would call standard solutions, i.e. solutions of known concentration. He was able to benefit by the French metric tradition and thought it natural to divide his scale into 100 parts, a procedure which greatly facilitated calculation. As his units he took grams and cubic centimetres, which were acceptable internationally in scientific circles. If Gay-Lussac had been British he would almost certainly have used a system of units with no interrelation and one which would have been less acceptable in other countries.[15] The very words 'titre', 'burette' and 'pipette' are all reminders that volumetric analysis began as a French science. As the standard history of analytical chemistry remarks, volumetric analysis grew in France but when foreign students studied in France and learned the new method, they took it back to their own countries. Szabadvary attributes the success of volumetric analysis largely to the contribution of Gay-Lussac.[16] Because a chemist as celebrated as he had devised new methods of working, it aroused the interest of academic chemists in its scientific possibilities.

According to Pelouze[17] Gay-Lussac in his later years had hoped to write a work on volumetric analysis in which he would have given within the covers of one book a general account of the method and a description of its many applications, so many of which had been developed or extended by himself. Such a work would have undoubtedly made his name the best known in the history of volumetric analysis. The fact that he never found the time and energy for such a synthetic work has meant that, at least in the English and German speaking worlds, Gay-Lussac's contributions to volumetric analysis are largely overlooked. Much better known is the name of Friedrich Mohr, who further improved the subject and who was the author of a synthetic work of a kind planned but never published by his French predecessor. In fact by 1850 volumetric analysis was in need of a fresh approach and Mohr's work, without detracting from Gay-Lussac's achievement, advanced the subject considerably.

### Apparatus

Gay-Lussac made important contributions to the apparatus of volumetric analysis, particularly by his burette. Primitive forms of the basic apparatus for volumetric analysis – a graduated flask, a pipette and a burette – were being used by Descroizilles by 1809. Descroizilles' 'burette' was essentially a graduated cylinder from which a reagent could be poured. It was Gay-Lussac who introduced the next important variation of the burette with a side tube, a basic design widely accepted in the nineteenth century and still included in some modern catalogues of scientific instruments.

The basic problem in pouring a liquid from a narrow graduated tube was that air had to enter in order to allow the smooth escape of the liquid. In the form used by Descroizilles it had a hole in the side of the tube which could be covered by the finger. The liquid could thus be poured out of the other side. Not the least of the difficulties of using this instrument, however, was that, unless it was manipulated very carefully, some liquid would escape through the air hole.

Gay-Lussac's burette (fig. 3a) represents a major transitional stage between the graduated cylinder and the modern burette. Although a part of it was a graduated tube, rather like a graduated cylinder, it had a thin parallel tube connected at the foot, so that when the burette was inclined the solution ran out at the top of the thin tube. The introduction

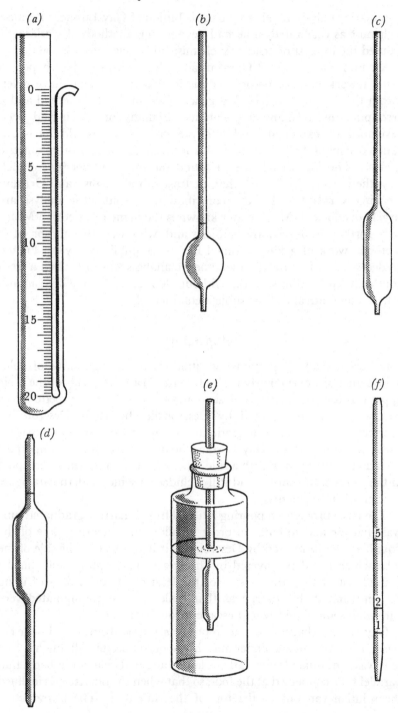

Fig. 3. Apparatus used by Gay-Lussac for volumetric analysis.

of this second tube allowed the air to replace the liquid poured out without risk of spilling and miscalculation of the amount used. The burette was first filled to above the zero mark and the solution was allowed to fill the side tube completely. A reading was taken at the beginning and end of the titration to determine the volume used.

Gay-Lussac's original burette was graduated in cubic centimetres. The orifice was slightly greased to prevent the liquid dribbling down the side. According to the diameter of the thin tube, Gay-Lussac found that 0.5 cubic centimetres was equivalent to six to ten drops. With a little practice the burette could be made to deliver a drop at a time. Thus the method could be made accurate to at least 1 part in 250 (assuming a volume of 25 cubic centimetres). For the estimation of bleaching powder Gay-Lussac at first claimed an accuracy of no more than 2 per cent[18] but a few years later he had improved this to 1 per cent, which, he said, for practical purposes was probably as great as one might wish.[19]

In making up his solutions Gay-Lussac took considerable precautions to ensure accuracy. Because a sample might not be completely homogeneous and also to make weighing out easier, he recommended weighing out ten times the amount of potash required, then making up the solution and diluting. Thorough stirring of solutions was recommended and all volumes were to be read from the bottom of the meniscus with the eye at that level. He recommended doing a titration more than once, pointing out how little time this took, especially once the approximate end point was known.

In the 1820s, 1830s and 1840s much ingenuity went into devising improvements to Gay-Lussac's burette[20] but the basic design persisted. In the 1850s this type was gradually replaced by a simpler burette, ending in a rubber tube with the flow of liquid controlled by a clip, invented by Friedrich Mohr. However, it was pointed out that the rubber was liable to perish or to react with the liquid in the burette (e.g. potassium permanganate). The alternative tap with a ground glass joint was difficult to make well and was expensive. Although of course this form has become standard in the twentieth century, in the second half of the nineteenth century Gay-Lussac's burette continued to be recommended[21] and widely used.

One can trace the evolution of the modern pipette in Gay-Lussac's successive publications on volumetric analysis.[22] In his work of 1824 he was using a primitive form of pipette with a round bulb (fig. 3b). By 1828 he had developed a pipette consisting of a thick part joined to a thin tube (fig. 3c); the thick tube contained an appreciable volume of the liquid, while the thin tube with a mark on the stem made it possible to control the precise amount of liquid. The pipette was filled by immersion in the liquid. Finally, by 1832, Gay-Lussac had developed the modern form of pipette with a thicker part at the centre (fig. 3d). The liquid was drawn up by suction until it reached the mark on the stem, when the pipette would contain a specified volume. Gay-Lussac instructed the operator, after sucking up the liquid, to place his moistened finger at the top to retain the liquid; when it was transferred to another vessel care had to be taken to include the last drop from the pipette. The problem, however, remained of dealing with poisonous solutions which might accidentally be drawn into the mouth. Gay-Lussac therefore went on to describe an immersion pipette which could be regarded as absolutely safe and foolproof (fig. 3e). He also developed a device for filling a pipette by means of an aspirator.[23] Finally Gay-Lussac described pipettes of variable content. According to which mark was used, the pipette contained, 5, 10, or 20 cubic centimetres, or in another instance, 1, 2, or 5 cubic centimetres (fig. 3f).

## 'Instructions'

An important part of Gay-Lussac's work in applied science was devoted to the rationalisation and simplification of procedures for testing a wide range of materials used in nineteenth-century France – whether silver or bleaching powder, alcohol or gunpowder. The products were of importance for commerce and industry, taxation and war, and represented different aspects of the life of a modern state. However the problem was not one of testing small quantities by a few leading scientists in Paris. The quantities and the variation of quality were considerable and the problem was a national one demanding testing throughout the country. What was therefore required was some kind of educational programme. Although in the crisis of the French Revolutionary wars young men had been recruited from all over France to be brought to Paris for a crash course in the preparation of saltpetre and gunpowder, there was a good case for the opposite approach. Instead of inviting an influx of potential students from all over France into Paris, it made more sense for an expert in the capital to send out information to anyone who might wish to have it. In fact the Revolutionary crash courses on gunpowder had been supplemented by the preparation of *Instructions*[24] which set out principles and procedures in simple language so that any literate person in the provinces could benefit. Where appropriate, plates were included, as in the *Instruction* on the separation of copper from church bells[25] or that by Monge on the casting of cannon.[26] Even before the Revolution occasional *Instruc-*

*tions* had been officially published, including some on practical problems such as agriculture and Lavoisier had written one on the establishment of nitre beds.[27] Probably the booklet of most far-reaching importance to be published by the government press was the *Instruction sur les mesures* (1794), which explained to Frenchmen the principles of the new metric system.

When the Revolutionary crisis was over the *Instruction* remained as a potentially valuable method of technical and scientific education. It was fitting that Gay-Lussac should continue a tradition to which the generation of Berthollet had contributed so much. There were, however, certain differences between the *Instructions* of the Revolutionary period and those of the Restoration. The first were obviously more restricted in their scope being concerned with the speedy production of materials needed in the war or cut off by economic blockade. During peace time obviously such restrictions were not relevant. However, even more important was the purpose of the *Instructions*. They were no longer concerned with the preparation of materials but with their evaluation. In a crisis almost any quality of saltpetre seemed better than none. In the calmer days of peace there was concern with the quality of the product. When it was discovered that chlorine could be applied to bleaching by absorption in lime, this seemed the end of a difficult technical problem – until it was discovered that the 'bleaching powder' produced could vary in quality from the very good to the very bad; obviously bleachers wanted a reliable method of distinguishing the good from the bad. The alcohol content of wines and spirits was not only of interest to the consumer but also to the state for purposes of taxation. If the taxation was to be directly proportional to alcohol content it was important that a method should be devised of estimating this with accuracy and that the method should be beyond dispute. The assaying of precious metals was of crucial importance to the economy and here again Gay-Lussac made his contribution. A final difference between the *Instructions* of the Revolutionary period and those drawn up by Gay-Lussac was the level at which they were written. Whereas the earlier *Instructions* had carefully avoided using scientific terms, Gay-Lussac was addressing an audience in which he could assume some basic knowledge of science. This helped to raise the standard of his pamphlets to something of value to science as well as technology.

Gay-Lussac's *Instruction* on the densities of alcohol/water mixtures has been described in the previous chapter, since it was closely related to his business interests. Other important *Instructions* making use of volumetric analysis were concerned with the respective estimation of saltpetre (1818), bleaching powder (1824, 1835), and silver (1830,

1832). We will look in turn at each of these before concluding with the *Instruction* on lightning conductors, not the least of Gay-Lussac's multifarious claims to fame.

## Estimation of saltpetre

The *Instruction* on bleaching powder of 1824 has been described as 'the first titrimetric work of Gay-Lussac'.[28] Actually, four years earlier he had published a joint paper with his friend Welter on the estimation of soda.[29] For Gay-Lussac's volumetric analysis to have had its beginnings in the rapidly expanding branch of French chemical industry may seem appropriate but further investigation of his research shows that his first successful application of the new method was in a military context.

One of the main problems in the production of gunpowder was the production of the saltpetre and the control of its quality. Even after refining, the saltpetre (potassium nitrate) often contained appreciable quantities of sodium, potassium, calcium and magnesium chlorides. The effect of these in 'diluting' the potassium nitrate might have been ignored but what could not be overlooked was that some of these impurities were deliquescent, that is that they readily absorbed moisture from the air, making the gunpowder damp. It was obviously a matter of concern to determine whether the chloride impurity was present and, if so, in what proportion. The use of a solution of silver nitrate solution to detect the presence of 'muriate' (i.e. chloride) was known in the eighteenth century and before Gay-Lussac's appointment at the Arsenal this test had been applied in a quantitative way to estimate the presence of common salt in saltpetre.[30] However it is important to note that the method used was gravimetric. A solution of silver nitrate was made up and added to a solution of the impure saltpetre until there was no further precipitation (of silver chloride). The flask containing the silver nitrate solution was then *re-weighed* in order to determine how much silver nitrate had been used.

The next directions for the determination of the purity of refined saltpetre were dated 7 December 1818 and published in June 1819 by the Imprimerie Royale. They were published as the work of the consultative committee attached to the General Direction of Gunpowder and Saltpetre and were signed by its president Count Ruty. However, we know from other evidence[31] that the *Instruction* was largely the work of Gay-Lussac, who had been appointed as a member of the consultative committee on 15 July 1818.

Gay-Lussac proposed that a standard silver nitrate solution should first be prepared to see if the concentration of chloride exceeded the acceptable limit of 1 part in 3000 by weight. If one started with a 10 gram sample of saltpetre, then 1 part of sodium chloride in 3000 would be 0.0033 gram, which was the weight which was found to react with 0.0096784 gram of fused silver nitrate. It would have been possible to add this quantity of silver nitrate dissolved in water to the solution of the saltpetre sample in one portion as a simple positive/negative test. However, the gunpowder administration was interested in determining the amount of impurity and it was here that Gay-Lussac introduced a volumetric approach. He first suggested that the silver nitrate solution should be added dropwise from a pipette controlled by a light pressure of a finger at the upper end. The drops could be taken as equal in volume and thus containing equal quantities of reagent. Gay-Lussac now turned to the question of concentration of the silver nitrate solution so that this could be measured in terms of volume. To decompose 0.0033 gram sodium chloride (the critical amount in a 10-gram sample of saltpetre) one would make up a solution in which say 10 grams contained 0.096784 gram silver nitrate.[32]

To test whether the saltpetre was acceptable, 10 grams was dissolved in a minimum of tepid water. Because of the sensitivity of the reagent, the test was carried out in two parts. One measure[33] of the test solution ('une mesure de la liqueur d'épreuve') was added and the solution filtered so that any further reaction would immediately be seen. The filtrate was then divided into two and a few drops of the test solution were added. If there was no precipitate one could be sure that the saltpetre did not have more than 1 part in 3000 of sodium chloride but if a further precipitate was produced the impurity was clearly above the acceptable limit.

To estimate the actual concentration of sodium chloride, the silver nitrate could be added not in one portion but several drops at a time, thus explaining the use of a pipette. If one measure of silver nitrate was equal to 30 drops from the pipette then the use of a half measure of 15 drops could determine the concentration of sodium chloride to 1 part in 6000 instead of 1 in 3000. If 10 drops were added at a time, the chloride would be estimated to 1 part in 9000. The obvious drawback in adding, say, 10 drops at a time is that one might add too much of the reagent. Instead of suggesting the addition of, say, only 1 or 2 drops, an alternative method is proposed for greater accuracy. A standard solution of sodium chloride is made up and added to the saltpetre solution. The titration is repeated and allowance made for the sodium chloride added.

## Estimation of bleaching powder

Although the principle of chlorine bleaching was established in France in the late-eighteenth century, the practice presented a number of major problems. The main problem was how to apply the chlorine to the cloth. It was found that a good method of 'storing' the chlorine was to absorb it in quicklime to make bleaching powder or 'chloride of lime'.[34] The cloth could then be treated with a suspension of this and the bleaching agent liberated by adding a dilute acid. The principal disadvantage of bleaching powder was its decomposition when exposed to the air. Thus, while bleaching powder was very useful if it had been freshly made or stored in an airtight container, other samples were practically useless and the product consequently acquired a bad reputation. Gay-Lussac considered that it was this uncertainty about the product which had prevented its wider adoption and in 1824 he consequently prepared an *Instruction* on the estimation of the bleaching powder. This was published both in the *Annales de chimie*[35] and as a separate pamphlet.

Gay-Lussac pointed first to the advantages of 'fixing' chlorine in this form. When combined with an alkali, chlorine could be made available in a form which was easily transportable and could be stored. However, if a concentrated solution of caustic soda or potash were used to absorb chlorine, some of the chlorine would be converted to chlorate and so lost for bleaching purposes.[36] There was no such problem with lime. The estimation was carried out by titration against a solution of indigo in sulphuric acid of given strength, as suggested by Descroizilles. The trouble was that the indigo available was of very variable quality so that it had been almost impossible to compare the qualities of bleaching powder on different occasions. To overcome this Gay-Lussac and Welter related their basic unit of bleaching power ('unité de force décolorante') to chlorine gas at 760 mm pressure and 0°C.

In 1835 Gay-Lussac published a revised method of estimating chlorine in bleaching powder.[37] Having mentioned previously the difficulty of storing a standard solution of indigo, he said that now after three years of trials he felt sufficiently confident in the new methods to publish them. He now introduced the use of a standard solution of arsenious oxide dissolved in hydrochloric acid. Two cubic centimetres of the arsenious acid was withdrawn by means of a pipette, transferred to a flask and a drop of indigo solution added as an indicator. The bleaching powder solution was added from a burette until the colour of the indigo disappeared. Alternatively he suggested titration with a 'normal' solution of potassium ferrocyanide. When this was acidified it turned yellow in contact with free chlorine. A drop of indigo made the mixture green and the end point, when all the ferrocyanide had reacted, was indicated by the sudden disappearance of this green colour. A third volumetric reagent which could be used to estimate chlorine was mercurous nitrate. When this was mixed with hydrochloric acid, a white precipitate of mercurous chloride was formed but this dissolved in the presence of free chlorine. Thus the end point was indicated by the formation of a permanent precipitate and this method had the advantage over the other two that it did not require indigo as an indicator. All of these new titrimetric reagents were reasonably stable. He had even kept a solution of arsenious acid in contact with oxygen for six months and had found very little change. Potassium ferrocyanide in the solid state was also stable.

The method could be applied to estimating samples of manganese dioxide by heating with excess hydrochloric acid and absorbing the chlorine evolved in an alkali. The resulting hypochlorite was then titrated. He suggests also using titrimetry to ascertain how much hydrochloric acid has reacted with the manganese dioxide. This memoir represents a high level of sophistication and when calculating volumes of chlorine he gives calculations to show how to convert a volume to $0\,^{\circ}C$ and 760 mm Hg.

### Estimation of silver

Mention has been made in the previous chapter of the *Instruction* by D'Arcet and Gay-Lussac which was ordered to be published in 1830. Because of the circumstances in which it was produced it was something of a compromise.

The *Instruction* gave a gravimetric method of determining silver as silver chloride but, instead of weighing the precipitate, the amount of silver was calculated from the *weight* of common salt (or hydrochloric acid) required for complete precipitation. The accompanying plate in the *Instruction* shows a pipette marked 10 grams and a burette reading up to 110 grams. However, Gay-Lussac added a note to the effect that they intended to give a more detailed description of the method in the next edition of the Mint's manual: *Manuel de l'essayeur*.[38] He went on to say:

We will describe these different methods which could simplify this kind of assaying and in particular the substitution of volumes for weights which makes it easier to carry out and thus it could be used more often.

Not only had Gay-Lussac been influenced by the traditional gravimetric methods used in assaying but he had been under pressure to produce a report quickly and the overriding consideration in this financial crisis was accuracy of assay.

It was not until 1832 that Gay-Lussac's *Instruction sur l'essai des matières d'argent* was ready for publication by the Imprimerie Royale. Gay-Lussac admitted in his preface that, having used the new process himself for more than two years and having spoken about it freely to anyone interested, he had not felt particularly inspired to write a full and detailed description of the process. The Count de Sussy, president of the Mint Commission had, however, persuaded him of the utility of the enterprise which resulted in the production of a quarto book of some ninety pages including twenty pages of tables showing the silver content of samples of alloy between 1 and 2 grams according to the quantity of common salt required to react with it. There were also folding plates with some fifty sketches of apparatus.

As opposed to the *Instruction* of 1830, Gay-Lussac was now writing on his own and we find therefore less tendency to compromise with the old method of cupellation since results from this new method were 'much more certain'.[39] He suggested that the silver to be assayed should be dissolved in nitric acid and the solution titrated against a standard solution of common salt. A standard solution of salt was to be made up of such strength that 100 cubic centimetres would precipitate exactly 1 gram of pure silver solution. If one was measuring to one tenth of a cubic centimetree this would give units of thousandths, and then the number of cubic centimetres of salt solution required to precipitate the silver contained in 1 gram of alloy would give directly the percentage assay of silver in the alloy.

Gay-Lussac now described two alternative methods, the first in which weights were used as much as possible. The burette was weighed before and after. In order to save time, if the approximate answer was known, then say 90 grams of the salt solution from the burette could be added at one time and then a few drops at a time, near the end point. But instead of counting the drops or noting the volume on the burette, the amount of additional solution used was read off from the position of a cursor on the balance arm. An allowance could be made if excess was added, i.e. if solution was added which produced no additional turbidity. Moreover for additional accuracy a solution only one tenth of the strength of a 'normal' solution of salt could be used. Gay-Lussac called this a 'decimal' ('décime') solution. To make a 'normal' solution of common salt 99.4573 kilograms pure water could be added to 0.5427 kilograms pure salt. However to use these weights would be to rely absolutely on the purity of salt which so easily absorbs moisture from the air.[40] Gay-Lussac therefore recommended making a large quantity of concentrated salt solution and determining the salt content by evaporating a sample to dryness. From the concentration one could work out how much water should be added to bring it to the right strength. He suggested that large quantities could be stored in jars of 50 litres capacity.

A large part of the *Instruction*[41] was devoted to a second method which was purely volumetric. Although Gay-Lussac agreed that the first method was more accurate in principle, the volumetric method had the advantage of speed and was sufficiently accurate for most purposes. He described how the standard solution was to be made up and standardised and what allowances were to be made for temperature differences. He appreciated the fact that a titration might well be carried out at a temperature different from that at which the standard solution had been made up and that this constituted a possible course of inaccuracy. He therefore devoted considerable space to explaining how temperature difference could be allowed for and he drew up a table from which corrections could be calculated.

On the practical side full details were given of the filling of the pipette either by suction with the mouth or by using an aspirator. Although the latter method seems an unnecessary refinement for solutions of sodium chloride it was of great potential value in titrations involving poisonous solutions.

A part of the booklet with important implications for industrial usage is a description of how to apply the new methods to a batch of say ten samples at a time. Gay-Lussac proposed that the whole determination should be carried out in small flasks supplied with ground glass stoppers, ten being handled at a time and each distinguished by a number on a label. Samples of the respective silver alloys were to be introduced into the flasks and a slight excess of standard nitric acid added to dissolve the silver. If this were done in a water bath with ten holes (reminiscent of a modern egg poacher), the flasks could be conveniently and simultaneously heated. When the reaction was complete each flask would contain a solution of silver nitrate. A standard amount of sodium chloride solution could be delivered from a pipette into each flask and if these were now stoppered and placed in an agitator suspended between two springs, thorough mixing could be ensured of the reagents before the resulting silver chloride precipitate was allowed to settle. Further standard sodium chloride solution[42] could then be added a few drops at a time to each sample in turn to determine the point at which no further precipitation of silver chloride occurred. The percentage of silver in the sample could be readily calculated from the total volume of sodium chloride solution added. Throughout this part of the *Instruction* Gay-Lussac showed not only his appreciation of the value of producing reliable data quickly but also his knowledge of the procedures of the Paris Mint. It was an eminently practical book yet one which also discussed all the theoretical implications. Liebig, who translated the book into German, said that the many new instruments and ideas contained in it would make it of special value to chemists:

It has not been written for the assayer alone, for in it Gay-Lussac has dealt with a scientific problem in all its aspects in such a way that the slightest criticism has been dealt with in advance.[43]

## Lightning conductors

Gay-Lussac became associated in a special way with advice on lightning conductors. The general principle of lightning conductors had been established in the eighteenth century by Benjamin Franklin but the problem continued to be raised from time to time in the nineteenth century. Gay-Lussac had been concerned with atmospheric electricity in his balloon ascent and this was his basic qualification to consider the subject of lightning conductors as a junior member of a commission appointed on this subject by the Academy in 1807.[44] In 1817 an eccentric memoir on lightning conductors was referred to Charles and himself as members of the physics section.[45] It was however, in 1822 that the effect of lightning became a matter of government concern. That year there were a number of instances of churches (including Rouen cathedral) being seriously damaged in thunderstorms and the Minister of the Interior decided that all churches should be protected by lightning conductors. An undertaking on such a scale called for the acceptance of certain general principles of construction and the Minister accordingly turned to the Academy of Sciences as the official body of experts to draw up specific instructions for the making and fitting of lightning conductors. At its meeting of 25 November 1822 the Academy decided on the rather unusual course of nominating all the members of one section, obviously in this case the physics section, to undertake this task. The physics section at this time consisted of Charles, Lefèvre-Gineau, Gay-Lussac, Poisson, Girard and Dulong. Charles as the senior member might have been expected to take responsibility for drawing up a report but he was already in his last illness and died shortly afterwards.[46] Of the remaining members of the commission Gay-Lussac was the one with the greatest experience and it was he who drew up the report finally presented to the Academy on 23 April 1823. It may have been on Gay-Lussac's initiative that the brief was extended from the specific problem of protecting churches to the general problem of protecting buildings. In the final report[47] a section of only two pages out of forty was specially concerned with churches. Even more pointedly, the report went out of its way to attack the old country custom of ringing church bells as a protection in thunderstorms. The commission pointed out that the fact that the bells were made of metal as well as the elevation of church towers made them particularly susceptible to being struck by lightning; ringing the bells might conceivably increase the chance of being struck. It was not even safe to assemble in a church without a lightning conductor during a storm. The report contains the reassuring statement that it was quite irrational to be afraid of the noise of thunder, since when one hears it the danger is past: 'Anyone struck [by lightning] neither sees the flash nor hears the clap'.

The memoir is worth mentioning not so much for its originality as its influence. In the second half of the eighteenth century Benjamin Franklin and others had established the basic theory of lightning conductors but they had addressed their finding specifically to men of science. This work came to be incorporated in general treatises of natural philosophy but these were still not very accessible to the public. Gay-Lussac's memoir was written as an *Instruction* and was therefore a compact illustrated work presented at an elementary level. It was divided into two parts, theoretical and practical. The basic facts of electrical conductivity were explained. The use of the term 'electric matter' reminds the modern reader that the members of the commission still accepted the eighteenth-century idea of physical explanation in terms of subtle fluids. It was stressed that it was essential to have a conductor which was thick, continuous and terminating in moist ground. To ensure that the conductor reached moist earth or water often meant digging very deep, but without special attention to earthing it was pointed out that the other precautions might be useless. The commission recommended that the pointed terminal rod should rise at least 7 metres above the highest part of the building and the very tip of the conductor should be gilded or made of platinum and this connected to a brass base. The major part of the conductor would be made of iron. Consideration was given in the report to the protection of powder magazines from lightning.

There had previously been some fear that lightning conductors were not only useless but actually harmful in so far as they attracted lightning. Of such beliefs the physicist Pouillet later remarked:

The *Instruction* of 1823 has contributed in no small measure to weaken them, not only on account of the authority which the sanction of the Academy has given it, but still more through the practical rules which it gives and explains in so clear and precise a manner that it is impossible to misunderstand them. The workmen themselves, with a little attention, have been able to understand what they were doing and on this account one had not to fear those errors in the placing of conductors which formerly were so frequent and often sufficient to neutralise their efficiency.[48]

In fact all problems were not solved immediately by the 1823 *Instruction*. In a thunderstorm at Bayonne on 23 February 1828 a powder magazine was struck by lightning despite the fact that it had been fitted with a lightning conductor. The Minister of War asked for an immediate local report on the incident and then turned to the Academy for an explanation. The Academy asked the physics section to investigate the matter, Gay-Lussac having the major responsibility. In reporting back on 30 March 1829[49] the physics section refused to accept any blame but ascribed the principal fault to poor earthing of the conductor.

Although the Gay-Lussac report was considered authoritative until at least the 1850s, occasional buildings continued to be struck by lightning despite their having lightning conductors. When a further investigation was called for in 1854 Pouillet attributed the fault to the belief that the area of protection of a lightning conductor extended to an area with a radius twice the height of the conductor. What Gay-Lussac had actually written on the subject was:

The distance to which a lightning conductor extends its efficient sphere of protection is not known exactly and indeed it depends on many circumstances which it would be difficult to evaluate; but in the time since buildings have been protected, several observations have shown that parts of buildings at a distance from the tip of the lightning conductor of more than three or four times its length have been struck. We estimate – and this was the opinion of Charles, who was much concerned with this question – that a lightning conductor can protect effectively from lightning a circular area of a radius double its height. Lightning conductors are erected on buildings according to this rule.[50]

Thus Gay-Lussac had characteristically been far from dogmatic. Pouillet said that the consequence of the literal observance of this rule had led to the construction of enormously high conductors. Yet in its illustrations the 1823 commission made it clear that for large buildings they recommended the erection of a series of conductors at various high points. If the 1823 commission was at fault, it was in accepting a traditional belief; anyone who read the above statement as a law of physics had himself to blame.

Gay-Lussac's *Instruction* continued to be used widely in the second half of the nineteenth century, sometimes with the addition of the 1854 report of Pouillet. The fact that metals were entering increasingly into the construction of buildings gave scientists additional problems but Pouillet suggested that the 1823 report required no alteration in its essential principles.

# 10

# Scientist and bourgeois in the political arena

'What society holds most dear and most sacred is the free
exercise of intelligence and respect for property'

Gay-Lussac[1]

There may be some scientists whose whole life is in their laboratories.
But, if a scientist applies his special expertise in commerce and industry,
he obviously relates to the wider world outside the laboratory. If he
accepts rewards for such work, he becomes involved in the society of his
time. This is all the more so if the rewards are substantial, enabling the
scientist to aspire to a superior social position. But science, through its
applications, can relate not only to private life but also to public life.
If a scientist stands as a member of parliament, he immediately becomes
a public figure. If in the course of his public office he expresses opinions
on government policy and legislation affecting science and technology,
then he is fully involved in contemporary society. Such a man was
Gay-Lussac. He did not divide his life into two separate compartments:
science and private life. In an alphabetical file of his papers we find not
only pure science and industrial tariffs together, but also papers on his
tax assessment next to papers on electrical instruments.

Gay-Lussac played an important part in the economic life of France
not only by his advice on the analysis of coinage but by his many
expert contributions to the system of taxation. In early nineteenth-
century France opinion had reacted against the imposition of high
taxes on property, and the idea of a tax related to income found little
favour except with Republicans. Most middle-class Frenchmen supported the idea of obtaining the major part of government income from
taxation of consumer goods, a system known as *contributions indirectes*.
The tax on alcoholic drinks was an example of this system, and obviously depended on some sophisticated quality control for which a
scientist could provide valuable advice. We have seen how Gay-Lussac
helped the government by improving the reliability of the means of
inferring alcoholic content of liquids from their specific gravity. The

assaying of gold and silver and the sale of tobacco were also under the control of this branch of the civil service. As if to complete Gay-Lussac's association he was appointed in 1843 professor at the school attached to the tobacco administration. Gay-Lussac also helped the Ministry of Finance as a member of the Academy of Sciences' commission appointed to give advice on preventing the fraudulent re-use of official stamped paper after the original had been bleached. But he was not simply a consultant in the background. As a member of the Chamber of Deputies and, later, of the Chamber of Peers, he was able to draw on his technical expertise in debates on taxation; and when he spoke on import duties, salt tax, or alcohol, he was listened to as an authority. A considerable proportion of his speeches concerned theoretical and practical issues relating to the *contributions indirectes*.

Besides making important contributions to science and technology, Gay-Lussac stood for the interests of a particular social group in the society of his day. Adeline Daumard, in her study of the middle classes in Paris in the period 1815–48,[2] divides the bourgeoisie into three categories: the petit bourgeoisie, including many small shopkeepers, the middle bourgeoisie with limited prospects, and the upper bourgeoisie. The last comprised many of those practising the liberal professions, and would have included Gay-Lussac. Above this class were the notables and the very rich, bankers, industrialists and landowners, a group with which Gay-Lussac had occasional contact (e.g. at Saint-Gobain) but which he never joined. Gay-Lussac was brought up in a middle-class family living in reduced circumstances because of the Revolution. By industry, thrift and application, he could aspire to the upper middle class. Money was never more relevant to social class than during the bourgeois monarchy of Louis Philippe. The accumulation of wealth became almost a social obsession, as some of Balzac's novels remind us. Any study of a savant of this period which attempts to place him in his social context must therefore include some consideration of finance.

As a professor, Gay-Lussac had little to contribute directly in the political arena of his time. As an editor he was a little more involved, considering occasionally such issues as censorship and copyright. But as a consultant, familiar with workshop and factory as much as laboratory, he was deeply involved with industry and felt strongly about the rights of the inventor and entrepreneur. Gay-Lussac was hardly a major political figure, but what political influence he exerted was in support of free enterprise. Although moved by humanitarian considerations, he was not much concerned with the problems of the ordinary working man. It was the larger issues of discipline, economy and stability which exercised him. Sometimes, when he appeared to be speaking on behalf of science, he was really representing the interests of chemical industry. This was all the more the case in so far as Gay-Lussac's period of political involvement came in the later part of his life, when he was most heavily involved in industry. It may be worth considering in a little more detail the extent of that involvement.

## Salaries and sympathies

In keeping with the tradition of heroic biography of scientists, it has been customary to list the positions held by a man of science. These are usually seen as titles to glory and only seldom as sources of income. To mention money in the traditional *éloge* would have been extremely bad taste. Yet in so far as an income enabled a scientist to undertake scientific research, and in so far as a salaried position may have guided a scientist to undertake research in one field rather than in another, it is not irrelevant to an understanding of his science as well as his private life.

In early nineteenth-century Britain a man of science was fortunate if he received any salary in connection with his scientific work. However, with the earlier professionalisation of science in France and its centralisation in Paris, the question about any recognised scientist was not whether he had a salary but rather how many salaries, since we have seen that it was possible to fulfil simultaneously the duties of several salaried posts. This system of *cumul*, as it was called, was a feature of French academic life. It was often attacked, since it reduced the number of people who could benefit from a salaried post. It was justified by the establishment as necessary to maintain a certain standard of living. Thus members of the Institute, men of international renown, received only a basic honorarium of 1500 francs which compared with the wages of a clerk and demanded supplementation by the scientist who saw himself as a member of the professional classes.

If Gay-Lussac not only benefited from several positions but actively sought out others, he was not unique in his concern for money. The old idealistic view of the scientist as someone above financial considerations has recently been eroded by reminders about Dalton[3] and Faraday,[4] to take only examples from the early nineteenth century. Gay-Lussac might have defended himself in two ways against the (palpably false) accusation that he was more concerned with money than with science. In the first place he was merely taking part in the recognised reward system of nineteenth-century French science. Most posts carried salaries which were not intended to be the total income of the holder of that post. As the duties might not amount to more than a few hours a week, for an industrious savant the duties of one post did not necessarily conflict with those of several others. Anyone with several posts was not greedy but enterprising, and the system could be defended as allowing the best people to be available in several capacities. Gay-Lussac could also have defended himself by pointing to the numerous committees on which he sat in an honorary capacity. His work for the Academy of Sciences is a good example of the expenditure of a great deal of time and effort in exchange for no more than a token salary. Having said this, however, it cannot be denied that Gay-Lussac went beyond a modest *cumul*, and particularly in his applied science consultancy posts he exceeded the bounds of some of his colleagues whose work was done entirely in an academic context.

One should not introduce a detailed record (table 1) of the accumulated salaries of a nineteenth-century French scientist without some explanation and warning. The attempt to produce a graphical record of this rather than, say, of Gay-Lussac's duties and responsibilities, is not to suggest the former is more important than the latter but rather that it can be more easily quantified. Indeed, with a few exceptions his annual income can be estimated fairly exactly.[5] A greater justification for adding up the different sources of income is that this is precisely what Gay-Lussac and his wife did. Few of Gay-Lussac's appointments were honorary and he tended to see each position as a further source of income. There is evidence that seekers as well as holders of positions tended to put a price tag on them.[6] On the other hand one should not forget that a scientific career often involved appreciable expenses. It was not unusual for Faculty professors to use their own apparatus to illustrate lecture courses and some research would be paid for by the researcher himself. Thus it is claimed that Gay-Lussac and Chevreul spent 40 000 francs trying to develop their stearic candles.[7]

The graph in fig. 4 does not necessarily represent Gay-Lussac's total income. For example his share of the profits of the flourishing scientific instrument company and his unearned income from property and shares are excluded. These would have significantly raised his total income for his middle and later years. Nor does fig. 4 include his fees for consultancy at the Charenton ironworks nor the large sum of 50 000 francs paid by the Saint-Gobain Company in 1844, 1845 and 1846 for the patent rights of the Gay-Lussac absorption tower. Fortunately for the ready interpretation of the data, the period covered does not start until some time after the severe inflation which followed the Revolution. A study of the cost of living in Paris in the period 1817–47 shows fluctuations both up and down of up to 10 per cent but does not reveal any overall rise.[8]

Table 1. *The accumulation of salaried appointments*

| Date | Position | Annual salary (francs) | Total salary (francs) |
|---|---|---|---|
| 22 Nov. 1800 | Graduate Student at Ecole des Ponts et Chaussées | 900 | 900 |
| 23 Sept. 1804 | *Répétiteur* at Ecole Polytechnique | 1 500 | 1 500 |
| 7 Feb. 1806 | Member of Bureau Consultatif des Arts et Manufactures | 2 400 | 3 900 |
| 8 Dec. 1806 | Member of Institute | 1 500 | 5 400 |
| 11 May 1809 | Professor at Paris Faculty of Science | 1 500 | 6 900 |
| 1 Jan. 1810 | Professor at Ecole Polytechnique (promotion from *répétiteur*) | 6 000 | 11 400 |
| 7 Aug. 1812 | (Professor at Paris Faculty of Science with duties at Ecole Normale–supplementary salary) | 3 000 | 14 400 |
| 1 Jan. 1816 | (Government economy–salary reduction for post at Ecole Polytechnique) | (− 1 000) | 13 400 |
| 19 Apr. 1816 | (Further reduction in salary for post at Ecole Polytechnique) | (− 2 500) | 10 900 |
| 5 Sept. 1816 | (Restoration of last salary cut) | +2 500 | 13 400 |
| 1817 | Joint editor of *Annales de chimie et de physique* | 3 000[a] | 16 400 |
| 15 July 1818 | Member of Comité Consultatif de la Direction Générale des Poudres et Salpêtres | 4 000 | 20 400 |
| 20 Nov. 1829 | Assay Master at Bureau de Garantie of the Mint | (est.) 20 000[b] | 40 400 |
| 16 June 1832 | Professor of Chemistry at Muséum d'Histoire Naturelle | 5 000 | 40 400 |
| 1832 | (Resigns professorship at Faculty of Science) | (− 5 000) | |
| 28 June 1832 | *Censeur* at Saint-Gobain | (est.) 5 000 | 45 400 |
| 21 May 1840 | Administrator at Saint-Gobain (instead of *censeur*) | (salary unknown) (est.) 10 000[c] | 55 400? |
| 18 Nov. 1840 | (Resigns professorship at Ecole Polytechnique) | (− 5 000) | 50 400? |
| 1843 | Professor at Ecole d'Application des Tabacs | 3 000 | 53 400? |
| 1847 | (back-dated to 1 Jan. 1846)[d] Senior consultant at Saint-Gobain (instead of administrator) | 6 000 | 49 400? |
| 1848 | (Resigns position at Mint) | (− 20 000) | |
| | (Resigns post at Muséum d'Histoire Naturelle) | (− 5 000) | |
| | (Resigns post at Ecole d'Application des Tabacs) | (− 3 000) | 21 400? |
| 13 May 1849 | Retirement pension in respect of service at Ecole Polytechnique | 3 333 | 24 733? |

[a] In 1818 this had risen to 4000 francs and it would be reasonable to estimate this in the 1820s as 5000 francs p.a.

[b] Although the salary was 10 000 francs, this was supplemented with other substantial payments. See Appendix, letter 11. The 25 000 francs mentioned in the letter is likely to be an over-estimate.

[c] This estimated figure is based on the assumption that an administrator or director received a high salary. Gay-Lussac must also have received a substantial salary in the years 1843–7 when he was company chairman but no information on this has been found.

[d] See Appendix, letter 14.

The most striking feature of the graph is the dramatic rise in income through the accumulation of salaried posts. Only twice in Gay-Lussac's life did he feel morally obliged to relinquish one source of income before accepting another. The first occasion was in 1804, when he relinquished his grant as a student at the Ecole des Ponts et Chaussées to take up the

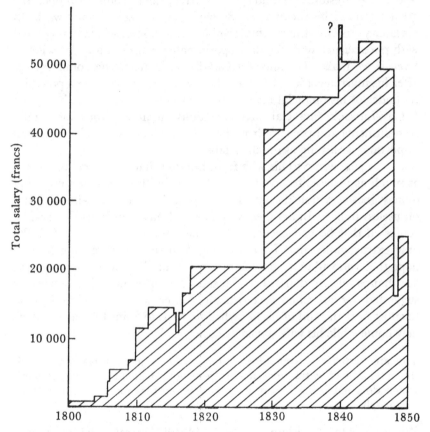

Fig. 4. Graph showing Gay-Lussac's combined income from salaried appointments listed.

post of *répétiteur* at the Ecole Polytechnique.[9] Then, nearly thirty years later, he effectively 'exchanged' his chair at the Faculty of Science for a chair at the Museum. On all other occasions a new appointment meant an additional income. The most striking (and most necessary) increase in Gay-Lussac's salary came at the beginning of his married life in 1809 and 1810. The salary cuts at the beginning of the Restoration affected all savants but, however alarming, we can see that they came at a time in his career when he was still left with a moderate income. The later stages of the graph reflect resignations. However, his resignation from his teaching post at the Ecole Polytechnique in 1840 was more than compensated financially by his directorship at Saint-Gobain. Finally the diagram shows that, whereas in the Napoleonic period Gay-Lussac's income was derived almost entirely from teaching, under the Restoration and particularly under Louis Philippe, the greater part of his income was derived from his consultancy work. In his later years Gay-Lussac thus combined prestige as a savant associated with pure science with the disproportionate rewards of applied science. Even if we make allowance for Gay-Lussac's increasing commitments after his marriage and the birth of his five children, we have a powerful reminder of his financial involvement with commerce and industry.

Gay-Lussac's financial preoccupations stemmed from his family commitments and pressure from his wife. Judicious expenditure was as important as earning as much as possible.

In 1817 he wrote to his wife from London that he was going on foot as much as possible to save money.[10] Even 'with the greatest economy money goes very quickly but,' he reassured Madame Gay-Lussac, 'not on useless things'. In another, when he had stayed in Paris to work, he confessed that he had been out to dinner and to the theatre. 'It is the only extravagance that I have indulged in since I have been here; I hope that you will not reproach me.'[11] The family purse strings were in the hands of Madame Gay-Lussac. The pressure she exerted is suggested by an uncharacteristically bitter letter[12] written by the scientist near the end of his life when he went to Saint Léonard to their new home while his wife stayed in Paris:

You are mistaken if you think that my solitary journey was made with the intention of extracting myself from your vigilant control for work which does not bring in any money; you say that money lost is trouble for oneself.

When his wife urged him that he could still 'be useful' he interpreted this as an attempt to get him to take up more paid work. For his part he had no regrets over the few thousand francs that he had lost by giving up some of his positions. Gay-Lussac now seventy years old felt that he had earned his retirement. He was no longer willing to fall in with his wife's concern for additional sources of income.

Throughout his professional life Gay-Lussac had been a skilful player in the French game of *cumul*. To his less worldly colleague Dulong, he appeared as someone who was too concerned with the financial implications of science, and when he told Berzelius that Gay-Lussac was taking over his lecture course, he added:

He is going to profit by this opportunity to write a treatise on chemistry in two or three volumes. In this country an elementary text put out by a university professor is a way to make considerable money and I believe that this consideration enters largely into his decision. . .[13]

We have seen that Gay-Lussac had no financial interest in the publication of his lectures, but Dulong's comments are of interest as they obviously reflect the opinion that Gay-Lussac's collegues had of him.

Although Gay-Lussac made a considerable amount of money out of science he did not work only for money. The case of the miners' safety lamp is often quoted as an example of the high principles of Humphry Davy, who refused to patent an invention which could benefit humanity. Gay-Lussac too made a humanitarian discovery when he investigated a means of making fabrics non-inflammable. He studied the effect of treating cloth (particularly cotton and linen) with solutions of volatile and non-combustible salts such as ammonium chloride or sulphate, since, not only would the vapours of such salts hinder combustion but 'in assuming the elastic state they would arrest combustion by absorbing much of the heat and thus lower the temperature below that necessary for combustion'.[14] He recommended immersing cloth in a solution of e.g. ammonium phosphate and then allowing it to dry. The discovery was reported in the *Moniteur*[15] as an example of disinterested science, since Gay-Lussac had communicated the results of his research to the government without thought of financial gain. His humanitarian concern is more explicit in another memoir in which he deplored fatal accidents involving children whose clothing had caught fire, and recommended work which might be used to reduce such risks.[16]

The accumulation of money over and above one's regular expenditure raises questions of investment and security. Gay-Lussac's connection with industry and his acceptance of the new investment pattern of nineteenth-century France is shown by his decision to buy shares. At the time of his death he had more than 100 000 francs in shares in the Banque de France and in US government bonds, not to mention lesser amounts invested in the Saint-Gobain Company and a coal mine. But Gay-Lussac was also attracted to the more traditional investment in property. In so far as his work provided him with a house in Paris – first at the Arsenal and then at the Jardin des Plantes – he had no strict need to own further property. But here economic considerations were reinforced by his love for his native health. Family reasons prompted him to invest much of his surplus wealth in property around Saint Léonard. His political role in the Chamber of Deputies in the 1830s was also rooted in such local attachments. In 1821 he had bought the property of Bassoleil, near Lussac, for 60 000 francs, and by the end of his life he owned eight properties near Saint Léonard with a total value of more than 200 000 francs.

Gay-Lussac also became a property owner in the Paris region. By the beginning of the July monarchy, his fame assured, he thought of some relaxation in a country estate. In 1832 he bought a house with some land at Chatillon, a short distance outside Paris. The new house provided a retreat which he could use for weekends or longer periods without losing contact with his various commitments in the capital. In October 1833 he wrote to his friend Dr Chamberet that he would probably be staying there until November if the weather was good.[17] However he went frequently to Paris on business.

At Chatillon, as earlier under Berthollet's roof at Arcueil, pleasure was often mixed with business. Pelouze, a frequent visitor, wrote to Liebig on 10 July 1834:

I lunched the day before yesterday at M. Gay-Lussac's country house and saw a proof [of a memoir by Liebig] which Jules had just corrected.[18]

Four years later[19] Pelouze mentioned to Liebig that he had dined the previous Saturday at Chatillon with Gay-Lussac and the guests included Thenard, Chevreul, Robiquet, Balard and Demarçay.

A year after Gay-Lussac bought the house at Chatillon he moved into the house at the Muséum d'Histoire Naturelle. Considering that the family went every summer to the Limousin and that they had three houses to keep up, this must have caused difficulties. Eventually in 1841 he sold the house at Chatillon and his life reverted to the simple pattern of alternating the academic year in Paris with long summers at Saint Léonard.

## The political arena

When the young Manchester physician and chemist William Charles Henry visited Paris in 1831 he was disappointed to find all the scientists he knew absorbed in politics. Henry's judgement was that of an Englishman and obviously made no allowance for a different tradition in France. As his father reported:

Even Gay-Lussac, to whose temper and habits one might think the tranquil pursuits of philosophy [i.e. science] more congenial, has been induced to take his seat in the national council.[20]

Gay-Lussac's political career is an integral part of the biography of his later life, although he was hardly politically minded. His experience of the Revolution with the arrest of his father gave him grounds for suspicion of the political left but this did not drive him to the opposite extreme. One of his rare political declarations in public was prompted by a threat to his friend Arago. At the second Restoration one of the professors of the Ecole Polytechnique at a staff meeting raised the question of politics.[21] Wishing to attack Arago but fearing his popularity, the professor of literature suggested that no Bonapartist should continue as a professor at the Ecole. He said this, knowing that Arago had signed his allegiance to Bonaparte's 'Additional Act to the Constitutions of the Empire', introduced when Napoleon returned from Elba in March 1815. Before Arago could reply, Gay-Lussac rose and declared with warmth that he too had signed the *Acte additionnel* and if this were to be considered as grounds for dismissal he should be the first to lose his job. He explained that under the circumstances in which France was threatened by attack by foreigners he would always support the government of the day, even a Jacobin one. He argued that patriotism was more important to him than politics and that in the circumstances his act had been a patriotic one. Thus the attack on Arago was thwarted and Gay-Lussac gave one more example of his personal courage, not this time in the cause of science but to protect a friend, one who, ironically enough, was to become well known for his extreme Republican views and beside whom Gay-Lussac would appear a reactionary.

No doubt it was also Gay-Lussac's friendship with Arago which prompted him to accept his brother Etienne as *répétiteur de chimie* at the Ecole Polytechnique. It did not help Gay-Lussac's standing with the Bourbon regime that Etienne Arago should have used the laboratory in 1821 as a focus of a politically subversive group.[22] This was all the more serious as the Ecole Polytechnique was always feared as a centre of political unrest, regardless of the government of the day.

Gay-Lussac was quite aware of the poverty of many people of his time. In a letter to his wife from England in 1817 he describes Brighton as 'an opulent town where the unfortunate do not trouble the pleasures of the rich'.[23] But perhaps Gay-Lussac exercised his critical faculties more clearly when he was in a foreign country. At home Gay-Lussac fitted in well with Guizot's injunction on the importance of building up one's personal wealth. In so far as he was involved in industrial enterprise he applied to the ordinary workers the same kind of benevolent paternalism which was traditional from the better landowners.

It has been said that Gay-Lussac's elevation to the upper house, the Chambre des Pairs, was delayed because as an assayer at the Mint he did manual work. However, practical science of any kind involves the use of one's hands and this did not prevent Thenard being nominated a baron in 1825. A more cogent reason was his continuing friendship with the liberal Arago, and there is some evidence that in government circles Arago and Gay-Lussac were classified together. The same issue of the *Moniteur* (4 June 1825) which carried the news of the elevation of two men of science, Thenard and Poisson[24] to the rank of baron, also reported that Arago and Gay-Lussac had been promoted to the Legion of Honour from ordinary members to the rank of officer.[25] This was obviously a lesser honour and may have been offered as an alternative to a barony for distinguished men of science who were not completely reliable politically. In view of Gay-Lussac's sympathies it was appropriate that he should have been elevated to the upper house not under the Bourbon Restoration, but under the bourgeois monarchy of Louis Philippe: in fact his whole political career falls into this period.

The revolution of 1830 has been called a negative revolution since its leaders, far from desiring important innovations, were fighting for the re-establishment of the principles of the charter of 1815, which had granted certain basic freedoms such as freedom of the press. The revolution has also and more appropriately been described as a 'bourgeois revolution' since it had the effect of driving out the nobility, which had become increasingly prominent in state positions and replacing them by middle-class Frenchmen. In the Chamber of Deputies civil servants, lawyers, writers, professors and intellectuals constituted a major part, a fact which was partly a reflection of the greater enfranchisement. Although universal male suffrage was not to come for another twenty years, a step in this direction was made by increasing the electorate from large property owners to all men owning a certain amount of property. In fact what was assessed was not income but property on which tax was paid. The tax qualification was decreased from 300 francs a year to 200 francs which had the effect of doubling the electorate.[26] Moreover a further concession was made to the middle class in recognising professional status as part of the necessary qualification. Members of the Institute, professors, physicians and lawyers could now vote if they paid only 100 francs tax. It was in this system of special favour for recognised professional attainment that Gay-Lussac was to rise to the peak of his fame, salary and achievement.

## The Chamber of Deputies

Under the constitutional monarchy of Louis Philippe candidates for the Chamber of Deputies had to be at least thirty years old and pay tax in excess of 500 francs. Gay-Lussac's tax return showed that in 1831, now in his fifties, he paid about 1100 francs, thus qualifying comfortably. What the law did not state was that the candidate had to be affluent enough to support himself in an unremunerative office, but we already know that Gay-Lussac's income was sufficient to solve this problem. On the positive side a successful candidate would be someone with an acceptable political stance and a good local reputation, for example, in his native town. It was thus that Gay-Lussac stood as a candidate in the elections of 1831, not soliciting the candidature but agreeing when approached by local people to stand for the district including his home town (officially: Limoges *extra muros*) in the Haute Vienne department. As a historian of the period has remarked:

The man of national eminence, the Great Name, was obviously the perfect candidate. Voters gave him their confidence in a burst of emotion where they would not have supported an obscure man of identical political views.[27]

Although there were no political parties it is clear that Gay-Lussac was standing for the new spirit of the bourgeois monarchy; his platform was patriotism, duty and impartiality and perhaps by implication the rights of the new middle-class meritocracy. A manifesto written on his behalf pointed out that he would not be one of those who used a place in the Chamber as a stepping stone in their careers. Gay-Lussac already had a European reputation by his contributions to science and industry.

An interesting claim was made about the effect Gay-Lussac's rational training as a scientist would have on his political judgement:

In the discussion of the questions which will be debated in the next session M. Gay-Lussac will display that great sagacity and depth of judgement which distinguish him.

His exact and positive mind ('son esprit juste et positif') accustomed to the search for truth and the rigorous study of facts, will not allow itself to be carried away by purely speculative abstractions and will demand wise application of a practical liberty.

In the defence of our rights he will show that unshakeable firmness and that generous devotion which have enabled him to confront the greatest dangers and even death itself to push back the frontiers of science.[28]

Gay-Lussac stood with three other candidates. At the first ballot on 6 July he obtained 124 votes out of 252 cast, his nearest rival having 77 votes.[29] At this election a successful candidate was required to have at least 50 per cent of the votes cast and a second ballot was therefore held. Gay-Lussac obtained 167 out of 236 votes and was declared elected, although the election was later challenged on the grounds that the polling station had been closed early.[30] Gay-Lussac took his seat as one of the five deputies for the Haute Vienne department. Of the 459 deputies elected nearly 200 had not previously belonged to any such previous parliament. Gay-Lussac, however, took some time to get used to his new role and remained a spectator in the debates.

Gay-Lussac's parliamentary career began slowly. On 25 March 1832 he wrote ruefully to Liebig that if he read the political reports from France he would not find a single mention of the name of Gay-Lussac.[31] And yet this was not for any lack of assiduity in attendance. Gay-Lussac told Liebig that he regularly sat for four hours at a time when the Chamber was in session and tried to follow the debates on national affairs for which he now felt some responsibility.[32] As a 'new boy' he took some time to overcome his natural shyness but an opportunity came a few days after writing to Liebig. Arago was ill in bed and had asked his friend to introduce an amendment on his behalf. Accordingly on 28 March Gay-Lussac rose in the Chamber of Deputies to submit a proposal on an annual chronometer prize to encourage French watch-makers.[33]

After this Gay-Lussac was happier about taking part in debates, although nearly all his interventions were in matters on which he had special expertise. Even then, however, he would begin typically by saying that he did not wish to take up more than the minimum of his colleagues' time. His speeches were therefore usually brief and to the point although we shall see that they were not always characterised by the supposed objectivity of the scientist.

In 1834 the Chamber was dissolved and elections held. Gay-Lussac stood once again for the second district of Limoges. He stood with three other candidates. At the first ballot on 22 June there was a slight majority for the legitimist candidate. To obtain a clear majority the candidate with the fewest votes was eliminated and a second ballot held. This provided Gay-Lussac with a majority of 1 vote but at a third ballot he polled 141 votes against 111 for his rival. In 1837 Gay-Lussac stood for the last time as a deputy. He was elected at the second ballot by 142 votes out of 271. Thus at the elections of 1834 and 1837 Gay-Lussac met with serious opposition. Perhaps the promise of 1831 of a new and better world under Louis Philippe had not been met. Certainly discontent was to grow both from the right and more seriously from the left. The Republican faction which had taken part in the Revolution of 1830 was not to attain power until 1848. But it is time that we examined Gay-Lussac's career in the Chamber of Deputies in the 1830s.

It is understandable that among the subjects on which Gay-Lussac, the professor, chose to speak in the Chamber of Deputies was higher scientific education. However most of his speeches on the subject were concerned not with general principles but rather the allocation of funds for institutions with which he was connected. In the budget debate of 1834 he asked successfully for an increase in the allocation for the Museum of Natural History in order to build up a chemistry collection.[34] If in this instance he was merely representing his own interests, his successful request in the following year for an additional 15 000 francs to buy a mineralogical collection which had come on the market was more clearly on behalf of the Museum.[35] In 1837 he rose in the Chamber to ask for further additions in the budget of the Museum, this time for the establishment of a new chair of physics as well as new acquisitions and improved facilities for display of specimens.[36] Gay-Lussac's colleagues at the Museum must have felt how much it was to their advantage to have one of their number in the assembly which controlled the purse strings of their institution.

Gay-Lussac was also able to speak on behalf of the Ecole Polytechnique, although when he contributed to the debate on this subject in 1835, it was not to obtain more money but rather to obtain a revision of the system of higher education in the physical sciences. Gay-Lussac regretted that so many graduates of the Ecole Polytechnique went into administration or other government service rather than into the teaching of science.[37] In this way, he said, they were 'lost for science'. Gay-Lussac, having seen the Polytechnique from the inside, was thus able to publicise a weakness in higher education in science, but his plea for a rethinking of the role of the Polytechnique fell on deaf ears.

In the decade before his former student Liebig wrote on agricultural chemistry, Gay-Lussac took part in a debate on agriculture. He braved the hostility of wine-growing interests in proclaiming viniculture a fundamental evil.[38] Thinking in terms of an agricultural cycle, he pointed out that nothing was returned to the soil by the products of the vine. Fertilisers had to be provided from outside. Turning to the nutritional aspect he remarked that if grapes were consumed, in that form, then their value might be comparable with cereals. However when the grapes were transformed into wine a large part of their nutritive value was lost. Gay-Lussac therefore deplored the extension of vine cultivation at the expense of cereals which provided grain for men, cattle feed and fertiliser. In the national interest cereals rather than wine should be encouraged.

## Protectionism

Gay-Lussac generally supported agricultural interests in France, and on these grounds he spoke against the importation on a large scale of tobacco from overseas.[39] He wished to support French tobacco growers and argued that if it was right to tax imported iron, thus encouraging the French iron and steel industry, it was also right to tax imported tobacco. Nevertheless, if he supported tariffs, he was not an extreme protectionist. Rather by supporting *low* tariffs he took a middle course. In a debate on tariffs on imported iron he remarked:

Gentlemen, I am in favour of a reasonable reduction in import duty because I am convinced that competition between one nation and another is always useful to them both and that a protective tariff should be calculated in such a way as to establish equilibrium and a useful competition.[40]

Gay-Lussac was fully aware that a tariff barrier was a double-edged sword. When it had been proposed that a high import duty should be charged on English iron Gay-Lussac saw that this would have the effect not of encouraging the production of iron in France by the same methods, but solely the effect of protecting French manufacturers against competition from a higher quality product. He clearly saw the potential evils of protectionism, and was only prepared to support tariffs as a means of allowing French industry or agriculture to grow. It is interesting to find Gay-Lussac a fervent supporter of *both* agriculture and industry, thus cutting across the division so deeply dividing his British contemporaries.

Gay-Lussac attacked tariffs which were based on distinctions not clearly recognisable by science. Thus there was a different tax on iron made with charcoal from that made with coke, and Gay-Lussac argued that it would be more rational to tax these at the same rate.[41] He also considered the difficulty of customs officials distinguishing between molasses from sugar cane and from sugar beet, and he supported the abolition of any tax on them.[42] He supported a law to levy the same import duty on all machines rather than charge more for steam engines. He felt that such a distinction was irrational. He also criticised a proposal to tax machines by weight – this was purely incidental. All import duties should therefore be rational and no tax should be introduced without the expertise necessary to impose the tax according to the law.

Gay-Lussac intervened in a debate in which it was suggested that the Ministry of Commerce might examine the claims of a secret formula for a pesticide.[43] It was not the government or its ministries which should carry out tests to satisfy themselves about the claims made for the pesticide. That was none of their business[44] but rather that of the appropriate scientific societies. The government should remain completely neutral, and, he implies, in the background. Only when the facts became clear was it proper for the government to take some action, as for example in awarding a prize.

In a debate in 1843 on patent law[45] Gay-Lussac took a moderate position. He agreed that the inventor must be protected but he should not be given a monopoly for any future invention. Referring to himself as an 'industrial inventor' he said that what was most dear, indeed most sacred, in society was 'the free exercise of the intelligence and respect for property'. Yet no invention should be viewed in isolation, and he quoted the saying that a discovery is the daughter of a discovery which in its turn may become the mother of a further discovery. He therefore opposed the granting of any additional classes of patents. Remarking 'I am a conservative', he supported the existing patent law which allowed for patents of five, ten and fifteen years.

In a debate on literary copyright[46] Gay-Lussac was angry with one of the speakers who had felt that 'material inventions' were hardly able to bear comparison with the fruits of literary genius. The Minister of Education pacified him by speaking of the 'sublime discoveries' of science which happened to have material expression. Gay-Lussac expressed publicly the feeling that there was too much snobbery in France about practical applications of science:

The English understand material interests much better than us, and that is the secret of their great prosperity. We, gentlemen, live always in abstractions. Certainly I do not deny the existence of an empire of ideas, but I say that people should not hold in contempt the application of ideas.

He defended the steam engine, which had been spoken of with contempt in the debate but 'which has rendered such great services to society'. It was industry which constituted the power of nations, and Gay-Lussac pointed to the prosperity of Britain and the United States. The English, he said, were 'a people who owed everything to their industry, who were very enlightened and who were leading [the world] in the industrial arts'.

One can sometimes see Gay-Lussac trying consciously to apply in the Chamber of Deputies the reasoning he had applied in his laboratory to chemical and physical matters. In a debate on silviculture in 1838 he remarked:

Like M. Arago I have doubts about many scientific matters; but this question has seemed to me impossible of solution precisely because of the difficulty of appreciating things which are divided into so many elements.[47]

In other words there were simply too many variables, not only in the human arena but even in matters of agriculture and meteorology.

As most of Gay-Lussac's interventions in the political assemblies had some relevance to science it might be tempting to see him as 'the voice of science' in this new arena. There are several reasons why such a picture would be a false one. In the first place Gay-Lussac was not the only scientist, not even the only distinguished scientist in politics. In the Chambre des Députés he had Arago as a colleague and, despite their lasting friendship, they did not always agree in debates.[48] When Gay-Lussac joined the Chambre des Pairs he could not always agree either with his friend Thenard.[49] There was, therefore, even among friends, no single voice representing science. If a chemical analysis was a matter of fact rather than of opinion, in any policy-making body it was opinions rather than facts which were debated.

## Applied science and industry

Although Gay-Lussac was listened to with respect in the political assemblies when he spoke as a scientific expert, there is considerable evidence in the reported debates that Gay-Lussac in the 1830s and 1840s was no longer concerned primarily with pure science. When he spoke it was usually as a representative of applied science and industry. Although in some of the early debates he spoke as a professor,[50] he came to identify himself increasingly with the interests of the manufacturer. Some debates make clear his concern not so much with manufacturers in general but big manufacturers whose interests might be prejudiced by hasty legislation.[51] In a debate on the Compagnie de Salines de l'Est, he defended the interests of the company in the problem of defining the limits of their concession for salt extraction.[52]

Gay-Lussac's contribution to the problem of the taxation of salt illustrates connections between government administration and chemical industry. In the 1830s Gay-Lussac was concerned with the problem both in the Chamber of Deputies and as a member of the Comité Consultatif des Arts et Manufactures.

The *gabelle*, the famous salt tax, had been a major source of revenue under the old regime. Although abolished in 1790 as one of the first reforms of the Revolution, in 1806 a salt tax was re-introduced. The tax was a substantial one and increased the price of salt from 10 francs to about 30 francs for 100 kilograms. This was a heavy blow to chemical manufacturers wishing to use salt as a cheap raw material to make soda by the new Leblanc process. The manufacturers immediately clamoured for exemption and an investigation was carried out to see if this was feasible, either by denaturing the salt[53] or by transferring it to locked stores in the factories. A decree of 1809 gave manufacturing interests favourable treatment (which continued under the Restoration) and agricultural interests also claimed they should have access to salt. Gay-Lussac's reply on this matter, in 1833 (one of his early contributions to debates in the Chamber), was effectively to use his authority to attack the popular view that salt was a fertiliser.[54]

Surviving manuscript notes in Gay-Lussac's handwriting show that he was also concerned with the problem as a member of the Comité Consultatif des Arts et Manufactures. In August 1836 the problem of the denaturing of salt for chemical industry was again discussed. The tax authorities considered that the addition of a large proportion of sodium sulphate was effective in stopping its use for ordinary consumption so that it could be supplied tax-free to chemical manufacturers. Gay-Lussac pointed out that at the same time as it constituted a nuisance to industry, it did not prevent the product being used illicitly, since the sodium sulphate could be separated out simply by crystallisation. He therefore recommended that the addition of this or any other denaturing agent should cease.

In the 1840s the subject of the salt tax again assumed importance. The government had the idea of reducing the tax but, by allowing very few exemptions, obtaining approximately the same revenue. After two sessions in the Chamber of Deputies it came to the Chamber of Peers in the summer of 1846, when its detailed examination was handed over to a commission of which Gay-Lussac was the spokesman. The government had hoped that salt would be used increasingly in agriculture, and Gay-Lussac now took a more moderate line in explaining its place in agriculture. One recognises his contribution in the explanation of the necessity of salt for livestock:

It cannot be denied that salt is useful for the animal economy in the sense that, if there were no other sources, it could provide the soda which exists [in combination] in the blood, bile, saliva, and albuminous liquids although in a very small quantity. The necessity of this alkaline base is even demonstrated by its constant presence in many similar secretions in very different animals...

Gay-Lussac went on to point out that this soda must come from foodstuffs of vegetable origin which in turn took them from the soil. Thus in this limited sense salt was necessary for the land. However, when Gay-Lussac turned from agriculture to industry, he was not prepared to be conciliatory. Salt had a marginal importance in agriculture but it was of major importance in chemical industry. Since the new law would result in the re-imposition of salt tax for chemical manufacturers, Gay-Lussac, on behalf of the commission, recommended its rejection. Gay-Lussac was no doubt congratulated by his friends in chemical industry, but in the countryside he was portrayed as the man who had opposed the lowering of the salt tax.[55]

Much can be learned about the interaction of government and industry in nineteenth-century France as well as about Gay-Lussac's own political views from a debate in the Chambre des Pairs in March 1840 on the employment of children in factories.[56] In the course of this one debate Gay-Lussac emerges as a great humanitarian and a champion of workers in occupations gravely injurious to their health. Yet at the same time he appears to be advocating unrestricted child labour in factories and he attacks any system of inspection or control. He began by suggesting that to lay down a maximum number of hours for factory work was largely irrelevant. The primary evil was unhealthy working conditions and he took cotton mills as an example:

The spinning mills are really unhealthy...because the workers live in an atmosphere which is always full of dust, continually filled with tiny filaments, which, when breathed by the workers are, I believe, the true cause or at least one of the principal causes of the state in which children are found and indeed workers in most factories and mills.

Gay-Lussac created a considerable stir when he went on to consider other industrial processes injurious to health and suggested that the average life of workers in lead extraction plants was only about two years. The manufacture of white lead, too, produced serious illness. Baron Mounier commented that if this were the case such industries should be banned. Given this encouragement Gay-Lussac continued:

Well, my God! I would not have to go far to find such factories. I have already mentioned lead works; I could add works where fulminating powder is made. Everyone knows the danger of handling flour. I could add the industries which work with siliceous stones, millstones or sandstone. When no care is taken to prevent the volatilisation of these materials by means of water the lives of the men who work there are usually very short and I know this by experience. In lead works they are obliged to send workers to the country because they lose their faculties and can only recover them after several months.

However, although Gay-Lussac helped to make his colleagues aware of the serious nature of many industrial diseases, on other industrial matters such as child labour he adopted a traditional stance.

The attempt to control the hours and conditions of children working in factories was interpreted by Gay-Lussac as a direct and shameful attack on manufacturers. His championing of the latter may be worth quoting:

The work done by children is considered as an immense source of wealth for the manufacturer, and the manufacturer himself is considered as animated with unlimited avarice which prompts him to exploit the work of children to a barbarous excess; it is also suggested that the law should assume the complete guardianship of children and that the manufacturer in return for the enormous profits of which he is assured, should be asked to undergo all kinds of sacrifices; that he should require less work; that he should no doubt pay more; that he should accept regulations for interference to ensure the salubrity of his works, food and clothing and the health of the children; that he should be liable to fines; and that finally his home should be permanently open to the village doctor, prefects, subprefects, mayors, public prosecutors, or their deputies, justices of the peace and policemen.

Gay-Lussac felt that a manufacturer was his own master and that government interference was not to be countenanced. A factory was 'a sanctuary which should be as sacred as the home'. Generally speaking the humanity and generosity of the manufacturer could be relied upon, and he assigned a position of benevolent paternalism to the manufacturer:

In the state the manufacturer is a true father of a family, naturally and by his position the supporter of order and peace.

Gay-Lussac was eventually convinced that some legislation to control child labour was desirable but he gave his colleagues a solemn warning:

Woe to the country if ever the government came to the point of meddling in the affairs of industry!

Gay-Lussac's advocacy of the rights of the entrepreneur against government interference can be seen in an entirely different context. His intervention in a debate on colonial affairs in 1845 provides a further insight into the political views of the mid-nineteenth-century middle-class Frenchman. He spoke in opposition to a proposal to legislate for conditions in sugar cane processing plants in the French colonies.[57] Such legislation, he said, would undermine the respect owed to masters by their slaves:

The authority of the master over the slave would be diminished. From this would come exaggerated pretensions on the part of the slave, less work and an increase in wages. Well, in my opinion, this would be a great mistake. I think it is better to allow to continue what is established in fact, what is established by a good agreement between master and slave and what is the result of sentiments of good will and humanity; for to transform acts of good will into rights is to destroy all feeling of gratitude.

A master, he said, was not likely to underfeed his slave 'for food is work'. Philanthropy should tread warily. Gay-Lussac saw legislation of this kind as having its logical conclusion in the emancipation of slaves and the ruin of the colonies. Yet if Gay-Lussac's support of the institution of slavery[58] seems lacking in humanitarian feelings, he was certainly no extremist in his period. Thus it was Gay-Lussac who proposed an amendment that an emancipated slave should only be required to stay with his master for a further two years instead of five.

Gay-Lussac's sympathies as a bourgeois did not often conflict with his sympathies as a scientist. His gospel of duty and hard work, freedom of the intellect and protection for the fruits of genius, all contributed to producing a disciplined science and a useful technology. The debate on the metric system in the Chambre des Députés in 1837 is therefore unusual in providing a field for conflict between Gay-Lussac, the representative of the people, and Gay-Lussac, the man of science. Knowing Gay-Lussac's advocacy of a scale of 100 for his alcoholometer and his use of a decimal scale in volumetric analysis, one would have expected him in the debate to have advocated the metric system as a system both rational and readily comprehensible. However, although Gay-Lussac the scientist used a decimal scale and metric units, Gay-Lussac as a middle-class Frenchman was conservative on the subject of weights and measures in commerce and industry. Although the metric system had been introduced after the Revolution, it was not until 1837 that the French government was prepared to take the decisive step of enforcing its use in ordinary life. It may well have expected support from scientists like Gay-Lussac but, far from acting in this problem as the representative of science, we find him supporting common usage against the impositions of the technocrats. He tried to introduce an amendment to allow the retention of the non-decimal fractions $\frac{1}{2}$, $\frac{1}{4}$, $\frac{1}{8}$, so that people could continue to ask for weights, for example, of 500, 250 and 125 grams.[59] The Minister told Gay-Lussac that if the government agreed to his amendment it would mean the destruction of the metric system. Eight years later in the Chambre des Pairs Gay-Lussac was criticised by the Marquis de Laplace (son of the great mathematician) for advocating the retention of non-decimal fractions of currency.[60]

Another occasion where there appeared to be a conflict between scientific and social considerations was in the debate on deforestation.[61] Gay-Lussac called attention to the continuous extraction of coal, which was used increasingly in industry and obviously (and unlike trees) not being replaced. He foresaw the time when coal supplies would be exhausted, but pointed out that the government did not dream of imposing restrictions and antagonising the commercial interests involved. So in the problem of deforestation there should not be government interference. As he told his colleagues in the Chambre des Députés:

Let it be the same with the plantations. Allow things to follow their ordinary course and be persuaded that private interest is much more able than you; be persuaded that every time this interest is interfered with there is a risk of serious error.

His colleague Arago had more faith in the predictive power of science and accepted the evidence that deforestation could have serious effects on drainage of the soil, etc. Gay-Lussac, however, liked to believe that if a wood was transformed into a meadow this would not affect the absorption of rain by the soil. How could a few trees, he asked, modify the climate? Here again Gay-Lussac, the landowner, could not be impartial.

Gay-Lussac's political career began with minor suggestions for the improvement of educational institutions, but his later political career shows that he was becoming less the professor and more the consultant to chemical industry. His influential voice representing the chemical manufacturer in political circles in the 1830s and 1840s invites comparison with Chaptal in previous decades, although Chaptal was more of a politician and administrator and less an original scientist. Gay-Lussac also spoke as a landed proprietor. He had never been an official representative of the scientific community, although throughout his political career his scientific expertise was respected. Although he spoke on a wide variety of subjects, nearly all had some aspect where Gay-Lussac could claim special knowledge.

In the Revolution of 1848 which brought his Republican friend Arago into the provisional government, Gay-Lussac (now aged 70), withdrew completely from the political arena. In the December 1848 presidential election he voted for General Cavaignac as the representative of 'order', and pointed out that in a time of national uncertainty, with many people out of work, he was giving employment to more than twenty workmen who were completing his new house at Lussac. 'What better proof', he wrote, 'of my political faith'.[62]

# I I

# The legacy

'We all teach...the chemistry of Lavoisier and Gay-Lussac'

Marcellin Berthelot (1877)[1]

A scientist's reputation long after his death depends not only on his achievement but also on a number of other factors. The relation of his work to later research obviously affects his posthumous influence. Gay-Lussac's law of combining volumes of gases was of great importance to nineteenth-century chemistry, although in the twentieth century it is likely to be mentioned only in an elementary exposition of the subject. The law of thermal expansion of gases is today more commonly associated with the name of Charles than that of Gay-Lussac. Posthumous fame so often depends on the practice of eponymy, misleading though it may be. A teacher's name tends to be kept alive by his students and we shall have to consider to what extent Gay-Lussac influenced students at the research level. The author of a text-book inevitably influences a wide circle of students but we have seen that the only text-books published under Gay-Lussac's name were extracts from lecture courses, compilations of which he disapproved.

After Gay-Lussac's death anyone who wished to consider his work would have to extract it for the most part from the appropriate journals, a method of publication more likely to bury than display any achievement. Of course the devotion of a family to the work of a scientist may help to perpetuate it but not so much as the more informed esteem of his profession and this brings us back to the question of students. Finally a reputation may depend on a national need for a hero. After the defeat in the Franco-Prussian war some French scientists were all the more determined to claim adherence to a great French tradition of science, going back to Lavoisier. Marcellin Berthelot in 1877 claimed that his chemistry was 'the chemistry of Lavoisier and Gay-Lussac'.[2] In fact Berthelot, the grand old man of science of the Third Republic, was, through Pelouze, the student of Gay-Lussac at one remove, a connection of which he was proud.[3]

*Students and research associates*

Gay-Lussac must have had several thousand 'students', using the term in its broadest sense. Many spoke appreciatively of his lectures and a special compliment was paid to him by a certain Auguste Chevalier, who, when he published a text-book on physics in 1833, entitled it *Traité élémentaire de physique d'après M. Gay-Lussac*, claiming as his authority that he was a former student. But we have seen how impersonal were the ordinary relations between professor and student in the French higher educational system. Although many of the students who attended his lectures would have increased his reputation both at home and abroad, it is the advanced students, those who had a closer contact with Gay-Lussac, who would have perpetuated his ideas and set the seal on his reputation.

We need, therefore, to turn from the lecture theatre to the laboratory and remember that, whereas anybody could attend Gay-Lussac's lectures, only the privileged few were allowed into his laboratory. Indeed the only regular method of collaboration with the master was to be his official assistant like Jean-Jacques Colin (1784–1865), appointed *répétiteur* for chemistry at the Ecole Polytechnique in 1809, the year in which Gay-Lussac was appointed to the chair. At the time of Gay-Lussac's research on iodine it was Colin who discovered the violet colour produced when the new element reacted with starch, a reaction which has ever since been used as a test. A particular protégé of Gay-Lussac was Pierre Robiquet (1780–1840), appointed *répétiteur* at the Ecole Polytechnique in 1813. Gay-Lussac several times acted as an intermediary at the Institute to present research carried out by him.[4]

When in 1817 Sertürner reported the discovery of morphine and its reaction with acids to form salts, Gay-Lussac asked Robiquet to repeat and confirm the experiments.[5] In the winter of 1817/18 Colin and Robiquet resigned their posts at the Ecole Polytechnique and César Despretz (1791–1863) became one of the new *répétiteurs*.[6] A paper published by Gay-Lussac at this time on the salinity of the Atlantic Ocean acknowledged that the experiments had been carried out in his laboratory with the greatest care by Despretz.[7] Some of the later research carried out by Despretz shows strong evidence of Gay-Lussac's influence.[8] When Despretz published a text-book of physics[9] he dedicated it to Gay-Lussac and Arago. Another associate of Gay-Lussac was Claude Pouillet (1790–1869), who shared with him the physics course at the Faculty of Science.[10]

Even after Gay-Lussac had given up teaching physics, he still remained officially a physicist according to the formal classification of the Academy of Sciences. Following the death of Charles in 1823 and Lefèvre-Gineau in 1829 he became the senior member of the physics section and he seems to have used this position to introduce his protégés into the Academy. It was fitting that, when Gay-Lussac died in 1850, both Pouillet and Despretz should have pronounced eulogies at the graveside of their friend and patron.

Among French scientists the one who stands out most clearly as a disciple of Gay-Lussac was Pelouze. Indeed I have even found the claim made that Pelouze was Gay-Lussac's only (advanced) student![11] Jules Pelouze, born in 1807, was some thirty years Gay-Lussac's junior. His father was employed at Saint-Gobain and Dumas[12] tells us that one day, when returning to Paris on foot after a visit to his family, Jules Pelouze was caught in heavy rain. He hailed a passing carriage which happened to contain Gay-Lussac! Gay-Lussac was favourably impressed by the younger man's scientific interests and knowledge and offered to find a place for him in his laboratory. After some time at the laboratory he was offered a teaching post. The town council of Lille had established a chemistry course and there was a vacancy for an assistant. Pelouze obtained this post on Gay-Lussac's recommendation, but Pelouze need not have feared this as banishment to the provinces. When the post of *répétiteur* at the Ecole Polytechnique became vacant in 1831 Pelouze was appointed. He had met Liebig when the latter was studying in Paris in 1822–4 and their subsequent correspondence throws some light on Pelouze's position in Gay-Lussac's laboratory. In 1832 Pelouze reported that he had prepared and purified specimens of organic compounds recently discovered by Liebig; the specimens were to be used in Gay-Lussac's lectures. Pelouze took special pride in presenting his master with fine samples.[13] When Liebig warned his young friend to avoid speculation in science, Pelouze replied touchingly:

The illustrious scientist with whom I have the good fortune to work thinks exactly as you do. I recall often having heard him say the same thing and I shall never forget it.[14]

Pelouze undertook some scientific collaboration with Jules Gay-Lussac. In 1830 they published a short joint memoir on salicin.[15] Gay-Lussac had himself hoped to do some research on lactic acid[16] but, with many other preoccupations, he passed the research on to his young collaborator and his son. They had hoped to isolate a new acid from the extract of sugar beet but careful examination of the acid and its salts showed that it was only lactic acid. The resulting paper was competent rather than brilliant but it is sometimes worth reporting purely negative conclusions in science and Gay-Lussac senior was happy to publish it.[17] When a vacancy occurred in the chemistry section of the Academy in 1837, Pelouze expected to be able to count on the powerful support of Gay-Lussac. The latter however refused to canvass for his assistant,[18] possibly thinking that at thirty he was not yet ready to be elected among the immortals. Pelouze was, however, successful and in his subsequent career he followed quite closely in the footsteps of Gay-Lussac, both at the Assay Bureau of the Mint and finally, after Gay-Lussac's death, as his successor as consultant at Saint-Gobain. In the context of his assay work Pelouze developed a titrimetric method of estimating copper, following the precedent established by Gay-Lussac for the estimation of silver.

Although Gay-Lussac's contribution to volumetric analysis owed something to his patron Berthollet and particularly to Descroizilles, there is no question but that during his lifetime volumetric analysis was a French science which was learned by the numerous foreign students who came to study in Paris. Some who were not able to learn the method from Gay-Lussac himself were initiated at second hand by Pelouze. According to the latter, Gay-Lussac had himself wished to write a treatise on the subject but he never found the time. His work is to be found in a handful of memoirs and it was from these that that other great figure in the history of volumetric analysis, Friedrich Mohr, learned the subject. Mohr's text-book on volumetric analysis was published in 1855 and 1856 and the author headed each section with the name of the person responsible for the introduction of that method. Thus under his own name he put the arsenious oxide–iodine titration but the whole of acidimetry and alkalimetry appeared under the name of Gay-Lussac. Although Mohr's important book became a classic, it was not the first to introduce volumetric analysis into Germany. This had been done in 1850 by a more modest compilation by K. H. Schwartz, who had learned the subject under Pelouze in Paris. When Mohr's book appeared in French translation eight years after Gay-Lussac's death, it was appropriate that the dedication should be to Pelouze as his legitimate heir. Pelouze had many students but probably the most famous was Claude Bernard. He also opened his laboratory to the young Marcellin Berthelot who was soon to collaborate in a study of chemical equilibria with another of Pelouze's students, Péan de Saint-Gilles, who would undoubtedly be better known in the history of science if he had not died in his late twenties.

An associate of Pelouze and a strong candidate for identification with the 'Gay-Lussac school' was Edmé Fremy (1814–1894). He began his chemical training in Gay-Lussac's laboratory at the Ecole Poly-technique.[19] It was Fremy whom Gay-Lussac asked to take over his lectures at the Muséum in the 1840s. Pelouze and Fremy collaborated in the publication of several text-books[20] in which the work of Gay-Lussac is prominently represented. Following the career pattern of Gay-Lussac and Pelouze, Fremy finally became a consultant at Saint-Gobain.

Gay-Lussac's tendency to weigh impartially the merits and appoint-ments of candidates for posts is suggested in the case of Regnault. When the post of *répétiteur* at the Ecole Polytechnique became vacant in 1836, Gay-Lussac at first felt that preference should be shown to Laurent, who had no job in Paris, rather than to Regnault, who was employed as assistant at the Ecole des Mines.[21] Only the unanimous preference of the Conseil d'Instruction of the Ecole Polytechnique for Regnault, a graduate of the school, managed to convince Gay-Lussac that he should also support him. Once on the staff of the Ecole Polytechnique, Regnault sometimes took over Gay-Lussac's lectures[22] and became one of his associates. He took further Gay-Lussac's work on the thermal expansion of gases, giving it greater precision;[23] indeed he became obsessed with problems of accurate measurement.

Probably the most influential of all Gay-Lussac's students was the German chemist Liebig, who had come to Paris when only eighteen and had begun his French scientific education by joining the throng of students at the Faculty of Science who attended the lectures of Gay-Lussac, Thenard and Dulong. But through his compatriot Hum-boldt, Liebig was fortunate enough to secure a personal introduction to Gay-Lussac who was so impressed by the ability and enthusiasm of the young German student that he took the exceptional step of inviting him to work in his laboratory.[24] Gay-Lussac's previous research on cyanogen made him particularly interested in certain fulminating compounds which the two chemists studied in the winter of 1823–4 in Gay-Lussac's laboratory at the Arsenal. When the research was com-pleted Gay-Lussac published it in the *Annales de chimie* under their joint names and their memoir, 'Analysis of silver fulminate',[25] remains in the published record as the joint work of two of the greatest chemists of the nineteenth century. A joint memoir is sometimes a matter of convenience, quickly forgotten but, although the content of the memoir is not particularly memorable, Liebig later described the process of collaboration as a turning point in his life. Although Liebig tended towards a quarrelsome disposition and later, as his power grew, attacked French chemists both individually and collectively, he never forgot his debt to Gay-Lussac, a debt which he continued to acknow-ledge both privately and publicly for the rest of his life.

His debt was in the first place that, in an educational system where postgraduate practical instruction had no official place, Gay-Lussac had given him special treatment. Gay-Lussac was one of the great authorities in organic analysis and the joint memoir was principally concerned with this. When Liebig returned to Hesse-Darmstadt he was to make analysis the starting point of his practical course. As Hofmann later put it,[26] Liebig hoped to be to his students what Gay-Lussac had been to him. In fact, Liebig took the method of laboratory instruction for advanced students much further than Gay-Lussac had dreamed of and much further than the rigidity of the French system would allow.

Students are influenced not only by their teachers but by their teacher's heroes who can in a way form a model over several generations. Perhaps some of Liebig's students were influenced indirectly by his French postgraduate training and the insistence on careful quantitative work. However good a lecturer Gay-Lussac was, however many thousands of students sat in his lecture theatre and benefited from his courses, Gay-Lussac was not the founder of a research school in the way that Liebig and Dumas were. If the Ecole Polytechnique had been differently structured Gay-Lussac might well have had such a research school but, unfortunately for most of his students, chemistry was a course requirement for a career somewhere else.

## The family

Gay-Lussac had five children: two daughters and three sons. The parents were concerned to see that the girls married well in society. The eldest, Virginie, married Baron Barham Colemergez, a cavalry colonel; the second daughter, Josephine, became Madame de Saint Paul by her marriage to a member of the Chamber of Deputies.

Gay-Lussac's ambitions were probably focussed on his sons. The eldest of them, Jules, was born on 18 June 1810. Jules had an auspicious start in the world of science since we find the names of Thenard and Humboldt as the witnesses who signed the birth certificate. At the age of 20 Jules Gay-Lussac might have had to go as one of 80 000 men whose names were drawn by ballot to undertake military service. Gay-Lussac senior, however, followed the middle-class French system of protecting his son from this ordeal by paying 800 francs 'insurance' – if his name were drawn the insurance company undertook to provide a substitute.

Gay-Lussac wanted his sons to make some use of science in their careers. Chemistry was a particularly useful branch of science because of its many applications in industry and he decided to train Jules as a chemist. Many young men who were to be scientists went to the Ecole Polytechnique but the training given was intensely mathematical and was more suited to engineers than chemists. Gay-Lussac solved this problem by sending Jules to study chemistry under Liebig. In 1831 when Jules arrived in Giessen, Liebig was not yet an internationally famous chemist. His best known foreign students: Gerhardt (1836/37), Lyon Playfair (1839/40), Wurtz (1842), James Sheridan Muspratt (1843) and many others had yet to come. Liebig was merely the ex-student and young collaborator of Gay-Lussac who was beginning to make his name by research in which he increasingly involved a handful of students. We might even claim that Gay-Lussac senior, far from conforming to the fashion to study with Liebig, helped to initiate it. What better testimonial to the young German chemist than that the most famous of French chemists chose to send his eldest son there? During the 1830s and early 1840s the stream of foreign students increased from a trickle to a flood. Gay-Lussac's motives, however, were not entirely based on the good chemical training he knew he would receive. It was about that time that Jules, now aged 21, should seek wider horizons than his father's house. Gay-Lussac believed in the importance of learning foreign languages:

I hope that Madame Liebig will not speak a single word of French to him...Jules asks me for French books to study mineralogy and botany. Not only do I not wish to send him any but I would even ask you to take away from him all the science books in French which he has already acquired.[27]

In fact, Liebig not only took Jules into his laboratory, he took him into his home and, as Gay-Lussac said in a 'thank-you' letter of 11 January 1833, treated him like a son. Gay-Lussac thanked the German chemist not only for the instruction, ending in the doctorate, but in the moral example he had set him. After obtaining his doctorate *summa cum laude* in September 1832, Jules went on a tour of northern Germany. His father told him to visit industrial enterprises of all kinds and advised him to take notes.[28] On his return to France, we find Jules and Louis visiting Saint-Gobain, but it was the younger son who was to find a career there. Jules began to help his father at the Assay office. As assayer with responsibility for the accuracy of the assays there was nothing to prevent him allowing his son to assist, although of course he would not be paid for this. However, Gay-Lussac saw to it that Jules had an appropriate qualification and on 9 February 1835 he took an examination in both the theory and practice of assaying. This examination was a professional requirement for anyone wishing to do assay work anywhere in France. Nevertheless, for the key post at the Paris Mint it was obviously a necessary rather than a sufficient condition. The second qualification of Jules was his practical experience at the Paris Mint where he came to do more and more of his father's work. We see from a letter to Jules, written on 22 December 1845, that Gay-Lussac was worried about how he could pass on his duties as assayer to his son. He could not have him nominated as his assistant since no such post was officially recognised. Finally the Revolution of 1848 provided Gay-Lussac with the occasion to tender his resignation on the understanding that his son would be appointed. On 18 June the mayor of Paris made this nomination, which was approved by the president of the Mint Commission and on 8 July Jules was formally installed with his father handing over the keys and the official hallmark dies.

Jules Gay-Lussac later applied his knowledge of chemistry to sugar, thus continuing the important French contribution to the sugar beet industry. In a text-book published in Paris in 1874 and translated from the German he is described as 'Inspecteur Général des Sucreries du Vice-Roi d'Egypte'.[29] He died in 1877.

We find Gay-Lussac's second son Louis (born on 8 September 1813) working in 1834 in Gay-Lussac's laboratory at the Ecole Polytechnique with Pelouze.[30] Later he too was given a job through his father's influence, this time at Saint-Gobain. He was interviewed by the administration of Saint-Gobain on 10 April 1838 and given the post of agent for the company in the United States where there was a potential market. Louis Gay-Lussac was based in New York but went later that year to New Orleans and again the following year 'for personal reasons'. He had in fact met his future wife, whom he married in December 1839. He was later to be given a managerial position in the company, that of assistant director of the factories at Saint-Gobain and Chauny. In his retirement he returned to Gay-Lussac country, near Saint Léonard, where he continued to live the life of a country gentleman until his death in 1903.

The third son Gabriel, born in 1820, did not conform so easily to his father's wishes. He was packed off to boarding school at the age of 12.[31] He eventually became a district collector of taxes.

### Conclusion

For at least a century after Newton even the most able physical scientists could hardly escape falling under the shadow of their illustrious predecessor. Similarly anyone who contributed to chemistry in the generation or two after Lavoisier is likely to suffer from a dwarfing effect, all the more so if he is also a Frenchman. But despite the danger of comparing scientists of different generations, the circumstances of Gay-Lussac's life positively invite comparison with those of Lavoisier. In his interests in pure and applied science, in his extension of Lavoisier's work, in his association with the Academy, his editing and his final position of authority, he might well be thought of as an early nineteenth-century Lavoisier. Although, unlike his predecessor, he lived unscathed through several political revolutions, he was as closely involved in the social and economic system of his country. Whereas in Lavoisier's time there was a distinction between the financial work for which he was paid and the science done in his leisure hours, Gay-Lussac, professional scientist, professor and consultant, did not have to divide his life into separate compartments. If we compare Gay-Lussac with other chemists of Lavoisier's generation, Fourcroy was a great lecturer, Chaptal made important contributions to chemical industry, Vauquelin excelled in preparative and analytical chemistry, Gay-Lussac's contemporary Chevreul was a great organic chemist, but Gay-Lussac was all of these things. Indeed Gay-Lussac's work covered such a wide range in the physical sciences that, when his eldest son died in 1877, it was said that the son was descended from *two* great scientists![32]

We have seen how Gay-Lussac as a young idealist abandoned prospects of a traditional career to follow the new banner of science. At first disinterested, his marriage and family commitments led to his increasing concern with money. Under the bourgeois monarchy this concern was a common one but since scientific consultancy was better paid than academic science it may have influenced his increasing concern with the application of science. Gay-Lussac's careful provision for his family provides a contrast with his colleague Dulong, whose family often went short when he used his salary to pay for his scientific research, or Laurent, who died leaving a wife and two young children practically destitute. But, money apart, Gay-Lussac had a special genius for solving practical problems by applying his physical and chemical knowledge. In purely utilitarian terms he was possibly more creative in the 1820s and 1830s than as a young man.

Yet Gay-Lussac's applied chemistry probably did more for the cause of science than if he had remained in a rigidly academic tradition. The success of his methods of volumetric analysis, particularly his estimation of soda, of bleaching powder and of silver made known to the manufacturer and the general public the economic value of a training in chemistry.[33] Manufacturers would not have been so willing to send their sons for abstract studies of the nature of matter. If, however, a knowledge of chemistry enabled one to evaluate and control one's reactants and products, then a chemical education could be seen as a valuable investment.

Gay-Lussac's influence was hardly less in physics than in chemistry. His report on lightning conductors was translated into English and published as late as 1881, the translator Richard Anderson considering it as evidence that in France things were organised better than in Britain. When, he asked, was the British government going to consult scientists as a matter of course on affairs of public concern such as the protection of buildings in thunderstorms?[34]

In so far as Gay-Lussac contributed to preparative chemistry: boron, cyanogen, a share of iodine and its compounds, his achievements stand without substantial modification. His record for a balloon ascent, which he had undertaken in 1804, stood for half a century. The value obtained in 1805 by Gay-Lussac and Humboldt for the volumetric composition of water[35] remained the standard one for nearly the whole of the nineteenth century – as was pointed out in 1893 by A. Scott when he undertook a re-evaluation. The classic 'modern' experimental determination was that carried out in 1895 by E. W. Morley, who gave the ratio hydrogen : oxygen as 2.00269 : 1. If Gay-Lussac's value for the coefficient of expansion of gases was challenged by Regnault during his lifetime, his total achievement is greater than that of Regnault, who became obsessed with the problem of accuracy. Gay-Lussac's law of combining volumes of gases remains one of the cornerstones of chemistry, despite the fact that all gas laws are only approximate under ordinary laboratory conditions.

Gay-Lussac had done so much important work over a wide area of chemistry that no one in post-Napoleonic France could teach the subject without frequent references to his contributions. He was referred to some forty-five times in a course of lectures given in 1828 by Laugier at the Museum of Natural History.[36] Ironically when Gay-Lussac himself took over these lectures he omitted to mention his part in various aspects of research – Gay-Lussac was not a man to blow his own trumpet. Going further than modesty, Gay-Lussac's reputation was affected by his unwillingness to make unqualified assertions. By excessive caution he left to others the glory of staking claims.

Although his family suffered during the Revolution, Gay-Lussac was strangely fortunate to have been a young man at the time. Had he been born ten years earlier he would presumably not have benefited from the education offered by the Polytechnique. But if Gay-Lussac's basic scientific education belongs to the later Revolutionary period his early scientific career was clearly in the Napoleonic period. Napoleonic science sharpened the appetites of young men by holding up prospects of recognition and reward but it also implied a certain conformity to orthodoxies which is inimical to originality. This was not political pressure – Napoleon's police agents were not concerned with scientific theories – but rather academic pressure. Young scientists were inspired by personal loyalties, and inducements to conformity took the form of a carrot rather than a stick. I am not suggesting that an overt system of bribery existed. Rather there was some subtle indoctrination and an 'atmosphere' or 'climate of opinion' which has been called 'Laplacian',[38] although an adjective derived from the name of Berthollet might be more appropriate in the case of Gay-Lussac.

A balanced view of Gay-Lussac must mention failures as well as successes. One consistent error was in treating the diverse problems of cholera and the preservation of food as if they were purely chemical. The generation before the advent of bacteriology tended to suffer from an over-ambitious reductionism. But in problems within the physical sciences Gay-Lussac showed a rare brilliance in devising experiments and in investigating new phenomena.

Gay-Lussac was a great theoretical scientist but he was also an outstanding experimentalist. This is shown by the wide range of apparatus he devised throughout his life. Thus the method of organic analysis by oxidation (1810) involved devising and constructing strong apparatus in which the gases evolved could be collected. His ideas on vapour density required apparatus (1811) in which the volume of a known weight of a compound could be measured. His portable barometer (1813–16) was widely used. Gay-Lussac in the development of volumetric analysis in the 1820s improved the pipette from its most rudimentary form to a recognisably modern instrument and his burette enabled volumes of liquid to be accurately measured. His work for the Mint also involved the construction of new apparatus to provide a quick and foolproof method of estimating silver (1829–32). His alcoholometer, although involving no new principle, is a further example of his practical interests.

In the late eighteenth century chemistry had been in need of some basic reorganisation. After the 'chemical revolution' associated with the name of Lavoisier there was a long period of consolidation and modification rather than revolution. Lavoisier had pointed out the elements as the building blocks of chemistry and in the next generation new elements isolated included sodium, potassium, boron, chlorine and iodine, all of which were prepared and studied by Gay-Lussac, Thenard and Davy. But Gay-Lussac contributed more by his interest in analysis:

the analysis of organic compounds by oxidation, the analysis by volumes of volatile compounds and the use of volumetric analysis in practical problems. No doubt the analytical approach agreed with Gay-Lussac's cautious and positivistic turn of mind, preferring limited knowledge with a high degree of certainty to bold theories which might be completely mistaken. Looking for correlations, he deserved to discover fundamental laws of nature but he regarded these as descriptive of events rather than causally.

Gay-Lussac reached his peak of achievement in the late Napoleonic period and the early Restoration; after the 1820s he made few contributions to pure science. By the late 1830s, when he attained the age of 60, he had reached the height of his public fame but he must have seemed to some of the younger and more ambitious of French scientists to represent a conservative influence. His adherence to caloric and his apparent lack of interest in optics and electricity and magnetism meant that he had become something of an anomaly as the Academy's senior physicist. There was something to be said for allowing Academicians to change to a section representing their current interests. But even in chemistry his reluctance to use the new atomic notation meant that he was no longer in the forefront of scientific advance. However, from lecture notes that have survived we can give him full credit for keeping fully up to date with current research, and this applies until a year or two before his death. Where Gay-Lussac continued to excel was in applied science. In his fifties and sixties he was able to solve a number of practical problems on a wide front by adapting his scientific knowledge to the occasion.

In some ways Gay-Lussac's work conforms to the programme of Saint-Simon, who wanted science to be directed to the needs of industry, and industry to develop under the direction of science. But in so far as the Saint-Simonians were early socialists they cannot have received much sympathy from Gay-Lussac. Their idea of converting private property to public property by the absorption of inheritance by the state was in direct conflict with his ideal of providing generously for his own family.

What we have said earlier about the professional scientist as a teacher required some modification in the case of Gay-Lussac in view of his later career. Whereas it is true for many of his contemporaries: Thenard, Biot, Dulong, and Poisson, Gay-Lussac's industrial involvement extends, although it does not contradict, the professorial model of the scientist. Gay-Lussac established himself as a scientist through government positions and, when his talents were widely recognised, he applied his scientific knowledge to problems of private industry. Here was another possible career pattern for the science graduate and one more immediately acceptable in Britain or the United States than a career depending on a centralised Education Ministry. Indeed it was the utility of science and especially chemistry which, by creating an economic demand for trained scientists, helped the scientific professions to develop later in the nineteenth century.

One explanation for the great achievement of Gay-Lussac was his concentration on one issue at a time; when asked what was his recipe for success, he replied, 'Il faut toujours y penser'. Yet he was no narrow specialist. Whereas a modern scientist usually does research within a small area representing one aspect of a branch of a major science, Gay-Lussac's interests ranged over the whole spectrum: physiology – chemistry – physics – meteorology – geology.[39] His public position obliged him to consider the place of science in the French state and to this extent he was interested in politics. He had no great interest in the arts. After his family his principal hobby was perhaps his concern with his property. This was particularly true in the later years of his life when his major energies went into planning a new house near his native Saint Léonard.

Gay-Lussac was a central figure in the French scientific establishment in the first half of the nineteenth century. Not only in the Academy of Sciences but in the Ecole Polytechnique and the Museum of Natural History he held an important and influential place. Although he had a position in the Paris Faculty of Science from its foundation in 1808 until 1832, he was not as influential in the politics of the university as, for example, his friend Thenard. On the other hand he is the prime example in his generation of a scientist who took his sciences beyond the academic world into commerce and industry. Although a man who adds to the number of elements and compounds has a niche in the history of chemistry and a man who discovers basic laws of nature has an assured place in the history of science, Gay-Lussac's influence was only partly within pure science. His readiness to apply physics and chemistry to practical affairs set an example if not a precedent which had broad repercussions in the nineteenth century. In an age when chemical industry took a recognisably modern form, Gay-Lussac made a significant contribution; his development of volumetric analysis was perhaps most important for the respectability he gave to this new method of testing, with its wide applications. The influence of Gay-Lussac in France on the wine and spirit trade became even stronger after his death with the enforcement of his alcoholometer as the only legal measure by the law of 1881.

As the editor of a key journal Gay-Lussac was in correspondence with scientists not only in France but internationally. His European orientation had begun with his tour of the Swiss, Italian and German states with Humboldt. As a professor at the Ecole Polytechnique he obtained authorisation for a number of foreign students to attend his lectures. This was in addition to the semi-public lectures given at the Faculty from which many foreign savants benefited.

In Britain Gay-Lussac's opposite number in the Napoleonic period was the brilliant Humphry Davy; they eventually met when Davy came to Paris in 1813. However, Davy had little original contribution to make to science after the Napoleonic wars and there was no obvious successor in British chemistry. Gay-Lussac's contact with British scientists was helped by his visits to England in 1816 and 1817. Gay-Lussac had a particular admiration for Wollaston, and Marcet kept him in touch with his ideas.[40] When Daniell invented his hygrometer he sent a model to Gay-Lussac.[41] It was apparently the height of ambition of the Belfast chemist, Thomas Andrews (1813–1885), a former student of Dumas, to be introduced to Gay-Lussac.[42] When he visited Paris in 1836 he wrote of being introduced to Jules Gay-Lussac, who 'promised to introduce me to his father when he returns to town'. The addresses of British correspondents kept by the French chemist in the later period of his life emphasise his industrial interests: I. K. Brunel, Percival Johnson, David Pollock, Andrew Ure (London), W. Charles Henry (Manchester), Henry Enderby (St Helens), William Gossage (Bromsgrove), Richard Phillips (Birmingham).

As far as the United States is concerned, Gay-Lussac was at the beginning of his career a close friend of the Irish–American diplomat David Bailie Warden.[43] Warden was in Paris from 1804–10 and he took advantage of this to follow a course at the Ecole Polytechnique.[44] Gay-Lussac mentions contact with several Americans in the early years of the nineteenth century and in 1818 when Benjamin Silliman launched his *American Journal of Science*, the French chemist received a copy.[45] When Joseph Henry visited Paris in 1837 he attended some of Gay-Lussac's lectures at the Muséum d'Histoire Naturelle.[46]

As an indication of his reputation abroad, we may mention that Gay-Lussac was made a foreign member of the respective scientific societies of Berlin, Bologna, Bonn, Edinburgh, Göttingen, Harlem, London (Royal Society), St Petersburg, Turin and Uppsala, among others.

We have seen that Gay-Lussac had great influence in his own country and he can be linked directly with those key figures in nineteenth-century French science, J. B. Dumas and P. E. M. Berthelot. In Britain he influenced William Prout and particularly Thomas Thomson, whose experimental work derived in several ways from his French contemporary.[47] In Germany it is enough that he helped to inspire Liebig. When we consider also his influence on the Swede Berzelius and the Italian Avogadro, Gay-Lussac emerges as a key figure in European science in the nineteenth century.

# Appendix: select correspondence

## LIST OF LETTERS*

| | | Source | Date |
|---|---|---|---|
| 1. | Abbé Dumonteil to Gay-Lussac's father | RGL | 27 March 1795 |
| 2. | Gay-Lussac to his father | RGL | '5 brumaire' (27 October 1795?) |
| 3. | Gay-Lussac to his mother | RGL | '17 floréal' (6 May 1798?) |
| 4. | Gay-Lussac to his father | RGL | 15 January 1803 |
| 5. | Gay-Lussac to his father | RGL | 26 May 1804 |
| 6. | Gay-Lussac to his father | RGL | 17 January 1805 |
| 7. | Gay-Lussac 'Notice historique de mes travaux sur l'iode' | Limoges | (1813/14) |
| 8. | Gay-Lussac to his wife | RGA | 6 August 1822 |
| 9. | Gay-Lussac to Berzelius | KVA | 25 May 1826 |
| 10. | Colonel Aubert to General [Valée?? Inspecteur général, Service d'Artillerie] | Vincennes | 18 May 1828 |
| 11. | Chabrol to Gay-Lussac | RGL | 20 November 1829 |
| 12. | Gay-Lussac to his son, Jules | RGL | 17 August 1832 |
| 13. | Gay-Lussac to his son, Jules | RGL | 15 September 1832 |
| 14. | Administrators of Saint-Gobain to Gay-Lussac | StG | 8 April 1847 |
| 15. | Commemorative speech by Liebig | RGL | 22 April 1867 |

*Abbreviations*

RGL   Collection of Roger Gay-Lussac
KVA   Archives of the Royal Swedish Academy of Science
StG   Saint-Gobain

* Two of these primary sources are not strictly letters but documents making interesting historical claims.

### 1. Abbé Dumonteil to Gay-Lussac's father at St Léonard

The abbé Dumonteil reassures Gay-Lussac senior on the fate of his son in the capital. There were some difficulties previously but Gay-Lussac is now settled in a *pension*.

[Paris] ce 7 germinal 3° an. rép.

[27 March 1795]

Les sollicitudes que vous montrez pour votre fils sont bien naturelles: elles sont aussi un témoignage de votre excessive tendresse pour lui et du vif intérêt que vous prenez à son avancement; mais rassurez-vous sur son compte. Si nous avons eu d'abord de la peine à trouver une pension qui pût répondre à vos intentions, nous sommes bien dédommagés pour celle qui a fixé notre choix. Votre fils, que je vis encore hier, s'y trouve à merveille et s'y plait infiniment. La nourriture du corps y est saine et en suffisante quantité. Celles de l'esprit et de l'âme n'y sont point négligées. La vie désoeuvrée et peu réglée qu'il avait menée jusqu'alors, son éloignement de vous me faisaient craindre pour les premiers jours quelque peu d'ennui: mais j'ai été heureusement trompé dans mes conjectures. Son heureux caractère, sa raison l'ont porté à dévorer avec ardeur toutes ces premières difficultés qui jettent souvent dans l'âme des jeunes gens une apathie et un dégoût fort préjudiciables. Il n'y a dans toute la pension qu'une voix sur son compte. Jugez, Citoyen, du plaisir que je goûte à vous faire part de ces détails. J'espère n'en avoir jamais de moins satisfaisant à vous donner, parce qu'il parait tout décidé à mettre à profit les sacrifices que vous avez faits en sa faveur, et dont il sent tout le prix. Je le surveillerai le plus qu'il me sera possible et l'entretiendrai dans ces bons sentiments afin de correspondre à l'opinion avantageuse que vous avez conçue de moi, qui vous honore infiniment. Daniel Monteil.

Veuillez bien faire agréer mes assurances à la C[itoyenn]e Gay.

### 2. Gay-Lussac to his father

Gay-Lussac had written home for his young brother's birth certificate because he thought that he would be conscripted in the army. He had learned some English and drawing with M. Sencier and he also requires a mathematics text-book. Finally he asks for a watch.

Passy 5 brumaire [an 4?]

[27 October 1795]

Il y a déjà longtemps mon cher père que je suis délivré de mes inquiétudes; elles étoient mal fondées et avoient été occasionnées par un faux bruit qu'on avait fait courir; suivant ce bruit on requerroit depuis l'âge de seize ans jusqu'à 40. C'étoit ce qui m'avoit engagé à vous demander l'extrait baptistaire de mon frère; de la mairie dont je vous ai parlé j'ai reçu cet extrait...je vous demandois un certificat pour l'employer seulement comme subterfuge. La crainte de me voir privé d'une éducation solide, m'avoit engagé à faire ces démarches. Vous me marquez qu'au milieu d'une grande ville comme Paris, on est à portée de recevoir des services de ses amis, mais nos amis où sont-ils? Tel se pare de ce nom, qui est votre ennemi; et celui qui vous fait bonne mine dans son adversité nous mit dans la nôtre. C'est une maxime que la révolution a bien prouvée. Je suis les conseils de Mrs [i.e. Monsieur] Sencier et me fais un vrai plaisir d'en recevoir; je me souciois peu d'apprendre l'anglois, mais Mrs Sencier me l'a conseillé et même l'a voulu, et après qu'il m'a eu fait ses remontrances j'ai reconnu que j'en tirerois une grande utilité. Vous avez reçu mes dessins d'architecture, mais vous ne me marquez pas s'ils vous ont plu. Je les soumets à votre jugement, ainsi veuillez, je vous prie, m'en dire votre façon de penser. Mrs Sencier vous avertira lorsque les 6 mois seront échus. Le livre mathématique que vous m'avez envoyé est la Caille, je suis donc obligé d'acheter Marie, et pour ce faire je vais demander de l'argent à Mrs Sencier. Je vous souhaite une santé pareille à la mienne, c'est à dire bonne...

Je joins ici un prospectus; je vous prie de le remettre à M[adam]e Basty.

J. Gay

[P.S.] Nous sommes mon cher papa dans une maison où nous n'entendons pas les heures à raison de notre éloignement de l'horloge, ce qui fait que je ne puis suivre aucun ordre dans mes études. Je me lève et je me couche sans savoir à quelle heure. Je vous prierai donc, si vous croyez que j[ai] l'âge mérité, de me donner votre montre d'argent. Ne croyez pas que ce soit par orgueil ou par ostentation que je vous le demande, c'est seulement par utilité; j'espère que vous n'aurez pas de répugnance à me confier une montre attendu qu'il n'y a pas ici de jeune homme au-dessus de 14 ans qui n'en ait une.

### 3. Gay-Lussac to his mother

Gay-Lussac has now been joined in Paris by his brother who is studying medicine. Both are managing to exist on their small allowances by making major economies. Gay-Lussac is borrowing money from his brother to buy a pair of trousers.

Paris 17 floréal [an 6?]<br>[6 May 1798?]

...je viens de voir mon frère dans son petit domicile, oui son petit domicile, car il l'est tellement que deux personnes ne sauroient s'y remuer: mais il a pris parti pour les moyens économiques et les a même portés au suprême degré. Hélas! Sur 10 écus qu'il a par mois, il emploie 7ff pour sa chambre, il achète un livre, celui qui lui est le plus nécessaire et un paire de souliers; il s'éclaire et se blanchit là-dessus et à peine lui reste-t-il 6ff pour sa nourriture qui se réduit à du pain sec depuis le matin jusqu'au soir. Je suis presque dans le même cas que lui; il m'est dû près de 100ff à mon école, et j'ai employé une partie de celui qu'on m'a donné à me procurer mes livres les plus nécessaires. Je croyais que j'aurais 60ff par mois comme je vous l'avois marqué mais le maximum n'a été porté qu'à 45ff. Je viens de quitter un traiteur où j'étois en pension à 24ff par mois: parce que avec la manière dont je suis payé, et la dépense qu'il faut que je fasse pour ce qui m'est nécessaire, je ne puis y suffire, et je me suis rangé au parti le plus économique. C'est là le cas où l'on pourroit nous appeler malheureux jeunes gens et nous le serions en effet si notre travail nous donnoit le temps de penser à notre situation. A l'égard de mes habits, je peux dire omnia mecum porto. Mon frère m'a remis 24ff et je vais acheter un pantalon car c'est ce dont j'ai le plus de besoin. Vous me marquez, maman que vous ne m'avez voulu rien acheter, parce que vous vouliez me laisser choisir à mon goût: mais je vous dirai que je n'en ai point de ce côté-là; et il m'importe peu que mon habit soit de telle ou telle couleur, de telle ou telle qualité pourvu que je sois couvert, voilà tout ce que je demande. Je ne veux rien de beau, je veux me conformer aux circonstances, et c'est pour cela que je vous avois écrit.

### 4. Gay-Lussac to his father

In this invaluable letter Gay-Lussac traces the educational circumstances which turned him from a career in law to one in science. The key lies in the abandonment of Latin for mathematics and the path thus opened to the Ecole Polytechnique. He has now opted for a scientific career although he does not pretend there is any money in science.

Paris 25 nivose an 11
[15 January 1803]

Mon bon ami. J'aurai eu le plus grand tort de ne pas vous consulter, si j'eusse été libre, ou mieux indifférent, sur le choix que j'avais à faire; mais depuis plus de trois ans il s'en fixe malgré moi, et il est aisé de sentir qu'il a entièrement dépendu du genre d'instruction que j'ai reçu. Lorsque je vins à Paris, je ne songeais qu'à devenir avocat, et mon premier soin fut de m'occuper de ce qui avait quelque rapport avec cette profession; mais bientôt vous voulutes que je me livrasse aux mathématiques, et je le fis avec d'autant plus de plaisir que les circonstances où se trouva Mr Sencier m'obligèrent de suspendre mon latin. N'étant pas plus avancé dans les mathématiques que dans le latin, je n'avais aucune raison de fixer mes goûts, et je tenais encore au projet de me faire avocat lorsque j'eus les espoirs d'entrer à l'Ecole Polytechnique. Je vous communiquai mon dessein et vous l'approuvates parce que outre que je pouvais m'instruire dans cette école, j'y étais à l'abri de la conscription qui me menaçait fortement. Dès lors, mon ami, je ne me suis occupé que de sciences exactes et j'ai conçu une très grande antipathie pour tous les états où on ne s'en occupât pas essentiellement. Chacun prépare la science dont il s'en occupe bien au-dessus de toutes les autres, et cela doit être. Si donc j'ai renoncé au barreau, c'est parce que je ne m'en étais pas assez occupé et qu'au contraire je m'étais beaucoup livré aux mathématiques. Pourquoi n'ai-je pas fait autrement? C'est parce que à l'Ecole Polytechnique on ne s'occupe que de sciences et point du tout de ce qui concerne le barreau. Notre état dépend donc de la direction qu'on donne à notre éducation et de la plus grande application qu'on donne à cette science plutôt qu'à telle autre. Les connaissances générales qu'on acquiert à l'Ecole Polytechnique permettent aux élèves de choisir entre six états différents suivant leurs goûts; mais comme l'étude des sciences m'avait déjà séduit, je ne songeais entrer dans aucune école d'application. Ce ne fut qu'un mois avant mon examen que je pensais à me faire recevoir, parce que mon amour propre aurait été blessé si on eût cru que je n'avais pas été en état de concourir, et que d'ailleurs étant reçu, je pourrais aisément continuer mes études sans vous être à charge plus longtemps. Je poursuivais ainsi mes projets, attendant qu'une heureuse circonstance les fit éclater d'eux-mêmes. Je vous les avais cachés, même à St. Léonard parce que j'étais très persuadé que vous ne les approuveriez pas, et à vous parler franchement, je regrette souvent pour votre bonheur d'avoir connu des sciences qui ne sont pas compatibles avec un état que j'aurais exercé auprès de vous. Mais aujourd'hui le mal est fait; il est irréparable. Je ne me dissimule pas que je n'ai pas pris un chemin qui puisse me conduire à une brillante fortune; mais ce n'est pas là ce que j'ambitionne le plus; le pur nécessaire me suffira toujours. Je sais que ce qui pourrait contribuer le plus à votre bonheur serait de me voir uni à la personne que vous avez prise d'avance en amitié parce qu'elle appartient à une famille qui vous a rendu beaucoup de services, et que de mon côté, j'estime infiniment. Cependant je me suis expliqué bien clairement dans le temps, et si j'avais cru que mon sort fût attaché au sien, je n'aurais pas laissé échapper une occasion que je savais ne devoir plus se présenter jamais. Rappellez-vous mon ami que vous m'avez dit plusieurs fois que mille écus étaient très bon à emporter à Paris, et que vous êtes très persuadé qu'étant dans les Ponts et Chaussées je ne pouvais aller habiter St. Léonard; qu'est-ce donc qui peut vous donner tant d'inquiétudes? Je n'ai pas encore d'état fixe, j'en conviens, mais suis-je tant à plaindre de ce côté là et ai-je passé l'âge d'en avoir un? Ne saurais-je être heureux que dans le mariage, moi qui en suis infiniment plus éloigné que je l'aurais désiré? Ce qui peut vous intéresser le plus et contribuer à votre bonheur est d'être persuadé que je ne suis point malheureux, que ma conduite est irréprochable et que je ne songe qu'à m'occuper et vous faire honneur. Je ne puis pas vous abuser en vous promettant de persister dans un dessein que je n'ai jamais eu, celui de devenir ingénieur, car je sais[?] que je ne pourrais jamais tenir ma promesse, et qu'il n'y a aucune considération qui puisse me déterminer à quitter ma manière de voir. Je suis assez jeune pour avoir un état, mais je suis trop vieux pour recommencer mes études.

### 5. Gay-Lussac to his father

Gay-Lussac is moving from Arcueil to Berthollet's house in Paris. Next year he may take over some of Berthollet's lectures on applied chemistry at the Ecole Polytechnique. He is already rehearsing his lectures in his room.

Arcueil 6 prairial an 12
[26 May 1804]

...Je vais aller demeurer à Paris dans la rue d'Enfer chez Mr B[erthollet] Je perds trop de temps par les courses que je suis obligé de faire tous les jours à Paris, pour ne pas y aller me fixer. Je resterai à la rue d'Enfer jusqu'en automne époque à laquelle j'aurai un logement à l'Ecole Polytechnique. Il ne faut pas croire que mon éloignement me mette mal avec Mr B[erthollet]; il veut mon bien et son intérêt et l'amitié qu'il a pour moi n'en sera point diminuée. Comme le cours de Chimie qu'il fait est particulièrement destiné aux arts, vous sentez combien il est important pour moi de bien me préparer et de fréquenter les plus célèbres manufactures. Il m'a d'ailleurs dit qu'il me feroit nommer à la fin de cette année son adjoint pour que à la fin de l'année prochaine il n'y ait plus aucune difficulté pour me faire nommer professeur. En conséquence il me laissera faire plusieurs leçons dans le courant de l'année prochaine, et je dois chercher à ne point me rendre indigne. Déjà mon ami je fais dans ma chambre des leçons à mes chaises: je recommence quand je n'ai pas bien fait[?]; je me suppose un auditoire très nombreux.

## 6. Gay-Lussac to his father

Long apologies for a delay in writing. Gay-Lussac is now working night and day to complete his joint memoir with Humboldt before leaving Paris on his European tour. Praise of Humboldt.

Paris 27 nivose [an 13]
[17 January 1805]

Je suis coupable mon ami d'avoir différé si longtemps de vous écrire parce que malgré mes occupations, si multipliées qu'on puisse les supposer, ne doivent pas m'empêcher de vous consacrer quelques instants. J'en suis honteux et très fâché, surtout à cause de l'inquiétude que vous pouvez avoir, vous ayant prévenu il y a près d'un mois que nous devions partir dans le peu[?]. Vous aurez bien pensé que je vous aime assez pour vous dire quand nous quitterons cette grande ville, mais j'ai auprès de vous une si mauvaise réputation de paresse qu'il est bien naturel que vous ayez quelquefois des doutes. Maintenant que je suis un homme qui a une activité singulière jointe à une très grande exactitude je vais me corriger et vous donner très souvent de mes nouvelles. C'est parce que le temps me paroit si court que je suis si inexact; j'écrirai demain me dis-je aujourd'hui, un jour est si peu de chose; et je ne pense pas que votre impatience vous fait compter tous les instants.

Nous sommes encore ici pour au moins un mois. A mesure que nous croyons finir il nous vient un surcroît de besogne. Ce qui nous a principalement arrêté c'est un travail que nous voulons publier avant notre départ et à la rédaction duquel je travaille nuit et jour. Nous comptons le lire lundi prochain à l'Institut, mais après j'aurai encore un autre travail à terminer. Vous pouvez compter que je vous écrirai avant notre départ.

Que je suis heureux, mon ami, d'avoir fait la connaissance de Mr Humboldt. C'est l'homme qui a le meilleur coeur que je connoisse, la plus grande sensibilité, le plus grand attachement pour ses amis. C'est un grand besoin pour nos coeurs de nous voir très souvent. Aussi M. H[umboldt] veut-il travailler tous les jours avec moi dans ma chambre très souvent depuis 9ʰ du matin jusqu'à après minuit [2ʰ du lendemain matin (crossed out)]. Nos goûts, nos sentiments sont absolument les mêmes; en un mot nous sommes de la plus grande intimité. Lui et M. Berthollet, voilà les deux vrais amis que j'ai ici. Ne soyez pas inquiet de moi pendant que je serai avec lui. Son existence ne lui est pas plus chère que le mienne; il a reçu votre lettre avec beaucoup de plaisir et il m'a prié de le rappeller à votre souvenir.

Ma lettre est de pièces et de morceaux. Après l'avoir interrompue deux fois, j'espère la finir celle-ci. J'arrive à l'instant de chez M. Bert[hollet] à Arcueil où j'ai dîné en grande société. Il avoit rassemblé plusieurs savants très distingués et d'opinions très différentes, en sorte qu'il y a eu beaucoup de discussions très intéressantes. Ce sont ces réunions, mon ami, qui font le bonheur de la vie. On ne les trouve qu'ici et je n'y renonce pas pour aller à Munich où on désire beaucoup m'attirer.

## 7. Notice historique de mes travaux sur l'iode

Gay-Lussac's letter to posterity setting out his claim for the discovery of hydriodic acid. The peculiar circumstances of Davy's communication to the Institute.

M. Clément a lu à l'Institut le 29 novembre, il a imprimé dans le moniteur le 2 décembre. Le 6 décembre j'ai lu à l'Institut sur l'acide hydriodique; ce jour-là j'ai montré chez M. Warden où nous avons déjeuné, de l'acide hydriodique à M. H. Davy et Warden; plusieurs jours avant j'avais com[muni]qué plusieurs faits à M. Davy sur les propriétés électriq[ues] de l'iode, sur sa combinaison avec l'a[cide] m[uriati]que oxygéné; je lui avois surtout dit qu'on formoit un acide particulier avec l'hydrogène. Le 12 dimanche ma note a paru dans le moniteur. Le samedi MM. Humboldt, Vauquelin et Chevreul dînaient chez Davy où il a été question de l'acide hydriodique parce que j'en avois montré à Chevreul au jardin des plantes. Le lundi 13 M. Davy a envoyé une lettre à l'Institut datée de 3 jours d'avance qui était adressée à M. Cuvier. Mais, dînant le soir chez M[adam]e. Lavoisier avec M. H. Davy, j'ai entendu, lorsque M. Cuvier y est venu après dîner, qu'il s'excusoit de n'avoir pas lu sa lettre à l'Institut au commencement de la séance, parce qu'il l'avait reçue très tard dans la journée; et en effet M. Cuvier ne parut à l'institut que vers la fin de la séance. La lettre fut la dernière pièce qu'on lut.

Le samedi 18 (veille de [illegible]) j'ai parlé à Davy de l'acide oxygéné de l'iode; il m'a déclaré n'en rien connaître et je lui ai dit que je l'avois obtenu.

Le lundi 20 j'ai lu ma note sur cet acide à l'institut et je l'ai portée de

suite à l'imprimeur qui m'a dit que le matin M. Guyton lui en avoit
envoyé une de M. Davy. J'en ai pris connoissance et j'ai vu qu'elle n'étoit
ni de la main de M. Davy ni de celle de M. Cuvier. Elle ne m'a pas paru
différer de celle que j'avois entendue à l'Institut excepté qu'en ce qui
regarde les sels détonnants.

### 8. Gay-Lussac to his wife

Gay-Lussac's wife has gone to their summer home and Gay-Lussac, left in
Paris, describes to her his round of duties, particularly in relation to his
membership of the gunpowder administration. This letter also shows his
relations with the families of scientists: Madame (Gay de) Vernon, wife of
the former deputy governor of the Ecole Polytechnique; Madame (Collet)
Descotils, widow of the metallurgist whose obituary Gay-Lussac had
written (*A.c.p.*, 4 (1817), 213–20); the astronomer Delambre who was
shortly to die (19 August 1822); and Madame Prony, wife of the director
of the Ecole des Ponts et Chaussées.

Paris 6 août 1822

Je vais essayer ma chère et bonne Joséphine de te donner des nouvelles de
ma conduite depuis ma dernière lettre. Le jour de mon arrivée j'ai fait des
visites à tout l'arsenal. Je suis allé à la direction des droits réunis pour mes
aréomètres et de là aux finances; mais tout cela n'avance pas. J'ai passé
une heure avec Humboldt; de là je suis allé porter le paquet de M[adam]e
Vernon que je n'ai pas trouvée, puis au comité consultatif et enfin chez
Arago où j'ai dîné. En me retirant j'ai fait une visite à M[adam]e Descotils;
nous irons demain chercher le Danal[?] à 3ʰ et je reviendrai dîner avec elle.

Le dimanche au matin je me suis occupé de ma petite fabrique; j'ai
déjeuné avec le Colonel, j'ai fait quelques arrangements et je suis allé
dîner à Arcueil.

Hier, lundi, je me suis occupé de la fabrique; déjeuner avec le Colonel;
institut et dîner chez Arago. Ce matin je m'occupe de toi; à midi j'irai
avec le comité consultatif à Charenton sur un bâtiment à vapeur qui nous
prendra au port d'Austerlitz: il est probable que le général Ruty et les
autres membres du comité des poudres viendront avec nous. Le soir je
dînerai chez M[adam]e D'Hérouville qui est encore dans la peine des
domestiques.

En arrivant ici j'ai appris l'affreux évènement arrivé à Colmar. La
poudrerie où se trouvaient 66 milliers de poudre a sauté. Sur 13 ouvriers 11
ont péri; les 2 autres sont très maltraités. Le commissionaire a perdu sa fille
aînée qui a été écrasée sous les décombres de sa maison; sa fille cadette par
suite d'une blessure grave a eu le bras droit amputé; lui-même et sa femme
ont reçu plusieurs blessures. Tout a été détruit par l'explosion ou par
l'incendie qui l'a accompagnée et il ne reste plus pierre sur pierre.

M[onsieu]r Delambre va de plus mal en plus mal; on a donné hier à
l'institut des nouvelles très peu rassurantes sur son état.

M[adam]e Prony est aussi très malade à Vichy.

### 9. Gay-Lussac to Berzelius

Gay-Lussac gives Berzelius a review of the data he has assembled on the density of alcohol/water mixtures. He wants one of his children to learn Swedish so that he can translate scientific memoirs.

Paris 25 mai 1826

Monsieur et très illustre ami,

Un voyage un peu précipité à la fonderie de Douay, pour assister à la fonte de quelques canons avec des alliages nouveaux, m'empêchera de pouvoir vous envoyer une petite boite de quelques alcoomètres et aréomètres; mais je le ferai plus tard. Voici quelques densités des divers mélanges d'eau et d'alcool à la temperature constante de 15° centigrade

| Alcool en volume | Densités | Alcool en vol. | Densités |
|---|---|---|---|
| 100 | 7947 | 70 | 8907 |
| 95 | 8168 | 65 | 9027 |
| 90 | 8346 | 60 | 9141 |
| 85 | 8502 | 55 | 9248 |
| 80 | 8645 | 50 | 9348 |
| 75 | 8779 | 45 | 9440 |
| | | 40 | 9523 |
| | | 35 | 9595 |

Ces densités doivent suffire pour la comparaison de l'alcoomètre Wilcke avec l'alcoomètre centésimal; mais si d'autres nombres étaient nécessaires je m'empresserais de vous les donner. L'adoption de l'alcoomètre centésimal rendra sûrement moins de services en Suède qu'en France, parce que sans connaître l'alcoomètre de Wilcke, il ne doit pas être aussi imparfait que celui de Cartier. L'administration des droits réunis avait souvent des procès avec les particuliers, et depuis l'adoption de l'alcoomètre centésimal elle n'en a pas eu un seul. Vous avez reconnu l'utilité de l'instrument, et s'il était besoin d'un nouveau titre à mon estime pour vous, j'en trouverais un dans cet amour du bien qui fait écarter préjugés et fausses rivalités nationales.

J'ai reçu votre mémoire sur le molybdène; Mr Pasch a bien voulu se charger de le traduire.

Nous sommes bien embarrassés pour les traductions, et c'est ce qui m'a déterminé à faire apprendre le Suédois à un de mes enfants. Mais le croirez-vous? Je n'ai pu me procurer ici les livres convenables; c'est-à-dire un dictionnaire et une grammaire de français en suédois. S'il en existait à Stockholm vous me rendriez service en prenant la peine de me faire parvenir ces livres.

J'ai remis à Mr Fresnel le paquet que vous m'avez adressé pour lui.

Recevez mon cher et illustre ami, l'assurance de ma profonde estime et de mon sincère dévouement.

Gay-Lussac

## 10. Colonel Aubert to General [Valée?? Inspecteur-général, Service d'Artillerie]

In view of the threat to suppress Gay-Lussac's position on the gunpowder administration Aubert describes the invaluable services he has performed, particularly in devising new methods for estimating the purity of saltpetre. He earns his salary of 4000 francs many times over in the savings he makes possible.

Paris le 18 mai 1828

Mon Général,

Vous m'avez permis de vous faire part de mes réflexions sur la suppression prévue ou probable de la place de M. Gay-Lussac près de la Direction du service des poudres. Je ne puis m'empêcher de vous faire connaître que mon opinion est que cette suppression paroit tout à fait contraire aux intérêts de l'etat et de la gloire des Corps de l'Artillerie.

Ses connaissances scientifiques positives ne sont pas et ne pourront jamais être tellement répandues dans le Corps, que les officiers des différents services n'aient encore souvent besoin de guide qui soit au courant de la science pour résoudre ou éclaircir une foule de questions qui se présentent journellement. Personne mieux que M. Gay-Lussac ne peut parmi les savants être ce guide. Outre les rares connaissances qu'il possède dans les sciences phisico-mathématiques et les arts chimiques, il est comme membre du comité des arts et manufactures placé de manière à connaître et être initié dans tous les procédés nouveaux et par conséquent il peut indiquer ceux qui offrent des perfectionnements pour les différentes branches de service et par là compenser et bien au-delà, le traitement annuel de 4000 f. qui lui était alloué comme membre du comité des poudres.

C'est déjà ce qu'il a fait en indiquant les moyens de distinguer et d'apprécier les quantités de *muriate de potasse* existant dans les sels provenant du raffinage de salpêtre. La Direction a pu ensuite donner une valeur à des quantités considérables de cette substance qui existaient depuis longtemps à Lille (plus de 130.000K) et cette valeur a été bien supérieure aux traitements que M. Gay-Lussac a pu toucher depuis dix ans qu'il est attaché aux poudres.

L'instruction sur les salpêtres publiée par la Direction est presqu'entièrement son ouvrage; il est aussi l'auteur du *mode d'épreuve* pour déterminer le *degré de pureté* du salpêtre raffiné; il a fait connaître un *nouveau moyen* d'analyser le *salpêtre brut*, moyen qui employé par l'administration de Paris sert de contrôle aux épreuves faites dans les divers commissariats et donne la certitude que les intérêts des salpétriers comme ceux des comptables ont été respectés. Il a encore indiqué un nouveau procédé pour analyser la poudre, etc., etc. Enfin ses conseils et son opinion ont servi plus d'une fois a rejeter les projets d'essais des procédés qui, avantageux au 1er aspect, péchaient par le principe et ces essais auraient occasionné des pertes réelles à l'etat si on les eût mis à exécution.

Il s'occupait depuis quelque temps d'un mode exact d'épreuve pour les alcalis et rassemblait les matériaux nécessaires pour former un corps d'ouvrage sur la vaporisation des eaux et sur la construction des fourneaux. C'est d'après ses principes que j'ai pu apporter quelques perfectionnements à plusieurs fourneaux de nos raffineries et particulièrement à ceux de Lille où l'on économise environ $\frac{1}{4}$ du combustible employé autrefois. C'est encore à lui qu'on doit l'indication des moyens à employer à la Raffinerie de soufre de Marseille, tant pour empêcher les vapeurs sulfureuses de nuire aux ouvriers et aux propriétés voisines de l'établissement que pour rendre le raffinage des soufres plus prompt et plus économique. Initié actuellement dans les détails pratiques du service des poudres de quelque autre branche de l'administration, il trouverait nécessairement l'occasion de faire faire de nouveaux perfectionnements dont la valeur ne peut pas être évaluée ni appréciée et dont l'etat tirerait seul tout le profit.

M. Gay-Lussac est connu de toute l'Europe savante et apprécié à un très haut degré tant pour son caractère privé et pour ses rares connaissances scientifiques. Les nombreux élèves et tous ceux qui l'estiment n'apprendraient peut-être pas sans quelque surprise que le gouvernement français pour une faible économie de 4000f. par an se prive des services d'un homme aussi distingué et aussi célèbre; qui sait même si dans le public on ne présentera pas la suppression de cette place comme une nouvelle disgrâce pour le sciences! D'un autre côté comme professeur de l'école polytechnique, les jeunes élèves voyaient avec plaisir et même avec orgueil M. Gay-Lussac faire partie d'un corps où plusieurs d'entre eux devaient entrer; peut-être c'est à cette idée que l'artillerie doit le choix de quelques bons sujets instruits qui, en entrant dans le corps, espéraient y retrouver pour soutien le professeur célèbre qui les avait distingués et déjà appréciés dans la foule de leurs camarades.

Voilà, mon général, quelques-unes des considérations, et je pourrais encore en ajouter d'autres, qui me font désirer que l'artillerie continue de s'attacher M. Gay-Lussac, soit en lui offrant un emploi continu et temporaire près le Comité Central de l'Artillerie et en lui conservant son logement à l'Arsenal où il sera mieux à portée d'utiliser particulièrement ses connaissances pour le service des poudres, soit de toute autre manière.

Dans ce que j'ai l'honneur de vous écrire, j'espère mon général, que vous ne verrez peu l'ami de M. Gay-Lussac mais bien l'officier d'artillerie jaloux de bien remplir ses devoirs et plus jaloux encore de voir s'accroître l'illustration du corps auquel j'ai l'honneur d'appartenir depuis longtemps.

Ennemi des abus, je ne serais. . .[illegible]. . .J'avais pû remarquer que la place de M. Gay-Lussac fut une sinécure mais j'ai reconnu qu'elle était profitable et avantageuse à un service du Roi et je le dis quoiqu'il en coûte. Je souhaite que l'artillerie ne soit pas dans le cas de regretter la perte qu'elle aurait faite en abandonnant M. Gay-Lussac, perte qui serait irréparable. Car on pourra bien si l'occasion se présentait offrir cette place à un autre savant, mais je pense qu'aucun autre n'est aussi propre à la bien remplir que lui.

Aubert

P.S. Lors de son entrée dans le service des poudres, M. Gay-Lussac était partisan de la libre fabrication des poudres par l'industrie particulière et par suite de la libre importation du salpêtre. Ce n'est qu'après [avoir] connu dans la discussion du Comité tous les détails de cette question qu'il a changé d'opinion et qu'il a éclairci celle de beaucoup de personnes appellées à la traiter.

### 11. Chabrol to Gay-Lussac

The prefect of the Seine department nominates Gay-Lussac director of the Assay office and explains how his salary will be calculated.

Préfecture du Département de la Seine, 5e Division, 1er Bureau
Paris 20 novembre 1829
Monsieur.

J'ai l'honneur de vous adresser une expédition de l'arrêté que j'ai signé pour vous nommer essayeur du Bureau de garantie en remplacement de M. Vauquelin.

En accomplissant par cette nomination un désir que vous m'avez témoigné, j'ai voulu tout à la fois rendre justice à un mérite que l'Europe entière reconnait, payer la dette de l'administration pour les services désintéressés que vous lui avez si souvent rendus, et vous donner un témoignage de la haute estime que je vous porte. J'aurais souhaité, Monsieur, pouvoir vous annoncer que vous jouirez de tous les avantages qui étaient attachés précédemment aux fonctions d'essayeur mais j'ai le regret d'avoir à vous prévenir qu'ils éprouveront une réduction.

La commission des Monnaies a fait, il y a plusieurs mois, à son excellence le Ministre des Finances des propositions par suite desquelles il seroit alloué à l'essayeur 1° le quart de ses recettes brutes, 2° un traitement fixe de douze mille francs, 3° enfin moitié des sommes excédantes, l'autre moitié devant être versée au fonds commun destiné à subvenir à l'insuffisance des traitements des essayeurs des Bureaux de garantie des départements.

M. le comte de Sussy, Président de la commission des Monnaies estime que, d'après cela, le traitement de l'essayeur du département pourra encore s'élever à vingt cinq mille francs par an.

Veuillez agréer, Monsieur, l'assurance de ma considération distinguée.

Le Conseiller d'Etat Préfet    Chabrol

### 12. Gay-Lussac to his son, Jules

Gay-Lussac wonders whether his son's progress justifies a longer stay at

Giessen with Liebig. He admonishes him, explaining again the importance of hard work and the need to learn German.

Paris 17 août 1832

Mon cher ami      j'ai trop de confiance en M. Liebig pour ne pas m'en rapporter à ses bons conseils sur la continuation de tes études à Giessen; je désire seulement qu'il ne consulte pas trop son coeur pour te retenir et qu'il juge froidement si tu as bien employé ton temps et si un plus long séjour te serait utile. Tu dois apprécier toi-même tous les soins bienveillants de M. Liebig et songer qu'ils doivent avoir un terme. J'avais pensé en t'envoyant auprès de lui, et tu le savais très bien aussi, qu'une année d'une application soutenue auprès d'un si bon maître devait suffire pour te familiariser avec la langue allemande, te faire connaître les principes de la chimie, te donner le goût du travail et t'apprendre à penser et marcher seul.

[added by Madame Gay-Lussac]
     ton père a acheté une jolie maison bien commode avec un jardin de 6 arpents qui est bien planté à Chatillon même distance de Paris qu'Arcueil, une vue délicieuse.

### 13. Gay-Lussac to his son, Jules

Gay-Lussac, while congratulating his son on graduation, points out sternly that this is only a beginning. He is to travel in northern Germany to visit factories and to perfect his German.

Paris 15 Sept. 1832

Mon cher ami nous avons appris avec plaisir ton élévation au Doctorat cum summa laude. Ta mère a déjà fait retentir bien de joie tes succès; pour moi je m'en applaudis sans rien dire. Je suis difficile et modeste pour toi comme pour moi. Ne t'imagine pas que ton enfantement soit fini; le titre de docteur n'est que ridicule quand il n'y a rien sous le bonnet. Ton voyage dans le nord de l'Allemagne achèvera de te rendre l'allemand familier; je mets à cela beaucoup de prix. Pour l'argent dont tu auras besoin pour ton voyage, tu pourras employer la même voie que la dernière, ou toute autre analogue, en ayant soin de m'avertir de l'époque de la traite. Je n'ai pas besoin de te recommander la plus stricte économie. Tu verras sans dout M. de Humboldt à Berlin. Je te recommande de ne point t'adresser à lui pour de l'argent et de refuser s'il t'en offrait. Aye bien soin de visiter autant que tu le pourras les établissements industriels de tout genre, de prendre des notes, de faire des croquis; ce n'est qu'ainsi que tu en conserveras d'utiles souvenirs.

### 14. Administrators of Saint-Gobain to Gay-Lussac

The members of the council of Saint-Gobain wish to retain Gay-Lussac's association with them without imposing the obligation of regular attendance at their meetings. They suggest that instead of the title president, with its implied obligation of chairing meetings, Gay-Lussac should have the title of senior consultant with a salary of 6000 francs.

[8 avril 1847]

Monsieur et honorable Président,

Le séjour que vous faites chaque année pendant plusieurs mois à la campagne a fait naître dans votre esprit des scrupules sur lesquels votre prochain départ nous fournit naturellement l'occasion de nous expliquer.

Nous avons cru reconnaître qu'à vos yeux les devoirs et la responsabilité de Président du Conseil d'Administration se conciliaient difficilement avec une absence aussi longue.

Nous redoutons la conséquence extrême à laquelle pourrait vous conduire une pensée dont nous ne pouvons que respecter la délicatesse mais qui nous parait aussi peu fondée que contraire aux intérêts bien entendus de la société.

Nous prenons donc la liberté de nous en ouvrir franchement avec vous.

A nos yeux votre nom n'est pas moins utile que le serait votre présence, à la bonne marche de nos affaires; la juste confiance qu'il inspire à nos directeurs au Conseil d'Administration, à nos actionnaires, est telle que nous sommes convaincus qu'aucune autre présidence ne pourrait être plus efficace et plus utile alors même qu'elle serait exercée avec des conditions de présence plus assidue.

C'est vous dire, Monsieur et honorable Président, que nous approuvons sans réserve et sans arrière pensée toutes les absences que pourront nécessiter le besoin de repos, ou le soin de vos intérêts; nous reconnaissons même que pendant le temps que vous passez à Paris, la multitude des affaires qui vous accablent, et les travaux de toute nature auxquels vous prenez part, rendant votre exactitude à nos séances extrêmement difficiles, il convient que vous puissiez vous considérer comme plus libre que par le passé dans l'exercice de vos fonctions de Président.

Mais à la [réflexion et tout??] bien établi et bien expliqué entre nous, nous insistons pour que vous restiez à notre tête; nous pensions unanimement que votre présidence réelle quand vos occupations vous le permettront, que votre présidence nominale en cas d'absence est préférable à toutes les autres combinaisons que nous pourrions rencontrer.

Nous ajoutons que si quelques scrupules subsistaient dans votre esprit, un moyen bien simple suffirait de les lever.

Déjà vous l'avez fait par le passé en donnant à la direction de nos fabrications des soins plus étendus et plus efficaces que ceux que comportent les fonctions ordinaires de l'administration.

En allant dans nos ateliers passer plus de temps que n'en passe jamais aucun des autres administrateurs, en consacrant à des expériences utiles à nos industries des moments que vous pourriez mieux employer dans l'intérêt de la science et dans votre propre intérêt, en nous faisant ainsi profiter des lumières et de l'expérience que vous ont données les longs travaux auxquels vous devez la position élevée que vous occupez, vous avez sans doute voulu suppléer, autant qu'il était en vous, à ce que vous trouviez d'incomplet dans votre présidence.

Permettez-nous de le considérer de la même manière, de vous demander de nous continuer pour l'avenir ces travaux extraordinaires et de vous assurer qu'ils seront accueillis par nous avec la reconnaissance qu'ils méritent, et appréciés comme un ample dédommagement au regret que nous avons de ne pas vous posséder toute l'année.

Mais en même temps et après avoir répondu aux si honorables scrupules qui vous agitaient, laissez-nous parler des nôtres et vous dire que nous nous considérons comme obligés à vous offrir une juste compensation pour les droits de présence dont vous priveraient votre séjour à la campagne et les absences que nécessitent les nombreux devoirs que vous avez à remplir.

Nous y trouvons notre honneur et notre délicatesse engagés; nous ne pouvons admettre que vous contribuiez de votre temps et de votre science à nos affaires dans une proportion plus grande que les autres administrateurs, et que vous y trouviez des avantages moindres et tellement minimes que le chiffre en parait en quelque sorte réduit et en conséquence nous avons unanimement décidé qu'il vous serait offert chaque année une indemnité dont le chiffre serait dès à présent fixé à six mille francs et que nous vous demanderons la permission de faire remonter jusqu'au 1er janvier 1846.

Il est bien entendu que cette indemnité serait indépendante de vos droits de présence toutes les fois que vous pourriez nous présider.

Nous ne pensons pas que cette résolution ait besoin d'autres justifications que celles que cette lettre explique suffisamment; nous voulons cependant la fortifier en vous offrant le titre de conseiller supérieur de la société pour la fabrication.

En distinguant ainsi en vous l'Administrateur et le savant, en attribuant aux travaux spéciaux du chimiste une faible indemnité, nous restons dans le vrai et nous acquittons une dette légitime.

Nous avons la confiance que cette lettre recevra de vous la réponse que nous sollicitons, elle est l'expression sincère et unanime de notre estime pour votre caractère et de notre reconnaissance pour les services que vous rendez à notre société.

Nous sommes heureux toutes les fois que nous trouvons l'occasion de vous témoigner ces sentiments et de resserrer les liens qui nous unissent.

Agréez Monsieur et cher Président l'assurance de tous nos sentiments d'attachement et de considération.

[Signed]   Gérard, Jouet, Hély d'Oissel, de Malezieu,
Marcellin DeFresne, Péan de St Giles


## 15. Commemorative speech by Liebig (1867)[1]

Liebig's speech at a banquet of French chemists in 1867 recalls his debt to Gay-Lussac and Thenard. He claims to have been Gay-Lussac's first (research) student and was indebted to Gay-Lussac not only for a place in his laboratory but also for training in organic analysis, a subject on which Liebig himself had built much of his reputation.

M[essieu]rs je suis vivement touché et profondément reconnaissant des sentiments que mon honorable confrère M. Balard vient de m'exprimer. Je le prie d'en accepter l'assurance bien sincère.

Appelé à mon tour à porter un toast, je vais vous en proposer un qui obtiendra, j'en suis sûr, toute votre approbation. Je vous propose un toast à la mémoire de deux des plus grands chimistes français, des deux fondateurs de notre science moderne, dont les admirables travaux n'ont jamais été surpassés et resteront toujours nos modèles – des deux savans qui, comme hommes, représentent les qualités les plus élevées de la Nation française – Vous devinez M[essieurs] que je veux parler de Gay-Lussac et de Thenard. Nous connaissons tous, M[essieurs], les grandes découvertes que nous devons aux efforts réunis de ces deux grands hommes liés par une amitié étroite et dont les travaux ont leur origine dans cette amitié même – Vous savez pourquoi leurs noms seront à jamais inséparables dans l'histoire de la Science.

Vous connaissez M[essieurs] leur mérite comme hommes de Science, mais il y a bien peu d'entre nous qui aient eu le bonheur de les connaître personnellement, et c'est pour moi le besoin de leur payer un tribut de reconnaissance qui me porte à vous dire quelques mots.

Ce qu'ils ont tous deux fait pour moi suffira pour vous faire comprendre ce qu'ils ont fait pour beaucoup d'autres.

J'arrivais à Paris il y a 44 ans – jeune étudiant, enfant de 19 ans sans recommandation aucune, si ce n'est celle de mon désir de m'instruire. J'avais apporté à Paris un petit travail sur les composés fulminans de mercure et d'argent et c'est à M. Thenard que je m'adressai pour le faire lire à l'académie. Le président de l'académie (car M. Thenard occupait alors cette position)[2] reçut le jeune étudiant avec la plus grande bienveillance. La note fut lue par Gay-Lussac et c'est Dulong qui en fit le rapport.[3] Dès ce moment, je possédai à Paris les amis les plus chaleureux.

Gay-Lussac me procura un laboratoire où je pouvais continuer mes travaux. Il m'admit dans sa maison et j'étais comblé de bonheur quand il me proposait de poursuivre et de terminer mon travail sur les fulminates de mercure et d'argent avec lui, dans son laboratoire à l'arsenal. Ce fut là ce qui donna la direction à tous mes travaux ultérieurs. 'Il faut vous occuper,' me disait-il, 'tous les jours de la Chimie organique, voilà ce qui nous manque.' Je fus, je crois, son premier élève; après moi c'était mon ami Pelouze qu'une maladie cruelle tient aujourd'hui séparé de nous.

Jamais je n'oublierai les heures passées dans le laboratoire de Gay-Lussac! Quand nous avions terminé une bonne analyse (vous savez, sans que je vous le dise, que la méthode et les appareils décrits dans notre mémoire commun sont de lui seul) quand nous avions terminé une analyse, il me disait 'Maintenant M. Liebig il faut que vous dansiez avec moi comme j'ai dansé avec Thenard, quand nous avions trouvé quelque chose de bon' et nous dansions!! Vous avez M[essieurs] souvent entendu appeler

Thenard le Père Thenard – et c'était véritablement notre père à nous, de nous tous qui tendait toujours et ne refusait jamais la main aux faibles pour les aider à monter l'échelle, et à vaincre les difficultés.

M. Dumas notre président aujourd'hui peut en dire quelque chose; il en était le favori et il l'était de droit; il était reconnu le premier et il est resté le premier.

Ainsi Mess[ieurs] à la mémoire de Gay-Lussac et Thenard, les fondateurs de notre science moderne les représentans des qualités les plus élevées du caractère français.

Prononcé le 22 avril 1867 par M. Liebig au banquet.

# Notes

## Preface

1  Unpublished notes, c. 1826, *Chymia*, 12 (1967), 131.
2  *Quarterly Journal of Science*, 18 (1825), 305.
3  Proposed by Boussingault, *A.c.p.*, 31 (1826), 278.
4  J. T. Merz, *A History of European Thought in the Nineteenth Century*, London, 1904, vol. 1, p. 279n.
5  Theodore Zeldin, *Times Literary Supplement*, no. 3, 774, 5 July 1974, p. 701.
6  The Academy of Sciences became the First Class of the Institute in the period 1795–1815.

## Chapter 1

1  Letter to his father, 15 January 1803 (Appendix, letter 4).
2  It must be remembered that the conceptual revolution in chemistry had occurred several years before the political revolution of 1789.
3  The Academy of Sciences became the First Class of the Institute in the period 1795–1815.
4  5–6 June 1787. *Travels in France*, ed. M. Betham-Edwards, London, 1889, pp. 21–3.
5  Abbé A. Lecler, *Martyrs et confesseurs de la foi du diocèse de Limoges pendant la révolution française*, 4 vols., Limoges, 1892–1904, vol. 2, pp. 480, 486–7.
6  The Concordat was an agreement between the Roman Catholic Church and the French government.
7  M. Berthomé, *L'Enseignement secondaire dans la Haute Vienne pendant la révolution (1789–1804)*, Paris and Limoges, 1913.
8  *Evxercies littéraires de Mrs. les écoliers du collège d'Eymoutiers*, Limoges, 4to, 32 pp. The date is torn off the title page but the examination questions refer to 1783.
9  J. le Francais de Lalande, *Bibliographie astronomique*, Paris, 1803, p. 757.
10  See Appendix, letter 1. In the Revolutionary period many people with the particle 'de' or 'du' indicating connection with the nobility, found it prudent to omit it from their names. Thus Dumonteil became Monteil.
11  L. Lacroix, *Notice historique sur l'Institution Savouré*, Paris, 1853.
12  See Appendix, letter 4.

13 C. Julii Caesaris, *Commentarii de bello gallico et civili*, Paris, 1788 (12 mo, vellum bound).

14 A. Fourcy, *Histoire de l'Ecole Polytechnique*, Paris, 1828, p. 120.

15 A uniform had been introduced for students at the Ecole Polytechnique as a means of strengthening discipline.

16 Letter from Gay-Lussac to his mother, 11 frimaire [year 4] (2 December 1795). 100 centimes = 1 franc; 1 sou = 5 centimes.

17 'Etat de distribution du traitement des élèves de l'Ecole Polytechnique pour les mois de floréal, prairial et messidor an 6°'. Archives de l'Ecole Polytechnique, carton 1797, dossier Etat des Elèves.

18 This 100 francs could refer to a sum given to students on enrolment as opposed to the monthly grant discussed in the next sentence of the letter.

19 The Ecole Polytechnique was obviously making economies in its budget.

20 Letter from Gay-Lussac to his mother, 17 floréal [year 6?] (6 May 1798). See Appendix, letter 3.

21 J. M. Thomson (ed.), *French Revolution Documents, 1789–94*, Oxford, 1933, pp. 255–6.

22 Letter from Gay-Lussac to his father, 5 brumaire [year 3?] (27 October 1795). See Appendix, letter 2.

23 Fourcy, however, mentions the calling up of ninety students in 1799 (*Histoire de l'Ecole Polytechnique*, p. 170).

24 Letter to his father, 15 January 1803. See Appendix, letter 4.

25 The lecture on physics given three times a month on the morning of the fifth day of the Republican week is not included in these statistics by Fourcy, *op. cit.*, pp. 378–9. (An explanation of the Republican week is given on pp. 13–14).

26 Law of 30 vendémiaire an 4, titre II, art. X.

27 In May 1798 Gay-Lussac wrote that he saw Vernon often.

28 *Abrégé de l'Histoire naturelle à l'usage des élèves de l'Ecole Royale Militaire*, vol 2, *Pour la seconde année de philosophie*, Paris, 1787.

29 Gay-Lussac's library has the third (1801) edition of Lavoisier's *Traité élémentaire de chimie*.

30 P. Rochon, *Recueil de mémoires sur la mécanique et la physique*, Paris, 1783.

31 This may possibly have been: Lacaille, *Leçons élémentaires d'optique* (Paris, 1764) which was in his library.

32 He refers to the author of this book as 'Marie' but there is no book of such a description in his library.

33 Quoted by Fourcy, *Histoire de l'Ecole Polytechnique*, p. 137.

34 *Journal de l'Ecole Polytechnique*, 4 (cahier 11, 1802), 319.

35 The *idéologues* were a group of French philosophers who flourished at the turn of the century and took as their starting point the philosophy of Condillac. They included Garat, Degerando and Cabanis.

36 D. F. J. Arago, *Oeuvres*, vol. 12, *Mélanges*, Paris, 1859, p. 683.

37 Letter of 11 February 1801.

38 Letter of 15 January 1803. See Appendix, letter 4.
39 Letter of 15 January 1803. See Appendix, letter 4. '...un dessin que je n'ai jamais eu, celui de devenir ingénieur'.
40 Letter of 11 February 1801.
41 The edition in Gay-Lussac's library was published in Paris in English and dated: 'The Ninth Year' (1801). He had begun to learn English in 1795.
42 Stretch, *The Beauties of History*, London, p. 24.
43 *Ibid.*, pp. 139-40.
44 Archives nationales 2AP5, *Leçons de physique*, M. Gay-Lussac professeur, 1820, f. 1.

## Chapter 2

1 Letter to his father, 17 January 1805 (Appendix, letter 6).
2 *Trona* is naturally occurring sodium sesquicarbonate, $Na_3H(CO_3)_2.2H_2O$.
3 *Leçons sur la philosophie chimique*, Paris, 1837, p. 379.
4 *Essai de statique chimique*, vol. 1, p. 25.
5 *Ibid.*, p. 48.
6 *Ibid.*, p. 73.
7 *Ibid.*, p. 94.
8 *Ibid.*, p. 30.
9 Berthollet sometimes explained chemical reaction in terms of cohesion between salts which had not yet been formed.
10 Arago, *Oeuvres*, vol. 3, *Notices biographiques*, p. 7.
11 Gay-Lussac to Jules Lussac, 6 October 1841.
12 Collection Madame Roger Gay-Lussac.
13 27 September 1804. Archives Ecole Polytechnique, dossier Gay-Lussac.
14 Wellcome MS Gay-Lussac No. 1567. All letters in the Wellcome Library are quoted by courtesy of the Trustees.
15 'Sur la dilatabilité de l'air et des gaz par la chaleur', *A.c.*, 1 (1789), 256-99.
16 'On the expansion of elastic fluids by heat', *Journal of Natural Philosophy*, ed. W. Nicholson, 3 (1802), 130-4. This article, published in October 1802 from the *Memoirs* of the Manchester Literary and Philosophical Society, was followed in November and December (pp. 207-16; 257-67) by an English translation of Gay-Lussac's memoir from *A.c.*, **43**, 137-75. The publication in the same volume provides a convenient opportunity for comparison.
17 *New System of Chemical Philosophy*, vol. 1, part 1, Manchester, 1808, p. 19. For subsequent evaluations by Joule and Partington see J. R. Partington, *History of Chemistry*, vol. 3, London, 1962, p. 770.
18 *A.c.p.*, 2 (1816), 235.
19 Dalton admitted differences in his readings of 6-8 parts in 325 (or 2 per cent) but was content to attribute them to the presence of water vapour. Gay-Lussac's greatest variation was between readings of

137.40 and 137.61 for the volume of air at 100°C which had been 100 at 0°C. Thus Gay-Lussac was measuring to 1 part in 13 000 as opposed to Dalton's 1 in 1000 and his maximum deviation was only 0.15 per cent. Dalton had been satisfied to begin his experiments at room temperature (55°F) rather than go down to the freezing point of water. Only after he knew of Gay-Lussac's work did he repeat his experiments and extend them to the lower temperature range.

20 Partington, *History of Chemistry*, vol. 3, p. 770.

21 P. G. Tait, *Sketch of Thermodynamics*, 2nd edn, Edinburgh, 1877, p. vii. C. G. Knott, *Life and Scientific Work of Peter Guthrie Tait*, Cambridge, 1911, p. 226.

22 I should like to thank Professor D. S. L. Cardwell for his insistence on this point. A number of nineteenth-century scientists, including the German physicist Clausius, accepted it as Gay-Lussac's law.

23 *J. de phys.*, 59 (1804), 320.

24 Brissot de Warville, *Un mot à l'oreille des Académiciens de Paris*, [1786], p. 11.

25 *J. de phys.*, 59 (1804), 314–20.

26 *Ibid.*, pp. 454–62.

27 The new record was claimed by Barral and Bixio. The altitudes are stated as they were recorded but it is doubtful if they are accurate to the nearest metre.

28 Berthollet described the accident in a letter to Blagden (Royal Society, BLA B133, 31 August 1808). Gay-Lussac went back to Saint Léonard to recuperate.

29 *Recherches physico-chimiques*, Paris, 1811, vol. 1, p. 46.

30 M. E. Weeks, *Discovery of the Elements*, 5th edn, Easton, Pa, 1948, p. 457.

31 *A.c.*, 91 (1814), 7n.

32 For a fuller account of the Arcueil group see the author's *The Society of Arcueil*, 1967.

33 Vol. 3 of the Arcueil *Mémoires* (1817) also lists Chaptal and Poisson as members but neither had a very long or a very close association with Arcueil.

34 For a list of these additional memoirs see the introduction to the modern reprint of the *Mémoires* of the Society of Arcueil (Johnson Reprint, New York and London, 1967, vol. 1, p. xlvi).

35 *M.S.A.* 1 (1807), 182.

36 For an account of this research see pp. 119–20 (chapter 6).

37 Laplace, *Système du monde*, 3rd edn, Paris, 1808, p. 369.

38 See chapter 3.

39 Ecole Polytechnique, *Conseil de l'Instruction et d'Administration l'an 10 à l'an 12*, f. 17, 21 vendémiaire an 11 (13 October 1802).

40 A *répétiteur* was so called because he had the duty of going over or *repeating* with students the work covered in lectures.

41 The delay in filling posts was probably due to the reorganisation of the school on a military basis in the summer of that year.

42 Archives de l'Ecole Polytechnique, 1794, dossier Berthollet.
43 Gay-Lussac to Chamberet, 30 April 1806. Wellcome MS No. 1568.
44 12 November 1806. Archives de l'Ecole Polytechnique, 1806, box 4. I owe this reference to Margaret Bradley
45 Thenard was also Fourcroy's demonstrator at the Athénée de Paris until 1808.
46 Cornell University.
47 *Dizionario portabile e di pronunzia Francese–Italiano e d'Italiano–Francese*, Lione, anno 11, 1802.
48 *Nouveau dictionnaire francois–allemand et allemand–francois*, 6th edn, 2 vols., Strasbourg and Paris, 1804.
49 E.g. *P.V. Inst.*, 5, *17* (17 February 1812).
50 *M.S.A.*, 1 (1807), 136.
51 Appendix, letter 9.
52 *M.S.A.*, 1 (1807), 1–22.
53 Wellcome MS No. 67799. Letter of 14 September 1808. The recipient is not named.
54 Ecole Polytechnique, *Registre du Conseil de Perfectionnement*, 1806–11, pp. 57–8.
55 Gay-Lussac's acceptance of multiple appointments is discussed on p. 228–33.
56 Letter from Gay-Lussac to his father, 15 January 1803 (Appendix, letter 4).
57 Letter from Gay-Lussac to his father, 17 January 1805 (Appendix, letter 6).
58 A Will of Madame Gay-Lussac, dated 22 January 1828, begins: 'N'ayant rien apporté en mariage'.
59 Arago, *Oeuvres, Notices biographiques*, vol. 3, p. 60.
60 Madame Gay-Lussac to J. B. Chamberet, 30 November 1809 (Wellcome MS Gay-Lussac No. 1559).
61 Gay-Lussac to his wife, 27 August [1818].
62 I.e. material for the *Annales de chimie et de physique*, of which Gay-Lussac was joint editor.
63 Gay-Lussac to his wife, 5 September [1818?].
64 Martha Somerville, *Personal Recollections...of Mary Somerville*, London, 1873, p. 187.

## Chapter 3

1 *A.c.p.*, 2 (1816), 130.
2 *Traité élémentaire de chimie*, 3rd edn, 2 vols., Paris, 1801. Also in Gay-Lussac's library were Lavoisier's *Opuscules physiques et chimiques*, 2nd edn, Paris, 1801, and Lavoisier's *Mémoires de chimie*, 2 vols., Paris, n.d. [1803?].
3 See chapter 7.
4 *Elements of Chemistry* [1790], trans. R. Kerr, Dover edn, New York, 1965, p. 245. *Traité élémentaire de chimie*, Paris, 1789, p. 267.

5 *A.c.*, 69 (1809), 204–20.
6 *M.S.A.*, 2 (1809), 317n.
7 Lavoisier insisted on both analysis and synthesis – see e.g. Kerr, *Elements*, p. 33.
8 *Ibid.*, p. 241.
9 *J. de phys.*, 70 (1810), 257–66. The outline above applies to vegetable substances. With animal substances, nitrogen was evolved and required separate estimation.
10 *M.S.A.*, 1 (1807), 379.
11 'Mémoire sur le rapport qui existe entre l'oxidation des métaux et leur capacité de saturation pour les acides', *M.S.A.*, 2 (1809), 159–75. See especially pp. 160, 166.
12 *Cours de chimie*, Paris, 1828, vol. 1, lecture 1, p. 18.
13 See pp. 131–3.
14 *A.c.p.*, 11 (1819), 297.
15 1801 edn, vol. 2, p. 39.
16 *A.c.p.*, 33 (1826), 22–3.
17 *A.c.p.*, 34 (1827), 93–4.
18 'Note sur l'acide prussique', *A.c.*, 77 (1811), 128–33.
19 'Recherches sur l'acide prussique', *A.c.*, 95 (1815), 136–231.
20 *M.S.A.*, 2 (1809), 357–8; *A.C.R.*, Edinburgh, No. 13, p. 48.
21 *A.c.*, 91 (1814), 97; *Annals of Philosophy*, 6 (1815), 124.
22 See pp. 78–9.
23 *A.c.p.*, 70 (1839), 427.
24 Lavoisier, *Oeuvres*, 6 vols., Paris, 1862–93, vol. 2, p. 550.
25 *Ibid.*, p. 197.
26 *Mécanique céleste*, vol. 4, Paris, 1805, 'Supplément à la théorie de l'action capillaire', p. 52.
27 *Ibid.*, p. 76.
28 *Ibid.*, p. 1.
29 *Ibid.*, pp. 56–9.
30 *Système du monde*, 3rd edn, 1808, p. 316.
31 *Ibid.*
32 *A.c.p.*, 91 (1814), 32–34n.
33 Laplace had recommended 'de s'attacher à déterminer par de nombreuses expériences, les loix des affinités; et pour y parvenir le moyen qui paroit le plus simple, est de comparer ces forces, à la force répulsive de la chaleur, que l'on peut comparer elle-meme à la pesanteur'. (*Système du monde*, 2nd edn, 1799, p. 287.)
34 This result can be understood from the following calculation:

Let $D$ = density of the earth, $d$ = density of Laplace's particle, $R$ = radius of earth = $6.37 \times 10^8$ cm, $r$ = radius of Laplace's particle = $10^{-4}$ cm, $M$ = mass of earth, $m$ = mass of Laplace's particle.

Force of attraction between a particle of mass $m'$ and earth is given by inverse square law of gravitation, i.e.

$$F = \frac{GMm'}{R^2}.$$

NOTES TO PAGES 52–7

Laplace thought it was reasonable to assume that his particle might exert the same force as the earth at its surface on a particle of mass $m'$, i.e.

$$F = \frac{Gmm'}{r^2}.$$

$$\therefore \frac{GMm'}{R^2} = \frac{Gm'm}{r^2}, \text{ by cancelling} \therefore \frac{M}{R^2} = \frac{m}{r^2} \text{ or } \frac{m}{M} = \frac{r^2}{R^2}.$$

Now consider the densities of the earth and of Laplace's particle:

$$\text{density} = \frac{\text{mass}}{\text{volume}} \qquad \therefore \text{ for the earth, } D = \frac{M}{4/3\pi R^3}$$

$$\text{and for the particle, } d = \frac{m}{4/3\pi r^3}$$

$$\therefore \frac{d}{D} = \frac{m}{M} \times \frac{R^3}{r^3} = \frac{r^2}{R^2} \times \frac{R^3}{r^3} = \frac{R}{r} = \frac{6.37 \times 10^8}{10^{-4}} = 6.37 \times 10^{12}$$

35 Berthollet insisted that chemical compounds, like alloys, could be formed by elements combining in a whole range of possible proportions.

36 *A.C.R.*, Edinburgh, No. 4, p. 8.

37 On 6 July 1807 Laplace had presented to the First Class of the Institute his 'Supplément à la théorie de l'action capillaire' (usually found bound at the end of vol. 4 of the *Mécanique céleste*) in which he described the latest work carried out on capillary attraction with the help of Gay-Lussac and announced his intention 'de mettre de plus en plus en évidence l'identité des forces attractives dont cette action dépend, avec celles qui produisent les affinités' (p. 1).

38 Thomas Thomson, *Système de chimie*, 9 vols., Paris, 1809, vol. 1, Introduction, p. 10.

39 *A.c.p.*, 2 (1816), 130.

40 *J. de phys.*, 59 (1804), 461; 60 (1805), 154.

41 *J. de phys.*, 59 (1804), 317.

42 *J. de phys.*, 60 (1805), 145.

43 *J. de phys.*, 60 (1805), 134 (my italics).

44 J. A. Chaptal, *Elements of chemistry*, 2nd edn, trans. W. Nicholson, London, 1795, vol. 1, p. xliii.

45 *M.S.A.*, 1 (1807), 181.

46 *M.S.A.*, 1 (1807), 200, 202.

47 Laplace, *Mécanique céleste*, vol. 4, 1805, 'Supplément à la théorie de l'action capillaire', [?1807], p. 39.

48 *J. de phys.*, 59 (1804), 456.

49 However he sometimes used the terms 'principe' and 'loi' as synonymous, e.g. *M.S.A.*, 2 (1809), 160.

50 'M. Gay-Lussac a lu à la Société, le 12 juin 1807, une note dans laquelle il annonce qu'en comparant la pesanteur spécifique des corps avec leur capacité de saturation, il a cru reconnoitre ce principe: que plus un corps a de pesanteur spécifique, moins il a de capacité de

saturation', etc. However, Gay-Lussac did not wish to present these 'principles' as 'verités incontestables'. Note, *M.S.A.*, 1 (1807), 379.

51 *M.S.A.*, 2 (1809), 166. This 'law' can be stated in various ways, see e.g. *ibid.*, 160.
52 *A.c.*, 91 (1814), 131ff.
53 See p. 132.
54 *Recherches physico-chimiques*, vol. 1, pp. 17–19.
55 *Ibid.*, p. 27 (my italics).
56 *Ibid.*, p. 31.
57 *A.c.p.*, 2 (1816), 130 (my italics).
58 *J. de phys.*, 60 (1805), 148.
59 *M.S.A.*, 1 (1807), 218.
60 I. Freund, *The Study of Chemical Composition* (1904), Dover edn, New York, 1968, p. 303.
61 'Mémoire sur la combinaison des substances gazeuses, les unes avec les autres', *M.S.A.*, 2 (1809), 207–34, translated in *Foundations of the Molecular Theory*, A.C.R., No. 4, Edinburgh, pp. 8–24 (p. 24).
62 Boyle's law too may be deduced from first principles but this modern exercise has nothing to do with the development of ideas on air in the seventeenth century which led Towneley, Boyle and Mariotte successively to the law relating the volume of air and its pressure.
63 'Mémoire sur l'acide fluorique' (read 23 January 1809), *A.c.*, 69 (1809), 204–24. Gay-Lussac and Thenard had hoped to isolate fluorine by preparing pure hydrofluoric acid and then decomposing it with potassium. Although this was not possible, a preliminary experiment in which calcium fluoride was heated with boracic acid produced their 'fluoric gas'.
64 W. H. Wollaston, 'On super-acid and sub-acid salts' (read 28 January 1808), *Phil. Trans.*, 98 (1808), 96, 102; Thomas Thomson, 'On oxalic acid' (read 14 January 1808), *ibid.*, 63–95.
65 Thomson, *Système de chimie*, 9 vols., Paris, 1809, vol. 1. Gay-Lussac refers to Berthollet's Introduction to Thomson's book in his memoir (*A.C.R.*, No. 4, Edinburgh, p. 9).
66 Berthollet received Dalton's *New System* in August. The Introduction was finished during that autumn but a further part was added after 15 November.
67 *A.C.R.*, No. 4, Edinburgh, p. 23.
68 Thomson, *Système de chimie*, vol. 1, p. 22 (my italics). In 1809 Berthollet corrected an obvious mistake in this passage. The reference to 'double the volume' should read: 'a volume equal to that' (*M.S.A.*, 2 (1809), 470). At the same time he repeated experiments which he had carried out as a direct outcome of Wollaston's work. This increases the probability that Gay-Lussac too was directly influenced by Wollaston's paper.

The combining volumes of ammonia and carbonic acid gas (given as 100, 275) were mentioned in a paper read by Berthollet to the

Institute in March 1806 (*Mémoires de la Classe des Sciences Mathématiques et Physiques de l'Institut National de France*, **7**, part 1 (1806), 236).

69 *Cours de chimie*, 1828, vol. 1, lecture 1, p. 12. This view of the theory was very similar to that current in England in the 1820s, e.g. as expressed by Humphry Davy.

70 *A.c.*, **74** (1810), 57; *Phil. Mag.*, **38** (1811), 64–5.

71 *Leçons de physique*, Paris, 1828, vol. 1, p. 7.

72 *A.c.p.*, **70** (1839), 444.

73 E. Verdet, *Notes et mémoires*, Paris, 1872, pp. xxv–xxvi.

74 M. Daumas, who remarked that Dulong was similar in temperament to Gay-Lussac, considered the latter generally superior – 'son intelligence scientifique semble plus achevée, plus accomplie' ('Gay-Lussac', *Revue d'histoire des sciences*, **3** (1950), 338).

75 *M.S.A.*, **2** (1809), 252–3. The related memoir is printed on pp. 207–34.

76 E.g. *Recherches physico-chimiques*, 1811, vol. 2, pp. 259–64.

77 *Leçons de physique*, vol. 1, lecture 26, p. 397.

78 *Cours de chimie*, vol. 1, lecture 4, p. 24.

79 *Ibid.*, p. 25.

80 *A.c.p.*, **11** (1819), 314–15; *Annals of Philosophy*, **15** (1820), 11 (my italics).

81 *Instruction pour l'usage de l'alcoomètre centesimal et des tables qui l'accompagnent*, Paris, 1824.

82 'Premier mémoire sur la dissolubilité des sels dans l'eau', *A.c.p.*, **11** (1819), 297.

83 *A.c.* **91** (1814), 128; *Annals of Philosophy*, **6** (1815), 186.

84 'Memoir on the combination of gaseous substances', [*M.S.A.*, **2** (1809), 230] *A.C.R.*, Edinburgh, No. 4, p. 22.

85 *A.c.*, **91** (1814), 96; *Annals of Philosophy*, **6** (1815), 124.

86 *Ibid.*, 128–9.

87 G. L. Armoire: 'Classification'.

88 G. L. Armoire: Corps simples, e.g. '3ᵉ leçon, 13 mars [18]47. Groupe des fluorides: fluore, chlore, brome, iode'. Some sketchy information on Gay-Lussac's early classification of elements is given by Ampère in his 'Essai d'une classification naturelle pour les corps simples'. See *A.c.p.*, **1** (1816), 296.

89 *Bulletin de la Société chimique*, **42** (1884), 134.

90 *A.c.p.*, **11** (1819), 297.

91 *A.c.p.*, **2** (1816), 130.

92 P. Achinstein, in a study using previous research by the present writer, characterised Gay-Lussac's reasoning as inductive (*Law and Explanation*, Oxford, 1971, p. 149).

93 *M.S.A.*, **1** (1807), 239.

94 *J. de phys.*, **60** (1805), 130.

95 This is a more sophisticated view of phenomena than is given by Lavoisier in the preface to his *Traité élémentaire*, where he seems to

consider facts as individual units, rather like, say, pebbles on a beach, which are waiting for the scientist to pick them up.

96 *Leçons de physique*, vol. 1, p. 7.

## Chapter 4

1 Speech in Chambre des Pairs, 27 March 1843 (*Moniteur*, 1843, p. 584.)

2 *M.S.A.*, 1 (1807), 21-2.

3 Another useful comparison is with John Dalton who was, however, some eight years Gay-Lussac's senior (1766-1844).

4 Correspondence is published in Berzelius, *Bref*, Uppsala, 1912, vol. 3, part 7, pp. 113-26.

5 *A.c.p.*, 2 (1816), 417.

6 Royal Institution, Davy notebook 69, box 14, 14i, p. 39.

7 The first chemist to prepare nitrous oxide was Priestley.

8 *Researches, chemical and philosophical chiefly concerning Nitrous Oxide*, 1800, p. xv.

9 *Ibid.*, pp. 56, 107.

10 *Ibid.*, pp. 140-1, 150-1.

11 *Bibliothèque britannique, Sciences et Arts*, 19 (year 10), 43-66, 141-54, 321-39; 20 (year 10), 27-48, 217-46, 346-76.

12 *A.c.*, 44 (1802), 43-4.

13 Royal Institution, Davy notebook 13i, p. 156.

14 *P.V. Inst.*, 3 (1808), 626-30.

15 *M.S.A.*, 2 (1809), 299-301.

16 H. Davy, *Collected Works*, ed. John Davy, 9 vols., London, 1839-40, vol. 5, p. 101n.

17 *M.S.A.*, 2 (1809), 311-16; trans. in M. E. Weeks, *Discovery of the Elements*, 5th edn, Easton, Pa, 1945, pp. 339-40.

18 'An account of some new analytical researches...', *Works*, vol. 5, pp. 140-204.

19 *Nouveau Bulletin de la Société Philomatique*, 1808, no. 10 (July), p. 173.

20 Davy, *op. cit.*, p. 177.

21 J. R. Partington, *History of Chemistry*, vol. 4, 1964, p. 49.

22 *Elements of chemical philosophy*, vol. 1, part 1, London, 1812, pp. 472-3.

23 Davy, *Works*, vol. 5, p. 285 (my italics). The full passage from which the above is taken also mentions 'muriatic acid' and makes the claim that Davy had previously referred to this experiment. I have been unable to find any such published reference.

24 *Ibid.*, p. 345.

25 *M.S.A.*, 3 (1817), 604.

26 *Traité de chimie*, 1st edn, vol. 1, 1813, p. 562n.

27 *A.c.*, 91 (1814), 96.

28 J. A. Paris, *The Life of Sir Humphry Davy*, London, 1831, vol. 1, p. 332.
29 Davy, *Works*, vol. 5, p. 503.
30 *Ibid.*, p. 3n.
31 Royal Institution, Davy, box 3, folder 2, lecture 5: 'Electrochemical science'.
32 Paris, *The Life of Sir Humphry Davy*, vol. 2, p. 18.
33 *Ibid.*, p. 22.
34 *Moniteur*, 2 December 1813, p. 1344.
35 For details of elections in the First Class of the Institute, see *P.V. Inst.*, 5, e.g. 270–1.
36 Despite the war with Britain Davy had been given special permission by the French government to travel through France.
37 H. Davy, *Fragmentary Remains*, ed. J. Davy, 1858, p. 186 (letter of 18 March 1814 to John Davy).
38 Paris, *The Life of Sir Humphry Davy*, vol. 2, p. 10.
39 *Ibid.*, p. 27.
40 M. Faraday, *Life and Letters*, ed. Bence Jones, London, 1870, vol. 1, p. 95.
41 In his published research Davy explained in some detail how he came to be doing the same work as Gay-Lussac (Davy, *Works*, vol. 5, pp. 439–40).
42 Davy, *Fragmentary Remains*, p. 186.
43 Most recently by June Z. Fullmer, 'Davy's priority in the iodine dispute: further documentary evidence', *Ambix*, 22 (1975), 39–51.
44 Jacques Payen suggests that (untypically) Berthollet 'harboured a lasting grudge against Clément'. See article 'Clément' in C. C. Gillispie (ed.), *Dictionary of Scientific Biography*, New York, 1970–, vol. 3, p. 316.
45 For a full discussion of the prize see R. Fox, *The Caloric Theory of Gases from Lavoisier to Regnault*, Oxford, 1971, pp. 134ff.
46 '...d'après l'invitation de M. Clément, son ami' (*Moniteur*, 1813, p. 1384).
47 Thus Gay-Lussac gave the substance a name before Davy. Davy's first use of the term 'iodine' occurs in his paper read to the Royal Society in January 1814 (Davy, *Works*, vol. 5, p. 454).
48 Davy, *Works*, vol. 5, p. 512. Davy seemed anxious to associate Gay-Lussac with Clément's early unsucessful work on 'X' and stated that they worked together (*ibid.*, pp. 439, 512n). The printed account in the *Moniteur* (1813, p. 1344) of Clément's research made no mention of Gay-Lussac.
49 'Je lui avois surtout dit qu'on formoit un acide particulier avec l'hydrogène.'
50 The full text of Davy's long letter was soon published in French journals: *J. de phys.*, 77 (1813), 456–60; *A.c.*, 88 (1813), 322–9. It is discussed in Mrs Fullmer's article, *Ambix*, 22 (1975), 39–51.
51 Originally the letter was dated 10 December but Davy changed this

to '11 December'. The date 10 December, however, remained in the version published in *Phil. Trans.*, 104 (1814), 74–93.

52 The original letter in the Institute archives is endorsed by Cuvier 'reçu le 13 décembre 1813'. The Academy of Sciences recognised the practice of depositing a sealed note with the secretary to secure priority. The date registered, however, was always the date of *receipt* by the secretary.

53 *Moniteur*, 1813, pp. 1384–5.

54 Gay-Lussac had actually suggested *iode* from the Greek ιον (violet) from the colour of its vapour. Davy's variation, *ione*, is closer to the Greek but it does not alter the historical point.

55 *A.c.*, 91 (1814), 252–72.

56 *A.c.*, 90 (1814), 87–100.

57 *Ostwald's Klassiker der Exakten Wissenschaften Nr. 4, Untersuchungen über das Jod von Gay-Lussac (1814)*, Leipzig, 1889, pp. 50–1.

58 *Annalen der Physik*, 49 (1815), 2.

59 John Davy, *Memoirs of the Life of Sir Humphry Davy*, London, 1836, vol. 1, pp. 464–5.

60 *A.c.*, 91 (1814), 8; *Annals of Philosophy*, 5 (1815), 102.

61 *Annals of Philosophy*, 5 (1815), 211.

62 *Ibid.*, p. 109.

63 None of the early values obtained by either Gay-Lussac or Davy for the equivalent of iodine was at all accurate.

64 The anhydride of iodic acid has the formula $I_2O_5$.

65 *M.S.A.*, 1 (1807), 202–3.

66 *A.C.R.*, No. 13, Edinburgh, p. 48 (my italics).

67 *Moniteur*, 12 December 1813, p. 1385 (my italics).

68 The work of Arago shows an interesting similarity to this attitude. Dr John Cawood has remarked on 'Arago's passion for [deliberate] unanalysed observation', e.g. in his meteorological studies. See A. J. Cawood, 'The scientific work of D. F. J. Arago (1786–1853)', Ph.D. thesis, University of Leeds, 1974.

69 Davy, *Works*, vol. 5, p. 57.

70 *Ibid.*, pp. 49–56, see especially p. 54.

71 'An account of some new analytical researches on the nature of certain bodies', *ibid.*, pp. 140–204. Gay-Lussac and Thenard were of course equally wrong in thinking that potassium contained hydrogen.

72 *A.c.p.*, 70 (1839), 427.

73 See D. M. Knight, *Studies in Romanticism*, 6 (1967), 74.

74 *A.c.*, 95 (1815), 156; *Annals of Philosophy*, 7 (1816), 357.

## Chapter 5

1 'Chemical observations on certain atomic weights as adopted by different authors', quoted by A. Thackray, *John Dalton*, Cambridge, Mass., 1972, p. 101.

2 The names Ampère and of Dumas must also be included in even the

briefest sketch of the history of the molecular theory. Ampère's key memoir, which takes Gay-Lussac's work as a starting point and which demands further study is in *A.c.*, 90 (1814), 43–86. An analysis of Dumas' atomic theory is given in: L. J. Klosterman, 'Studies in the life and work of J. B. Dumas', Ph.D thesis, University of Kent, 1976.

3 *Lettres de M. Euler à une Princesse d'Allemagne*, 3 vols., Paris, 1787. See especially letters 45, 69.

4 Euler, *op. cit.*, 2nd edn, 2 vols., 1806, vol. 1, p. 2.

5 *Leçons de physique*, vol. 1, p. 10.

6 *Ibid.*, p. 29.

7 *Essai de statique chimique*, vol. 1, p. 18.

8 *Ibid.*, pp. 245–7.

9 *Ibid.*, pp. 522–3.

10 *A.c.*, 43 (1802), 160.

11 'Expériences sur les moyens eudiométriques et sur la proportion des principes constituans de l'atmosphère', *J. de phys.*, 60 (1805), 129–68.

12 *Phil. Mag.*, 38 (1811), 60–8.

13 *Cours de chimie*, lecture 22, p. 26.

14 *Phil. Mag.*, 38 (1811), 64.

15 *M.S.A.*, 2 (1809), 218; *A.C.R.*, No. 4, Edinburgh, p. 15.

16 *M.S.A.*, 2 (1809), 233; *A.C.R.*, No. 4, Edinburgh, p. 24.

17 *M.S.A.*, 2 (1809), 207–8; *A.C.R.*, No. 4, Edinburgh, p. 8.

18 M. P. Crosland, 'The origins of Gay-Lussac's law of combining volumes of gases', *Annals of Science*, 17 (1961), 1–26.

19 *Annals of Philosophy*, 6 (1815), 188.

20 *A.C.R.*, Edinburgh, No. 4, p. 24 (my italics).

21 An advocate of the atomic theory would have referred to the high atomic weight of iodine (I = 127).

22 *A.c.*, 91 (1814), 17–18; *Annals of Philosophy*, 5 (1815), 105.

23 Marcel Oswald, *L'Evolution de la chimie au XIXᵉ siècle*, Paris, 1913, p. 39. Any statement about a 'first' is usually open to challenge. In this case we see Gay-Lussac's work as an extension of that of Berthollet and Laplace.

24 *A.C.R.*, Edinburgh, No. 4, p. 15.

25 'Proportion de plusieurs composés dont les élémens sont gazeux', *M.S.A.*, 2 (1809), 253.

26 The modern value is 0.9670 at 0°C (Leduc).

27 The modern value is 0.5971 (Leduc).

28 *A.C.R.*, No. 4, Edinburgh, p. 16. The modern equation is:
$$2C + O_2 = 2CO$$
1 volume   2 volumes

29 E.g. in the formation of water vapour 3 volumes are reduced to 2; in the formation of ammonia the contraction is 50 per cent.

30 Partington, *History of Chemistry*, vol. 4, pp. 160–5.

31 *A.C.R.*, Edinburgh, No. 4, p. 22 (my italics).

32 Lavoisier, *Elements of Chemistry*, trans. R. Kerr, p. 27.

33 *A.c.*, 91 (1814), 16–17; *Annals of Philosophy*, 5 (1815), 105.

34 *A.c.*, 91 (1814), 24–5; *Annals of Philosophy*, 5 (1815), 108.
35 *A.c.*, 91 (1814), 30; *Annals of Philosophy*, 5 (1815), 208. The term 'nitrogen' has been substituted for 'azote' in the original translation.
36 Since

$$2 \text{ volumes H} + 1 \text{ volume O} \rightarrow \text{water below } 100°\text{C}$$
$$50 \qquad\qquad 25 \qquad\qquad \text{(contraction of 75;}$$
$$\text{of negligible volume, therefore)}$$

37 A modern representation of the reaction would be:

$$2HCN + 5CuO \rightarrow 2CO_2 + H_2O + N_2 + 5Cu$$
$$2 \text{ vols.} \qquad\qquad 1 \text{ vol.}$$

38 *Annals of Philosophy*, 2 (1813), 430.
39 The difference between such compounds was later explained in terms of two quite different types of bond: sodium iodide and ethyl iodide, ionic and covalent.
40 *Traité de physique*, Paris 1816, vol. 1, pp. 291–9, also plate III.
41 *Ibid.*, p. 297.
42 This calculation was checked by Dr Klosterman who kindly showed me that, from the data supplied, Gay-Lussac should have obtained a value for the vapour density of 2.488. Perhaps, after seeing the close agreement with theory, Gay-Lussac did not bother to check his calculation.
43 *A.c.*, 95 (1815), 315. See Partington, *History of Chemistry*, vol. 4, p. 341.
44 *Annals of Philosophy*, 2 (1813), 359n.
45 *Annals of Philosophy*, 2 (1813), 450.
46 *A.c.p.*, 50 (1832), 172.
47 'Leur composition réelle', *A.c.p.*, 33 (1826), 339.
48 J. B. Dumas, *Leçons sur la philosophie chimique*, Paris, 1837, lecture 7, pp. 258–90.
49 Liebig, *Introduction to the First Elements of Chemistry*, translated from German by Thomas Richardson, London, 1837, p. 72. The translator said that his book 'had been submitted to the kind inspection of Professor Liebig'.
50 *Ibid.*, pp. 81–2.
51 H. Sainte-Claire Deville and L. Troost, 'Mémoire sur les densités de vapeurs à des temperatures très-élevées', *A.c.p.*, 58 (1860), 257–99.
52 *Ibid.*, p. 282.
53 *A.c.p.*, 60 (1835), 113–51.
54 *Précis de chimie organique*, Paris, 1844, vol. 1, p. 47.
55 For some misuse in France of 'the great law of Gay-Lussac' see P. Colmant, 'Querelle à l'Institut entre équivalentistes et atomistes', *Revue des questions scientifiques*, 143 (1972), 493–519, e.g. p. 150.
56 Fox, *The Caloric Theory of Gases*, p. 319.
57 *A.C.R.*, No. 18, Edinburgh.
58 *A.C.R.*, No. 20, Edinburgh, p. 25.
59 *New System of Chemical Philosophy*, vol. 1, Part 2, Manchester, 1810, p. 559 (my italics).
60 *Quarterly Journal of Science*, 16 (1823), 340.

## Chapter 6

1 *A.c.p.*, 26 (1824), 174.

2 On fig. 2 joint papers by Gay-Lussac and another person are counted as one half. Numbers are based on the list published in the *Royal Society Catalogue of Scientific Periodicals*, 1868–72, vol. 2, pp. 800–7.

3 Archives nationales, F¹⁷, 1933.

4 Lavoisier, *Oeuvres*, vol. 2, p. 285. Lavoisier, however, had decided by 1789 that caloric was a substance since he included it in his list of elements.

5 *Essai de statique chimique*, vol. 1, pp. 171–2.

6 *Leçons de physique*, vol. 1, p. 240.

7 'Premier essai pour déterminer les variations de température qu'-éprouvent les gaz en changeant de densité, et considérations sur leur capacité pour le calorique', *M.S.A.*, 1 (1807), 180–203.

8 This research was cited by Julius Robert Mayer in 1845. It was, however, unknown to J. P. Joule who did a similar experiment in 1845. See D. S. L. Cardwell, *From Watt to Clausius*, London, 1971, pp. 230–4.

9 'avec la plus grande réserve' (*M.S.A.*, 1 (1807), p. 202).

10 *Ibid.*, p. 190.

11 Fox, *The Caloric Theory of Gases*, p. 132.

12 Fox (*ibid.*, p. 135) suggests the subject as something of a possible crucial experiment.

13 'Extrait d'un mémoire sur la capacité des gaz pour le calorique', *A.c.*, 81 (1812), 98–108.

14 'Note sur la capacité des fluides élastiques pour le calorique', *A.c.*, 83 (1812), 106–8.

15 See Laplace, *Mécanique céleste*, vol. 4, pp. xxi–xxii.

16 *P.V. Inst.*, 5, 514–17 (5 June 1815).

17 'Extrait d'un mémoire sur le froid produit par l'évaporation des fluides', *A.c.p.*, 21 (1822), 82–92.

18 'Sur le froid produit par la dilatation des gaz', *A.c.p.*, 9 (1818), 305–10.

19 'Détermination expérimentale du zéro absolu de la chaleur et du calorique spécifique des gaz', *J. de phys.*, 89 (1818), 321–46.

20 For a full discussion of the caloric theory of gases see Fox, *The Caloric Theory of Gases*.

21 *M.S.A.*, 1 (1807), 184–5.

22 'Sur le calorique du vide', *A.c.p.*, 13 (1820), 304–8.

23 P. Costabel, 'Le "calorique du vide" de Clément et Desormes (1812–1819)', *Archives internationales d'histoire des sciences*, 21 année, 82–3 (1968), 1–14.

24 *A.c.p.*, 19 (1821), 436–7.

25 *A.c.p.*, 20 (1822), 267–8. The modern value is 1.42. This research was

quoted by Sadi Carnot in his famous *Réflexions sur la puissance motrice du feu* (1824). See *Reflections on the Motive Power of Fire*, ed. E. Mendoza, New York, 1960, e.g. pp. 16n, 30n.

26 For further scientific information see B. S. Finn, 'Laplace and the speed of sound', *Isis*, 55 (1964), 7–19.

27 *A.c.p.*, 13 (1820), 304 (my italics).

28 'Hypothèses sur la nature du calorique', *Leçons de physique*, vol. 1, pp. 242–3.

29 Theodore M. Brown, 'The electric current in early nineteenth-century French physics', *Historical Studies in the Physical Sciences*, 1 (1967), 61–103.

30 'Expériences sur la torpille', *A.c.*, 56 (1805), 15–23.

31 'De l'action chimique du fluide galvanique', *Nouveau bulletin des sciences par la société philomathique*, 1, no. 4 (January 1808), 71–6.

32 *Ibid.*, No. 6 (March 1808), 110.

33 Gay-Lussac and Thenard, *Recherches physico-chimiques*, vol. 1, p. 12.

34 'Sur les piles sèches voltaïques', *A.c.p.*, 2 (1816), 76–86.

35 'Gay-Lussac fut physicien ingénieux, chimiste hors de ligne.' Arago, *Oeuvres, Notices biographiques*, vol. 3, p. 69.

36 Biot, *Mélanges scientifiques et litteraires*, 1858, vol. 3, p. 130.

37 *A.c.p.*, 2 (1816), 130.

38 *A.c.p.*, 1 (1816), 114.

39 W. E. K. Middleton, *The History of the Barometer*, London, 1964, pp. 140–2.

40 *A.c.p.*, 8 (1818), 159.

41 *Cours de chimie*, vol. 1, 1828, lecture 3, p. 6.

42 *A.c.p.*, 70 (1839), 431.

43 *M.S.A.*, 2 (1809), 348–51.

44 Partington, *History of Chemistry*, vol. 4, p. 85.

45 E. von Meyer, *History of Chemistry*, trans. G. McGowan, 2nd edn, London, 1898, p. 249.

46 *Journal of Science and the Arts*, 1 (1816), p. 289 (Davy, *Works*, vol. 5, pp. 517–18).

47 'Recherches sur l'acide prussique', *A.c.*, 95 (1815), 136–231; translated in *Annals of Philosophy*, 7 (1816), 350–64; 8 (1816), 37–52, 108–15.

48 *Essai de statique chimique*, vol. 2, p. 267.

49 *A.c.*, 77 (1811), 128–33. In contemporary language this was the action of muriatic acid on common prussiate of mercury.

50 Gay-Lussac gave $C = 44.39\%$, $N = 51.71\%$, $H = 3.90\%$;
    Porrett gave $C = 24.80\%$, $N = 40.70\%$, $H = 34.50\%$.

51 *A.c.*, 95 (1815), 161; *Annals of Philosophy*, 7 (1816), 360 (my italics).

52 *A.c.*, 95 (1815), 227; *Annals of Philosophy*, 8 (1816), 113 (my italics).

53 *A.c.p.*, 2 (1816), 12.

54 *Traité de chimie organique*, vol. 1, Paris, 1850, p. 123.

55 *C.R.*, 5 (1837), 567–72.
56 *A.c.*, 91 (1814), 130–52: 'Sur l'acidité et sur l'alcalinité', trans. in *Annals of Philosophy*, 6 (1815), 187–95.
57 *A.c.*, 91 (1814), p. 117 (my italics).
58 *A.c.*, 88 (1813), 318.
59 *A.c.*, 91 (1814), 147–8.
60 *Ibid.*, p. 145.
61 *A.c.*, 91 (1814), 149n. For Davy's use of a structural theory of acidity see M. Crosland, 'Theories of acidity in the nineteenth century', *Proceedings of XIIIth International Congress of the History of Science*, Moscow, 1971.
62 *A.c.p.*, 2 (1816), 130.
63 *A.c.p.*, 1 (1816), 166.
64 *A.c.p.*, 27 (1824), 200n.
65 *A.c.*, 91 (1814), 149n. W. V. Farrar, 'Dalton and structural chemistry', in *John Dalton and the Progress of Science*, ed. D. S. L. Cardwell, Manchester, 1968, pp. 290–9.
66 *Recherches physico-chimiques*, vol. 2, pp. 289–90, 340.
67 '[Ils] ont leurs molécules dans des situations diverses' (*Cours de chimie*, vol. 1, lecture 16, p. 26).
68 *Ibid.*, vol. 2, lecture 24, pp. 23, 16.
69 *Ibid.*, vol. 2, lecture 33, p. 13.
70 *Ibid.* vol. 2, lecture 32, pp. 16–17.
71 A. Thackray, *Atoms and Powers*, Cambridge, Mass., 1970, p. 276.
72 *Essai de statique chimique*, vol. 1, p. 499.
73 Thomson, *Système de chimie*, vol. 1, pp. 20–1.
74 *Ibid.*, p. 27.
75 *A.C.R.*, No. 4, Edinburgh, p. 9; *M.S.A.*, 2 (1809), 209.
76 *M.S.A.*, 2 (1809), 209n.
77 *Ibid.*, p. 231.
78 *Ibid.*, p. 232.
79 *Cours de chimie*, vol. 1, lecture 1, pp. 10–15.
80 *Cours de chimie*, vol. 2, lecture 24, pp. 9–10.
81 Thackray, *John Dalton*, p. 101.
82 H. Guerlac, 'Some Daltonian doubts', *Isis*, 52 (1961), 544–54.
83 *A.c.p.*, 70 (1839), 416–17.
84 *Cours de chimie*, vol. 1, lecture 3, p. 23.
85 *Ibid.*, lecture 1, p. 13.
86 *Ibid.*, p. 23.
87 *P.V. Inst.*, 8, 485–7.
88 'M. Dalton a été conduit a cette idée par des considerations systématiques' (*M.S.A.*, 2 (1809), 209n).
89 C. Appert, *L'Art de conserver pendant plusieurs années toutes les substances animales et végétales*, Paris, 1810.
90 'Sur la fermentation', *A.c.*, 76 (1810), 245–60. The translation used is from Nicholson's *Journal of Natural Philosophy and the Arts*, 31 (1812), 249–56 (252).

91 Nicholson's *Journal*, 31 (1812), 251.

92 *Ibid.*, p. 254.

93 *Etudes sur le vin*, 1866, part 2. 'De l'oxygène de l'air dans la vinification', *Oeuvres de Pasteur*, Paris, 1922–39, vol. 3, pp. 170ff. 'Sur l'expérience de Gay-Lussac relatif au départ de la fermentation du mout de raisin par l'action de l'oxygène de l'air' (1874), *Oeuvres*, vol. 2, pp. 427–9.

94 *C.R.*, 18 (1844), 546–54. This was the culmination of work started in 1842 in collaboration with Magendie and the young Claude Bernard. See J. M. D. Olmsted, *Claude Bernard Physiologist*, London, 1939, p. 46.

95 For other examples of Gay-Lussac's work in physiology see *P.V. Inst.*, 9, 500–2 (6 September 1830, a report with Serullas on a study of blood) and *ibid.*, pp. 502–6.

## Chapter 7

1 *Petite promenade physique*, Paris, 1818, p. 7.

2 *Biographie des hommes vivants*, vol. 3, October 1817.

3 Letter from Cuvier to Fontanes, 18 April 1809, Archives nationales $F^{17}$ 1933.

4 A. J. Ladd, *Ecole Normale Supérieure. An Historical Sketch*, Grand Falls, North Dakota, 1907, pp. 31–2.

5 Archives nationales $F^{17}$ 13794.

6 J. B. Piobetta, *Le baccalauréat*, Paris, 1937, pp. 73–5.

7 O. Gréard, *Education et instruction*, vol. 1, *Enseignement superieur*, 2nd edn, Paris, 1889, pp. 284–5.

8 A highly competitive examination. It was introduced in 1821 with three sections: letters, grammar and science.

9 'Rapport présenté par M. J. A. Dumas pour être adressé au Ministre de l'Instruction publique au nom de la Faculté des Sciences', O. Gréard, *op. cit.*, vol. 1, pp. 241–2.

10 Dora B. Weiner, *Raspail, Scientist and Reformer*, New York, 1968, p. 57.

11 Archives nationales $F^{17}$ 1933, dossier 1.

12 Lectures normally lasted $1\frac{1}{2}$ hours but students would arrive early in the hope of getting a good place.

13 Archives nationales $F^{17}$ 1933, dossier 3.

14 Bodelio, *Petite promenade physique*, pp. 372, 437.

15 Archives nationales $F^{17}$ 1933, dossier 2. (Biot continued to share the physics course with Gay-Lussac until 1826, when he took over the chair of physical astronomy.)

16 E. W. A. Kahlbaum and E. Schaer, *Christian Friedrich Schönbein, 1799–1868*, Leipzig, 1899, p. 81.

17 *Ibid.*, p. 82.

18 *The Life of Sir Robert Christison, Bart.*, edited by his sons, vol. 1, *Autobiography*, Edinburgh, 1885, pp. 207, 239.
19 Charles de Remusat, *Mémoires de ma vie*, ed. C. H. Pouthas, vol. 1 (1797–1820), Paris, 1958, p. 243.
20 Funeral *éloge* delivered by Pouillet as representative of the Faculty of Science, 1850.
21 Bodelio, *Petite promenade physique*, p. 75.
22 'Observation de M. Gay-Lussac sur la publication de ses leçons de physique et de chimie par les stenographes', *A.c.p.*, **37** (1828), 441–3. For reasons for the exchange of chairs with Dulong see Berzelius, *Bref*, vol. 2, part 1, p. 74.
23 O. Gréard, *Education et instruction*, vol. 1, p. 49.
24 'Les auditeurs externes', *Ecole polytechnique: Livre du centenaire*, Paris, 1895, vol. 3, p. 584n.
25 Faraday, *Life and Letters*, vol. 1, p. 99.
26 *Galerie des contemporains illustres*, vol. 6, Paris, 1843, pp. 25–6.
27 Arago, *Oeuvres*, vol. 12, *Mélanges*, p. 686.
28 A. D. Bache, *Report on education in Europe*, Philadelphia, 1839, pp. 551–2.
29 *Moniteur*, 1835, p. 1226.
30 Arago, *Oeuvres*, vol. 12, *Mélanges*, p. 678.
31 Ecole Polytechnique, *Conseil de Perfectionnement. Procès-verbaux*, vol. 6 (1832–50), p. 144.
32 *Ibid.*, p. 155.
33 Ecole Polytechnique, *Conseil d'Administration. Procès-verbaux*, vol. 17, p. 462.
34 Pouillet, funeral *éloge* of Gay-Lussac, 1850.
35 I.e. in the new recommendation the name of Gay-Lussac had been substituted for that of Serullas.
36 Dumas may have withdrawn from the contest in the hope of succeeding to one of the other of Gay-Lussac's chairs.
37 Archives Nationales, AJ 15* 130, p. 98.
38 Liebigiana 58, Letter No. 3 Jules Gay-Lussac to Liebig.
39 Thenard, *Traité de chimie*, 6th edn, 1834–6, vol. 5, pp. 409–519.
40 Student notes: Chimie organique, 17 avril 1838, mardi et samedi, Professeur Gay-Lussac. Amphithéatre du Jardin des Plantes.
41 *Sketches from the Life of Edward Frankland [1825–1899] edited...by his two Daughters M.N.W. and S.J.C.*, London, 1902, p. 80.
42 *Moniteur*, 1834, p. 1200.
43 This was re-named the Académie des Sciences in 1816.
44 Printed in full in *P.V. Inst.*, **3** (1807), 626–30.
45 *Ibid.*, **7**, 83 (4 September 1820).
46 *Ibid.*, **8**, 414.
47 *Ibid.*, **3**, 612–14. The other members of the commission were Rochon, Charles and Montgolfier.
48 *Ibid.*, **4**, 345–7.
49 *Ibid.*, **5**, 530.

50 *Ibid.*, **7**, 9; *ibid.*, **7**, 547.

51 *Ibid.*, **8**, 14–24, 29–30. Gay-Lussac presented a supplementary report on 23 February.

52 I should like to thank John Perkins who kindly drew my attention to this appointment of Gay-Lussac, which is referred to in obituaries of Pelouze. See J. B. A. Chevallier, *Journal de chimie médicale*, 5th series, **3** (1867), 444–7, A. A. T. Cahours, *Moniteur scientifique*, **10** (1868), 502–8.

53 *P.V.Inst.*, **7**, 217.

54 *Ibid.*, p. 467.

55 *Ibid.*, p. 468.

56 *Ibid.*, pp. 469–79.

57 *Ibid.*, **5**, 192.

58 *Ibid.*, **6**, 373, 374.

59 Report by Gay-Lussac and Thenard, *ibid.*, p. 185.

60 *Ibid.*, **10**, 142–5.

61 *Ibid.*, **4**, 399. The other members of the commission were Desfontaines, Cuvier, Berthollet and Bosc. See also Fox, *The Caloric Theory of Gases*, p. 134.

62 The voting on 7 January 1833 for vice-president was: Gay-Lussac 30 votes, Biot 11, Poinsot 3, Bouvard 1, Ampère 1 (*P.V. Inst.*, **10**, 184). There are no details of the 1821 election.

63 *P.V. Inst.*, **9**, 447.

64 *Ibid.*, **9**, 544 (22 November 1830).

65 A. Moquin-Tandon, *Un naturaliste à Paris sous Louis-Philippe. Journal de voyage inédit* (1834), ed. M. Roland, Paris, 1944, pp. 79, 104–5.

66 The March 1833 issue of the *Annales de chimie et de physique* (**52**, 190–222) devoted more than thirty pages to obituaries of scientists of various degrees of eminence. Of these eight were victims of cholera.

67 J. M. D. Olmsted, *François Magendie*, New York, 1944, pp. 181–94.

68 *Moniteur*, 1832, p. 564.

69 The above summary will be supplemented by the author in a detailed article on the history of the two journals. For an analysis of the *Annales de chimie* based on published sources see Mrs Court's M.Phil. dissertation: 'The *Annales de chimie*, 1789–1815', University of Leeds, 1969.

70 This was the number sold by 1 January 1818 (G.L.5). However, a further fifty were sold in 1819, suggesting that purchasers were completing their sets of the journal.

71 The chemistry had been supplemented with a certain amount of pharmacy under the influence of Fourcroy (d. 1809).

72 *A.c.p.*, **3** (1816), 335. Each issue contained 112 pages.

73 *A.c.p.*, **1** (1816), 416; **2** (1816), 54.

74 The exact total is 5769 francs 97 centimes.

75 The annual subscription was increased during 1816 to 24 francs

(*A.c.p.*, **3** (1816), 355). The benefit derived from this is reflected in the 1818 figures which show the editors received 14 556 francs from the publisher and a profit for each of the two editors of 4289 francs 50 centimes.

76  *A.c.p.*, **47** (1831), 198.
77  E.g. William Henry, *ibid.*
78  *A.c.p.*, **48** (1831), 402–3. (This was included in the December 1831 issue.)
79  *A.c.p.*, **51** (1832), 404–34.
80  *A.c.p.*, **50** (1832), 5–69, 113–63.
81  M. Faraday, *Experimental Researches*, 3 vols., London, 1839–55, vol. 2, 1844, pp. 164–165n.
82  *A.c.p.*, **4** (1817), 79n.
83  *A.c.p.*, **38** (1828), 426–32.
84  *A.c.p.*, **33** (1826), 5–29. (For a discussion of the Longchamp affair see pp. 175–6.)
85  *A.c.p.*, **2** (1816), 204–5.
86  *A.c.p.*, **3** (1816), 170–4.
87  *Ibid.*, pp. 174–7.
88  *A.c.p.*, **27** (1824), 199–200n.
89  *A.c.p.*, **5** (1817), 41–42n.
90  *Ibid.*, 275–88 (278).
91  *A.c.p.*, **23** (1823), 440–3.
92  *A.c.p.*, **8** (1818), 306–12.
93  Chevillot and Edwards, 'Sur le caméleon mineral'.
94  *A.c.p.*, **9** (1818), 441–3.
95  *Mémoires de l'Académie Royale des Sciences de l'Institut de France*, **3**, year 1818 (1820), 385–488.
96  *A.c.*, **95** (1815), 227–8.
97  Berzelius, *Bref*, vol. 3, part 7, p. 114.
98  *A.c.p.*, **1** (1816), 32–45.
99  Swedish Academy of Sciences, Berzelius letterbook: Correspondence with Gay-Lussac, f. 2.
100  *A.c.p.*, **5** (1817), 149–60. Berzelius' Swedish friend d'Ohsson, then in Paris, made some slight alterations to the text.
101  *Ibid.*, p. 166.
102  Gay-Lussac to Berzelius, 2 July 1817 (Berzelius, *Bref*, vol. 3, part 7, pp. 121–2).
103  By 1818–19 Davy had given up scientific research and Dumas and Liebig were only in their teens.
104  'Lettre à M. Pelouze', *A.c.p.*, **67** (1838), 303–26.
105  E.g. *A.c.p.*, **5** (1817), 174–81.
106  E. Millon and J. Reiset, *Annuaire de chimie*, **1** (1845), Avertissement, p. vi.
107  Thomas Graham's letter to his sister Margaret, 1 September 1836. R. A. Smith, *Life and Work of Thomas Graham illustrated by 64 Unpublished Letters*, Glasgow, 1884, p. 37.

108 *Phil. Mag.*, 5 (1834), 401–15.
109 *A.c.p.*, 66 (1837), 366–87.
110 'Théorie nouvelle de la nitrification', *A.c.p.*, 33 (1826), 5–29.
111 'Lettre de M. Gay-Lussac à M. Longchamp sur sa théorie de la nitrification', *A.c.p.*, 34 (1827), 86–95.
112 'Lettre de M. Longchamp à M. Gay-Lussac', *A.c.p.*, 34 (1827), 215–20. (Longchamp's letter is dated 18 March 1827.)
113 *Lettre de M. Longchamp, ex commissaire-en-chef des poudres à M. Gay-Lussac, Membre du Comité consultatif de la Direction des Poudres, l'un des Rédacteurs des Annales de Chimie et de Physique, etc., etc.*, Paris, April 1827.
114 'les administrateurs qu'il défend aujourd'hui avec tant de chaleur' (*Ibid.*, p. 9n).
115 *Ibid.*, p. 7.
116 *Ibid.*, pp. 27–9.
117 *Petite promenade physique*, Paris, 1818.
118 *P.V. Inst.*, 5, 360 records the submission by Bodelio of several memoirs on 11 July 1814. Bodelio thus took several years to compose his attack and get it published.
119 *Petite promenade physique*, pp. 9–13.
120 *Ibid.*, p. 162.
121 *Ibid.*, p. 256.
122 *Ibid.*, p. 68.

## Chapter 8

1 Speech in Chambre des Députés, 27 May 1839, *Moniteur*, 1839, p. 796.
2 *Mémoires de J. B. Boussingault*, 5 vols., Paris, 1892–1903, vol. 1, p. 153.
3 A. Mitscherlich, *Gesammelte Schriften von Eilhard Mitscherlich. Lebensbild, Briefwechsel und Abhandlungen*, Berlin, 1896, p. 22. For information about the business see pp. 190–3.
4 S. Pierson, 'Gay-Lussac and Berthollet's Theory', *XII^e Congrès Internationale d'Histoire des Sciences* (1968), Actes, vol. 6, 1971, pp. 83–6.
5 *History of Chemistry*, 1830–1, vol. 2, pp. 231–2 (my italics).
6 See chapter 9.
7 16 December 1818: E. Sarrau, 'Poudres et salpêtres', *Ecole Polytechnique, Livre du centenaire*, 1894, vol. 2, pp. 349–73.
8 Gay-Lussac was particularly concerned that the gunpowder should not become damp when exposed to the air. He thought the potassium and sodium chlorides were liable to absorb moisture from the air but it was later found that deliquescence was due to traces of magnesium and calcium chlorides associated with the sodium chloride.
9 'Analyse d'un mélange de chlorure de potassium et de chlorure de sodium' (Extrait des *Archives du Comité consultatif de la Direction des Poudres et Salpêtres*), *A.c.p.*, 12 (1819), 41–5.

10 'Procédé pour analyser la poudre à tirer' *A.c.p.*, **16** (1821), 434–9. See also Select Bibliography, p. 316.
11 J. J. Berzelius, *Autobiographical Notes*, trans. O. Larsell, Baltimore, 1934, p. 103.
12 A *millier* is 500 kilograms.
13 Letter to Madame Gay-Lussac, 6 August 1822.
14 G.L. 18.
15 Liebig and Gay-Lussac, 'Analyse du fulminate d'argent', *A.c.p.*, **25** (1824), 285–311.
16 *Rapport au Comité Consultatif des Poudres et Salpêtres sur les moyens que l'on emploie pour faire l'épreuve de la poudre et sur celui auquel on doit donner la préférence* (G.L. 16). *Résultats des expériences faites avec le fusil pendule par Mrs Gay-Lussac et Aubert, Lt.Col. de l'Artillerie* (G.L. 17).
17 A letter from Poisson is filed in G.L. 15.
18 G.L. 14, p. 5.
19 *Ibid.*, ff. 19, 27, 37.
20 *Alliage Ternaire. Documents relatifs aux opérations de la commission faites à la fonderie de Douai, 1825, 1826.* Archives de l'Artillerie, Register 2138.
21 *Moniteur*, 25 June 1828 (reporting debate of 23 June), p. 944.
22 *Lettre à M. le Lieutenant-Général Sébastiani par M. Gay-Lussac, membre de l'Académie Royale des Sciences, professeur de chimie à l'Ecole Royale Polytechnique* (8pp.)
23 *Réponse du Général Sébastiani, Député de l'Aisne, à M. Gay-Lussac, membre de l'Académie Royale des Sciences, etc.*
24 Letter to the Dauphin, 29 June 1828, Wellcome MS. 69222.
25 Ministère des Armes, Service Historique, Xd 245, 247.
26 Artillerie, Direction des Poudres et Salpêtres, dossier Gay-Lussac.
27 *Moniteur*, supplement 3 No. 160, 8 June 1836, p. 1349.
28 *P.V. Inst.*, 9, 330.
29 Decree by Chabrol, 20 November 1829.
30 Letter from Chabrol to Gay-Lussac, 20 November 1829 (see Appendix, letter 11).
31 N. Richardson, *The French Prefectorial Corps, 1814–1830*, Cambridge, 1966, p. 143.
32 Chabrol had been a member of the Commission of Sciences and the Arts in Bonaparte's Egyptian expedition and was therefore known to Berthollet.
33 Unknown to the Minister, Vauquelin had just died; his place in the commission was taken by Dulong.
34 *Instruction relative aux procédés suivis au laboratoire des essais de la Commission des Monnaies pour les essais, contre-essais, et verification des espèces et des matières d'or et d'argent.*
35 In Britain a Parliamentary enquiry into the Mint included in the report Gay-Lussac's *Instruction* as an appendix: *Report from Select Committee on the Royal Mint*, Parliamentary Papers 1837, vol. 16,

Appendix H, pp. 339–67. However the London Mint considered its own accuracy, using the traditional cupellation method, to be superior to that in continental Mints and it was not until 1871 that Gay-Lussac's volumetric method came to be adopted there (Sir John Craig, *The Mint*, Cambridge, 1953, p. 237).

36 'Recherches sur les moyens les plus surs, les plus exactes et les plus commodes de déterminer la pesanteur spécifique des fluides, soit pour la physique, soit pour le commerce', *Oeuvres*, vol. 3, pp. 427–50.

37 *Phil. Trans.* (1794), 275–382.

38 It was also an improvement on tables drawn up in 1811 for the Prussian government by Trallès (Liebig and Poggendorff 'Alkoholometrie', *Handwörterbuch der...Chemie*, vol. 1, part 1, Braunschweig, 1837, pp. 212–63).

39 *Instruction sur l'usage de l'alcoomètre centésimal*, 1824.

40 'Rapport sur les pèse-liqueurs par MM. Gay-Lussac, Benoist et Francoeur', Arago, *Oeuvres*, vol. 12, *Mélanges*, pp. 136–145.

41 *Moniteur*, 17 June 1824, p. 839.

42 Letter to Madame Gay-Lussac, 6 August 1822 (Appendix, letter 8).

43 Letter to Madame Gay-Lussac, 9 August 1822.

44 Charles Felix Collardeau du Heaume had entered the Ecole Polytechnique in 1815 but had left the following year without graduating.

45 Letter to Madame Gay-Lussac, 6 August 1822 (Appendix, letter 8).

46 *A.c.p.*, 39 (1828), 340n.

47 E.g. Letter to M. Berenger, Paris, 6 February 1833, Wellcome MS., Gay-Lussac, No. 1150; letter to M. Lewesky, Paris, 31 December 1843, Wellcome MS., Gay-Lussac, No. 69222.

48 Further collaboration between Chevreul and Gay-Lussac took place in 1826; see *A.c.p.*, 33 (1826), 335.

49 'Pour l'emploi des acides stéariques et margariques dans l'éclairage', *Description des machines...*, ed. C. P. Molard, 93 vols., Paris, 1811–63, vol. 41, p. 392, no. 4323.

50 Patent No. 5183, A.D. 1825.

51 M. Berthelot, *Science et libre pensée*, Paris, 1905, p. 256. Louis Figuier, *Les applications nouvelles de la science à l'industrie et aux arts en 1855*, 2nd edn, Paris, 1857, pp. 314–17.

52 *Compagnie de Saint-Gobain, 1665–1965*, [1965], p. 47.

53 In the 1780s Nicholas Leblanc worked out a process by which soda (sodium carbonate) could be made from common salt (sodium chloride). In the first stage the salt was heated with concentrated sulphuric acid to produce sodium sulphate. This was then calcined with charcoal and ground limestone (calcium carbonate) to obtain crude soda. Leblanc began manufacture under the patronage of the Duke of Orleans but the factory was sequestrated in the Revolution. Leblanc never succeeded in raising enough capital to resume the process and he committed suicide in 1806. Dr J. G. Smith has shown how the Leblanc process was developed by other French manufacturers during the Napoleonic period.

54 Some historians regard the manufacture of sulphuric acid in lead chambers earlier in the eighteenth century as the effective beginning of modern chemical industry.

55 The Saint-Gobain Company has recently sold its chemical interests and has reverted to its original specialisation in glass.

56 J. G. Smith, 'Studies of certain chemical industries in Revolutionary and Napoleonic France', Ph.D. thesis, University of Leeds, 1970, p. 180n.

57 G. P. Palmade, *French Capitalism in the Nineteenth Century*, trans. G. M. Holmes, Newton Abbot, 1972.

58 The minuted account suggests some previous activity behind the scenes. On the announcement of the vacancy, the shareholders retired for a few minutes and then stated that they had chosen Gay-Lussac. *Registre des procès-verbaux des séances du Conseil d'Administration de la Société Anonyme de la Manufacture Royale des Glaces de Saint-Gobain. Conseil extraordinaire*, vol. 1, 4B25, p. 49.

59 The *censeurs* had the duty of inspecting the accounts and confirming that the company had followed the statutes. *Statuts de la Société Anonyme de la Manufacture Royale des Glaces de Saint-Gobain*, 1830, art. 28. The Banque de France, founded 1800, also had *censeurs*.

60 For such visits they were given an allowance of 25 francs per day plus all expenses.

61 *Registre des procès-verbaux des séances du Conseil d'Administration...*, 4B5, p. 236.

62 *Registre* 4B6, p. 195, 30 October 1833.

63 *Registre* 4B7, pp. 45–51.

64 Meeting of 22 June 1843.

65 See Appendix, letter 14.

66 *Registre* 4B16, p. 62.

67 So-called since when the reaction was carried out on more than a laboratory scale it was transferred from glass vessels to large lead chambers, lead not being attacked by the acid. The alternative 'contact process' was developed in the second half of the nineteenth century.

68 *A.c.p.*, 1 (1816), 408. J. B. Dumas, *Traité de chimie appliquée aux arts*, vol. 1, 1828, pp. 192–4.

69 Patent No. 1096, Molard, *Description des machines*, vol. 12, p. 92. This at least produced potassium nitrate which was itself used in the lead chambers.

70 *Registre* 4B7, p. 49.

71 See p. 202.

72 Molard, *Description des machines*, vol. 90, pp. 463–9. A third patent is dated 11 September [=November?] 1842.

73 No. 12262. 'Brevet d'invention de quinze ans', Molard, *Description des machines*, vol. 90, 463–9.

74 No. 9558, A.D. 1842.

75 *Registre* 4B12, 26 May 1843.
76 Gay-Lussac had ceded his rights to Jules on 20 July 1842. However, Saint-Gobain considered that 'dans la réalité Gay-Lussac père ne cessait pas d'etre propriétaire du procédé dont il etait l'inventeur'. *Conseil Extraordinaire*, 4B24, p. 46.
77 *Ibid.*, p. 47.

## Chapter 9

1 *Rapport sur le mode d'essai des matières d'argent*, 1 March 1830.
2 *A.c.p.*, 26 (1824), 174.
3 E. Rancke Madsen, *The Development of Titrimetric Analysis till 1806*, Copenhagen, 1958.
4 *A.c.*, 2 (1789), 151–90.
5 *Eléments de l'art de la teinture*, 2nd edn, Paris, 1804, vol. 1, p. 235.
6 'Observations sur l'essai des soudes et des sels de soude de commerce', *A.c.p.*, 13 (1820), 212–21.
7 L. Laurens, *Observations sur les essais alcalimétriques des sels de soude du commerce et moyen simple pour apprecier le carbonate de soude et le sulfite qui s'y trouvent contenus*, Marseille, 1819.
8 *A.c.p.*, 13 (1820), 221.
9 *A.c.p.*, 39 (1828), 338–41.
10 66° Baumé, but Descroizilles did not specify the temperature.
11 *A.c.p.*, 39 (1828), 337.
12 *A.c.p.*, 60 (1835), 226. This is not, if understood in context, a contradiction of his general volumetric approach.
13 *A.c.p.*, 26 (1824), 164.
14 *A.c.p.*, 39 (1828), 340n: 'titres'.
15 Gay-Lussac's only paper on volumetric analysis to be translated in a contemporary British journal had an appendix translating grams and litres into grains and cubic inches: *Annals of Philosophy*, 24 (1824), 225. See also the first edition of what was to become a standard text-book of volumetric analysis: F. Sutton, *Systematic Handbook of Volumetric Analysis*, London, 1863, pp. 12–16, where Sutton urged the advantages of 'the French decimal system'.
16 F. Szabadvary, *History of Analytical Chemistry*, Oxford, 1966, p. 227.
17 'Lettre de M. Pelouze à M. Forthomme', in Friedrich Mohr, *Traité d'analyse chimique à l'ordre de liqueurs titrées*, trans. C. Forthomme, Paris and Nancy, 1857, pp. viii–x.
18 *A.c.p.*, 26 (1824), 174.
19 *A.c.p.*, 39 (1828), 349.
20 Variations of Gay-Lussac's burette included those by Rosenthal, Mohr, Deschamps, Geisler and Binks. R. Christophe, 'L'analyse volumétrique de 1790 à 1860. Caractéristiques et importance industrielle. Evolution des instruments', *Revue d'histoire des sciences*, 24 (1971), 25–44.
21 'La burette de Gay-Lussac nous semble plus commode et plus exacte

que les autres': A. B. Poggiale, *Traité d'analyse chimique par la méthode des volumes*, Paris, 1858, p. 38.

22 A modern style pipette was, however, devised by Achard in Berlin as early as 1785, but for purely physical experiments. His pipette seems to have been unknown to chemists and Gay-Lussac's work in a chemical context was done quite independently. (Rancke Madsen, *The Development of Titrimetric Analysis till 1806*, pp. 199–200.)

23 *Instruction sur l'essai des matières d'argent*, 1832, plate III.

24 *Instruction pour tous les citoyens qui voudront exploiter eux-mêmes du salpêtre* (by Committee of Public Safety) Paris, year 2.

25 *Instruction sur l'art de séparer le cuivre du métal des cloches*, Paris, year 2.

26 *Description de l'art de fabriquer les canons*, Paris, year 2.

27 *Instruction sur l'établissement du nitrières et sur la fabrication du salpêtre*, 1777.

28 Szabadvary, *History of Analytical Chemistry*, p. 220.

29 'Observations sur l'essai des soudes et des sels de soude du commerce', *A.c.p.*, 13 (1820), 190–221.

30 *Instructions sur la manière de procéder à l'essai du salpêtre raffiné pour en constater le degré de pureté*, les Régisseurs-Généraux des Poudres et Salpêtres, Paris, 1817. A solution of lead nitrate had earlier been used for this purpose by Guyton de Morveau (C. C. Gillispie (ed.), *Dictionary of Scientific Biography*, vol. 5, New York, 1970, p. 602).

31 Letter by Aubert, 18 May 1828 (see Appendix, letter 10).

32 The *Instruction* says 'On prendra une quantité quelconque de nitrate d'argent fondu, exprimée en grammes; on la divisera par 0.0096784, et le quotient sera le nombre de grammes d'eau dans lequel il faudra dissoudre ce sel.' (p. 5) Thus to take simple numbers, 9.6784 grams would be dissolved in 1000 grams of water with an allowance made for the volume of silver nitrate: 'Le nitrate d'argent augmentant un peu le volume de l'eau, il y aurait une petite correction à faire...'

33 This might be 30 drops, as in the next example given.

34 The composition of bleaching powder is complex but modern analysis suggests that it is a mixture of basic calcium hypochlorite and basic calcium chloride. Gay-Lussac's name for it was 'chlorure de chaux'.

35 'Instruction sur l'essai du chlorure de chaux', *A.c.p.*, 26 (1824), 162–76.

36 Balard showed in 1834 that the formation of chlorate only takes place if the solution is warm.

37 'Nouvelle instruction sur la chlorométrie', *A.c.p.*, 60 (1835), 225–61.

38 Early editions of the manual were approved by the Mint Administration but the next edition (1836) simply appeared under the names of Vauquelin (who had died in 1829), Gay-Lussac and D'Arcet.

39 *Instruction sur l'essai des matières d'argent*, 1832, p. 10.

40 This comment suggests that Gay-Lussac's salt was contaminated with

small amounts of magnesium or calcium chloride, a factor which would slightly impair the accuracy of his results.

41 *Ibid.*, pp. 23–63.

42 If the standard amount of sodium chloride already added from the pipette had proved to be an excess, standard silver nitrate solution would be added instead and a suitable subtraction made.

43 *Vollständiger Unterricht über das Verfahren Silber auf nassen Wege zu Probieren*, Braunschweig, 1833, p. iii.

44 28 September 1807.

45 At the time a verbal report was considered sufficient but a written report was finally made by Gay-Lussac in 1820 (*P.V. Inst.*, 7, 77–8).

46 Charles was absent from meetings of the Academy from September 1822 until his death on 7 April 1823.

47 'Instruction sur les paratonnères', *A.c.p.*, 26 (1824), 258–98.

48 Pouillet, Report to the Academy of Sciences, 1854, in Gay-Lussac, *Information about Lightning Conductors issued by the Academy of Sciences of France*, trans. Richard Anderson, London, 1881.

49 *A.c.p.*, 40 (1829), 386–98.

50 *A.c.p.*, 26 (1824), 268–71.

## Chapter 10

1 Speech in Chambre des Pairs, 27 March 1843, *Moniteur*, 1843, p. 584.

2 *La bourgeoisie parisienne de 1815 à 1848*, Paris, 1963.

3 Thackray, *John Dalton*, p. 47.

4 L. P. Williams (ed.), *Selected Correspondence of Michael Faraday*, Cambridge, 1971, e.g. vol. 1, pp. 175–7, 229.

5 However, the salary shown is not always precisely the salary paid because of pension deductions. A deduction of 5 per cent was made from the Faculty of Science salary for a pension fund and a similar 3 per cent was deducted from salaries at the Ecole Polytechnique.

6 See e.g. E. Grimaux and C. Gerhardt, *Charles Gerhardt. Sa vie, son oeuvre, sa correspondance*, 1900, p. 175.

7 F. Larivière, *Expertise d'aréomètres par Jacquelin, Barreswil et Troost. Examen critique du rapport des experts*, Paris, 1861, p. 12n.

8 J. Rougerie, 'Remarques sur l'histoire des salaires à Paris au XIXe siècle', *Le mouvement social*, April–June 1968, pp. 106–7.

9 Even in this case there was a short overlap period when Gay-Lussac drew both salaries.

10 Gay-Lussac to his wife, 31 August 1817.

11 Gay-Lussac to his wife, 4 September 1822.

12 Gay-Lussac to his wife, 14 March 1849.

13 Berzelius, *Bref*, vol. 2, part 1, p. 75.

14 *A.c.p.*, 18 (1821), 211–16.

15 *Moniteur*, 12 May 1821, p. 332.

16 *A.c.p.*, 42 (1829), 214–21.

17 Wellcome MS. Gay-Lussac No. 1572, 8 October 1833. See also No. 1574, 17 June 1841.
18 Liebigiana 58. Pelouze to Liebig No. 15.
19 Liebigiana 58. Pelouze to Liebig No. 48 (3 July 1838).
20 W. Henry to C. G. B. Daubeny, 30 August 1831. Letter in library of Magdalen College, Oxford. I owe this reference to Wilfred Farrar.
21 Arago, *Oeuvres, Notices biographiques*, vol. 3, pp. 62-3.
22 J. P. Callot, *Histoire de l'Ecole Polytechnique*, Paris, 1958, p. 65.
23 Gay-Lussac to wife, London, 11 August 1817.
24 Thenard was one year older than Gay-Lussac but Poisson was three years younger.
25 In 1837 Gay-Lussac was promoted to *Commandeur* and in 1845 he became one of a small number to reach the rank of *Grande Officier*.
26 In 1831 there were 166 000 registered voters, about $2\frac{1}{2}$ per cent of the adult male population.
27 S. Kent, *Electoral Procedure under Louis Philippe*, New Haven, 1937, p. 166.
28 *Le Contribuable* quoted in Delhoume *La vie...de Gay-Lussac*, Limoges, 1950, pp. 144-5.
29 Archives Nationales C.1319.
30 *Moniteur*, 1831, p. 1276.
31 Letter from Gay-Lussac to Liebig, 25 March 1832 (Liebigiana 58).
32 'il est impossible de s'occuper d'autre chose que des affaires du pays'.
33 *Moniteur*, 1832, p. 908.
34 Ch.D., 9 May 1834, *Moniteur*, 1834, p. 1200.
35 Ch.D., 1 June 1835, *Moniteur*, 1835, p. 1379.
36 Ch.D., 7 June 1837, *Moniteur*, 1837, p. 1455.
37 Ch.D., 18 May 1835, *Moniteur*, 1835, pp. 1226-7.
38 18 April 1833, *Moniteur*, 1833, p. 1104.
39 7 January 1835, *Moniteur*, 1835, p. 65.
40 Ch.D., 20 April 1836, *Moniteur*, 1836, p. 822.
41 *Ibid.*
42 Ch.D., 28 April 1836, *Moniteur*, 1836, p. 926.
43 Ch.D., 4 February 1837, *Moniteur*, 1837, p. 245.
44 'cela ne le regarde en aucune manière'.
45 Ch.P., 27 March 1843, *Moniteur*, 1843, pp. 584-5, 602, 607, 615.
46 *Moniteur*, 1839, pp. 795-6, 815.
47 Ch.D., 5 March 1838, *Moniteur*, 1838, p. 486.
48 E.g. *Moniteur*, 1836, p. 989 (import duty on steam engines).
49 E.g. *Moniteur*, 1839, p. 815 (copyright law).
50 Debates on Muséum: *Moniteur*, 1834, p. 1200; 1835, p. 1379; 1837, p. 1455.
51 E.g. Ch.D., 27 May 1837, *Moniteur*, 1837, p. 1346.
52 Ch.D., 9 May 1837, *Moniteur*, 1837, pp. 1139, 1162, 1178.
53 The salt could be made unsuitable for culinary purposes by adding some obnoxious substance which would not, however, interfere with its use for industrial purposes.

54 *Moniteur*, 1833, p. 1095.
55 *Opinions des hommes politiques des savants, des agronomes et des agriculteurs sur l'utilité du sel pour les plantes et pour les animaux, publiés par M. A. Demesmay, Deputé du Doubs, en réponse au Rapport présenté à la chambre des pairs par M. Gay-Lussac, sur le projet de loi réduisant à 10c par Kilogramme l'impôt du sel; Adopté le 23 avril dernier par la Chambre des Députés.* Paris, 1846.
56 Ch.P., 5, 6, 7, 9 March 1840. Gay-Lussac's most interesting contributions can be found in *Moniteur*, 1840, pp. 429–30, 458–9.
57 Ch.P., 10 April 1845, *Moniteur*, pp. 929, 955, see also p. 945.
58 Slavery was finally abolished in the French colonies in 1848. It had been ended in British colonies in 1833 and was to be finally abolished in the United States in 1865.
59 Ch.D., 20 May 1837, *Moniteur*, 1837, p. 1264.
60 Ch.P., 6 June 1845, *Moniteur*, 1845, p. 1602.
61 Ch.D., 5 March 1838, *Moniteur*, 1838, pp. 484–6.
62 Letter to Jules Gay-Lussac, 15 November 1848.

*Chapter 11*

1 *C.R.*, **84** (1877), 1194.
2 *C.R.*, **84** (1877), 1194.
3 *Traité pratique de l'analyse des gaz*, Paris, 1906, p. vi.
4 E.g. *P.V. Inst.*, 6, 85 (14 October 1816), 200 (23, 30 June 1817).
5 *A.c.p.*, 5 (1817), 275. Partington, *History of Chemistry*, vol. 4, p. 241.
6 Fourcy, *Histoire de l'Ecole Polytechnique*, p. 360.
7 *A.c.p.*, 6 (1817), 426–36.
8 Notably that on vapour densities, *A.c.p.*, 21 (1822), 143.
9 *Traité élémentaire de physique*, 4th edn, 1836.
10 See *Cours de physique par M. Gay-Lussac à la Faculté des Sciences de l'Académie de Paris*, 2 vols., 1828. Vol. 1 is by Gay-Lussac, vol. 2 by Pouillet.
11 The circumstance was a meeting at Saint-Gobain after Gay-Lussac's death to present Pelouze as his successor to a post of consultant (*Procès-Verbaux. Conseil d'Administration*, 4 June 1850, 4B16, p. 54).
12 J. B. Dumas, *Discours et éloges scientifiques*, vol. 1, 1885, pp. 127ff. Different biographers of Pelouze provide conflicting accounts of how he first met Gay-Lussac.
13 'afin de les présenter bien purs et bien beaux à M. Gay-Lussac' Pelouze to Liebig, 8 March 1832, Liebigiana 58, No. 2.
14 Pelouze to Liebig, 19 June 1832, Liebigiana 58, No. 1.
15 *A.c.p.*, 44 (1830), 220–1.
16 Pelouze to Liebig, 14 June 1833, Liebigiana No. 5.
17 *A.c.p.*, 52 (1833), 410–24.
18 Gay-Lussac declared himself 'neutral' in the contest between Pelouze and Joseph Pelletier, who was twenty years older than Pelouze (Pelouze to Liebig, 11 May 1837, Liebigiana 58, No. 32).

19 *Revue générale des sciences pures et appliques*, 5 (1894), 139.
20 E.g. *Cours de chimie générale*, 4 vols., Paris, 1848–50.
21 Regnault to Liebig, 2 December 1836, Liebigiana 58, No. 9.
22 Regnault to Liebig (n.d.), Liebigiana 58, No. 22.
23 'Recherches sur la dilation des gaz', *A.c.p.*, [3], 4 (1842), 5–63.
24 See Appendix, document 15.
25 *A.c.p.*, 25 (1824), 285–311.
26 A. W. von Hofmann 'The life work of Liebig', *Nature*, Extra number, 6 February 1880, i–xl.
27 Gay-Lussac to Liebig, Paris, 25 March 1832, Liebigiana 58.
28 Gay-Lussac to Jules, 15 September 1832.
29 L. Walkhoff, *Traité complet de fabrication et raffinage du sucre de betteraves*, 2nd French edn, 2 vols., Paris, 1874.
30 Pelouze to Liebig, 11 February 1834, Liebigiana 58, No. 11.
31 Letter from Gay-Lussac to Pelouze, 5 October 1832, Académie des Sciences, dossier Pelouze.
32 'Il descendait des deux grands physiciens qui ont porté ce nom' (*Revue scientifique*, 20 (1877), 504).
33 Some discussion of volumetric analysis was introduced by Gay-Lussac into the academic lectures of the Paris Faculty of Science (*Cours de chimie*, vol. 2, lecture 30, p. 25).
34 Gay-Lussac, *Information about Lightning Conductors issued by the Academy of Sciences in France*, trans. Richard Anderson, 1881, p. vi.
35 Hydrogen : oxygen = 2 : 1.
36 *Cours de chimie générale*, 4 vols., Paris, 1829.
37 M. P. Crosland, 'Humphry Davy – an alleged case of suppressed publication', *British Journal for the History of Science*, 6 (1973), 304–10.
38 R. Fox, 'The rise and fall of Laplacian physics', *Historical Studies in the Physical Sciences*, 4 (1974–5), 89–136.
39 In Gay-Lussac's only full memoir on a geological subject ('Réflexions sur les volcans', *A.c.p.*, 22 (1823), 415–29) he tried to use his knowledge of chemistry and physics to throw light on the subject but he was the first to admit that he had no specialised geological knowledge.
40 *P.V. Inst.*, 7, 306.
41 *Ibid.*, p. 365.
42 *The Scientific Papers of Thomas Andrews with a Memoir by P. G. Tait and A. Crum Brown*, London, 1889, pp. xiii, xix.
43 *Dictionary of American Biography*, London, 1928–36, vol. 19, p. 443.
44 Fourcy, *Histoire de l'Ecole Polytechnique*, p. 387. Warden was a book collector who presented Gay-Lussac with several books, including Newton's *Opticks*.
45 *A.c.p.*, 10 (1818), 124.
46 Private communication, M. Rothenburg, N. Reingold, 1976.
47 Partington *History of Chemistry*, vol. 3, p. 721.

## *Appendix*

1 An English translation from another copy of this speech is given in *Journal of Chemical Education*, 15 (1938), 256–7. The document is not, however, commonly known even to historians of nineteenth-century chemistry.
2 Gay-Lussac's presidency of the Académie des Sciences in 1822 was followed by that of Thenard in 1823, the year in question.
3 'M. Gay-Lussac lit un mémoire de M. le docteur Liebig sur *l'Argent et le mercure fulminant*, M. Gay-Lussac et Dulong Commissaires': Monday 28 July 1823, *P.V. Inst.*, 7, 519.

# Select bibliography

*Manuscript sources*

Paris           Académie des Sciences

Archives de l'Artillerie

Archives nationales

Archives de la Seine

Ecole Polytechnique

Ecole des Ponts et Chaussées

Madame Roger Gay-Lussac

Institut de France

Ministre des Armes. Service Historique

Muséum d'Histoire Naturelle. Bibiothèque Central

Saint-Gobain Industries

Haute Vienne      Archives départementales, Limoges

Musée, Saint Léonard

London         Royal Society

Royal Institution

Wellcome Historical Medical Library

Munich         Bayerische Staatsbibliothek

Stockholm      Bibliotek K. Vetenskaps Akademiens

Ithaca, N.Y., U.S.A.   Cornell University Archives

## Publications by Gay-Lussac

(a) *Memoirs*

Gay-Lussac published altogether more than 180 memoirs on pure and applied science. Some of the more important of those on pure science (with translations where available) are listed below; for applied science see section (c). For a fairly complete list of Gay-Lussac's papers in scientific journals (but which omits some brief notes by Gay-Lussac in the *Annales de chimie et de physique*) see the *Royal Society Catalogue of Scientific Papers 1800–1863*, 6 vols., 1863–72, vol. pp. 800–7.

'Sur la dilatation des gaz et des vapeurs' *A.c.*, 43 (1802), 137–75.
  *Nicholson's Journal of Science and the Arts*, 3 (1802), 207–16, 257–67.
  'Researches upon the rate of expansion of gases and vapours', in W. W. Randall (ed.), *The Expansion of Gases by Heat*, 1902.
  *Untersuchungen über die Ausdehnung der Gasarten und der Dämpfe durch die Wärme* (Ostwald's Klassiker der exacten Wissenschaften, No. 44), Liepzig, 1894.

'Relation d'un voyage aero-statique', *J. de phys.*, 59 (1804), 314–20 [with J. B. Biot]; 454–61.
  *Philosophical Magazine*, 19 (1804), 371–9; 21 (1805), 220–7.

[with Humboldt]: 'Expériences sur les moyens eudiométriques et sur la proportion des principes constituants de l'atmosphère', *A.c.*, 53 (1805), 239–59.

[with Humboldt]: 'Observations sur l'intensité et l'inclinaison des forces magnétiques, faites en France, en Suisse, en Italie, etc.', *M.S.A.*, 1 (1807), 1–22.

'Premier essai pour déterminer les variations de température qu'éprouvent les gaz en changeant de densité, et considérations sur leur capacité pour le calorique', *M.S.A.*, 1 (1807), 180–204.

[with Thenard]: 'Sur les métaux de la potasse et de la soude', *A.c.*, 66 (1808), 205–17.

[with Thenard]: 'Sur la décomposition et la recomposition de l'acide boracique', *A.c.*, 68 (1808), 169–74.
  *Nicholson's Journal*, 23 (1809), 260–3.

'Mémoire sur la combinaison des substances gazeuses, les unes avec les autres', *M.S.A.*, 2 (1809), 207–34.
  'Memoir on the combination of gaseous substances with each other', in *Foundations of the Molecular Theory*, A.C.R. No. 4.
  *Das Volumgesetz gasförmiger Verbindungen, Abhandlungen von Alex. von Humboldt and J.F. [sic] Gay-Lussac...Herausgegeben von W. Ostwald* (Ostwald's Klassiker der exakten Wissenschaften, No. 42), Leipzig, 1893.

[with Thenard]: 'Sur l'acide fluorique', *A.c.*, 69 (1809), 204–20.
*Nicholson's Journal*, 24 (1809), 29–37.

[with Thenard]: 'De la nature et des propriétés de l'acide muriatique et de l'acide muriatique oxigéné', *M.S.A.*, 2 (1809), 339–58.

[with Thenard]: 'Sur l'analyse végétale et animale', *A.c.*, 74 (1810), 47–64.
*Philosophical Magazine*, 38 (1811), 60–8.

'Mémoire sur l'iode', *A.c.*, 91 (1814), 5–160.
*Annals of Philosophy*, 5 (1815), 101–9, 207–14, 296–302, 401–13; 6 (1815), 124–32, 183–98.
*Untersuchungen über das Iod von Gay-Lussac*...Herausgegeben von W. Ostwald (Ostwald's Klassiker der exakten Wissenschaften, No. 4), Leipzig, 1889.

'Recherches sur l'acide prussique', *A.c.*, 95 (1815), 136–231.
*Annals of Philosophy*, 7 (1816), 350–64; 8 37–52, 108–15.

'Premier mémoire sur la dissolubilité des sels dans l'eau', *A.c.p.*, 11 (1819), 296–315.
*Annals of Philosophy*, 15 (1820), 1–12.

'Extrait d'un mémoire sur le froid produit par l'évaporation des fluides' *A.c.p.*, 21 (1822), 82–92.
*Quarterly Journal of Science*, 15 (1823), 294–7.

'Considérations sur les forces chimiques', *A.c.p.*, 70 (1839), 407–34.

*(b) Books (lecture-courses and research)*

*Cours de chimie, par M. Gay-Lussac, comprenant l'histoire des sels, la chimie végétale et animale*, 2 vols., Paris, 1828.
(The text was checked by Gaultier de Claubry to minimise errors.)
Another edn, 2 vols., Brussels, 1829.
Another edn, 2 vols., Paris, 1833.
Vol. 1 has lessons 1–18, vol. 2 lessons 19–33. Each lecture is separately paginated.

*Leçons de chimie de M. Gay-Lussac à la Sorbonne, comprenant l'histoire des sels, la chimie animale et végétale. Recueillies et publiées par M. Marmet, sténographe*, Paris, 1828 (13pp.).
(This rare booklet was the first attempt of a shorthand writer to reproduce Gay-Lussac's lectures.)

*Leçons de physique de la Faculté des Sciences de Paris, recueillies et rédigées par M. Grosselin*, 2 vols., Paris, 1828.
Ire partie, professé par M. Gay-Lussac.
IIe partie, professé par M. Pouillet.

[with Thenard] *Recherches physico-chimiques, faites sur la pile, sur la préparation chimique et les propriétés du potassium et du sodium...*, 2 vols., Paris 1811.

Vol. 1, 386pp. + table of contents + 5 folding plates; vol. 2, 424pp. + table of contents + 1 folding plate.

(c) *Instructions*

The following *Instructions* appeared under Gay-Lussac's name.

*Essai des potasses du commerce,* par M. Gay-Lussac, Paris, n.d.
Also in *A.c.p.*, 39 (1818), 337–68.
Another edn, Paris [1847].

*Instruction sur l'essai du chlorure de chaux,* Paris, 1824, 16pp.
Also in *A.c.p.*, 26 (1824), 162–76.

*Instruction pour l'usage de l'Alcoomètre centesimal et des tables qui l'accompagnent,* par M. Gay-Lussac de l'Académie royale des sciences, Professeur à l'Ecole royale Polytechnique et à la Faculté des sciences de Paris, etc., Paris 1824, 12mo (pages not numbered).

*Instruction sur l'essai des matières d'argent par la voie humide,* Paris, 1832, 4to, 88pp. + plates.

*Manuel complet de l'essayeur, par Vauquelin. Suivi de l'instruction de M. Gay-Lussac sur l'essai des matières d'argent par la voie humide, et des dispositions du laboratoire de la Monnaie de Paris; par M. d'Arcet.* Nouvelle édition...par A. D. Vergnaud. Paris, 1836, 18mo, 248pp. + plates.

*Vollständiger Unterricht über das Verfarhren Silber auf nassem Wege zu probieren,* Braunschweig 1833.

*Instruction sur les paratonnerres adopté par l'Académie Royale des Sciences le 23 juin 1823 et publié par ordre du ministre de l'intérieur* [signed Gay Lussac, rapporteur], Paris, 1824, 4to, 31pp. + plates.
Also in *A.c.p.*, 26 (1824), 258–98.
Another edn, Paris, 1824, 8vo, 51pp. + plates.

*Instruction sur les paratonnerres adopté par l'Académie des Sciences. 1re partie 1823, M. Gay-Lussac rapporteur. 2e partie 1854, M. Pouillet rapporteur,* Paris, 1855, 18mo, 130pp. + plates.

*Instruction sur les paratonnerres adopté par l'Académie des Sciences. 1re partie, 1823, M. Gay-Lussac rapporteur. 2e partie, 1854, M. Pouillet rapporteur. 3e partie, 1867, M. Pouillet rapporteur,* Paris, 1874, 12mo, vi. + 143pp. + fig.

*Die Anlegung der Blitzableiter, zu sicheren Schutze von Thürmen, Kirchen, Schlössern, öffentlichen Gebaüden, Pulvermagazinen und Pulvermühlen, Telegraphenleitungen, Seeschiffen und Privatwoh-nungen; nach den in Jahre 1823 von Gay-Lussac, ferner in Jahre 1854 und 1855 von Pouillet ausgearbeiten und von der französischen Akademie der Wissenschaften genehmigten und bekannt gemachten Instructionen,* trans. Dr Christ. Hein Schmidt, Weimar, 1856, in *Neuer Schauplatz der Künste und Handwerke,* vol. 221, pp. xii + 96.

*Information about Lightning Conductors issued by the Academy of*

*Sciences of France.* 'Instruction sur les Paratonnerres...', trans. Richard Anderson, London, 1881, vii + 63pp.

*Nouvelle instruction sur la chlorometrie*, Paris, 1835, 41pp.
   Also in *A.c.p.*, 40 (1835), 225–61.

[with J. J. Welter] 'Observations sur l'essai des soudes et des sels de soude du commerce', *A.c.p.*, 13 (1820), 190–221.

In addition, Gay-Lussac played a major part in the composition of several *Instructions* for the artillery and the Mint.

In a private notebook, Gay-Lussac claimed authorship of the following papers which were published anonymously in the *Annales de chimie et de physique* as 'Extrait des archives du Comité Consultatif de la Direction des Poudres et Salpêtres':

   'Analyse d'un mélange de chlorure de potassium et de chlorure de sodium', *A.c.p.*, 12 (1819), 41–5.
   'Procédé pour analyser la poudre à tirer', *A.c.p.*, 16 (1821), 434–9.
   'Inflammation de la poudre déterminé par la chaleur qui se dégage pendant l'extinction de la chaux', *A.c.p.*, 23 (1823), 217–19.

## Biographies and eulogies of Gay-Lussac

Entries are in chronological order.

Anon. 'Gay-Lussac', *Biographie des hommes vivants ou histoire par ordre alphabétique de la vie publique de tous les hommes qui se sont fait remarquer par leurs écrits*, vol. 3, Paris, October 1817, p. 239.

Loménie, L. L. de 'M. Gay-Lussac', *Galéries des contemporains illustres par un homme de rien*, vol. 6, Paris, 1843, 36pp.

'Discours prononcé sur la tombe de Gay-Lussac, le 11 mai 1850', par MM. Arago, Thenard, Chevreul, Becquerel, Pouillet et Despretz.

Gardeur le Brun, A. *Notice sur M. Gay-Lussac*, Société d'Agriculture, Commerce, Sciences et Arts du Département de la Marne, Chalons [sur Marne], 1850, 23pp.

Biot, J. B. 'Notice sur Gay-Lussac', (Read at the Royal Society of London, 30 November 1850 and published in *Journal des Savans*, December 1850), *Mélanges scientifiques et littéraires*, 1858, vol. 3, pp. 125–42.

Massoulard 'Notice sur Gay-Lussac', *Journal de Limoges* [not seen].

Arago, D. F. J. 'Gay-Lussac. Biographie lue en séance de l'Académie des Sciences le 20 décembre 1852', *Oeuvres, Notices biographiques*, vol. 3, pp. 1–112.

   There is a note extant written by Arago to Gay-Lussac's widow inviting her to attend the meeting at which the *éloge* was to be read. Under such conditions one should hardly expect the *éloge* to be an objective biography.

Hasemann, J. 'Gay-Lussac', *Allgemeine Encyclopädie*, ed. J. S. Ersch & J. E. Gruber, vol. 55, Leipzig, 1852, pp. 121–8.

F[argeaud], A. 'Gay-Lussac', *Biographie universelle*, (Michaud), new edn, vol. 16, 1856, pp. 71–89.

Fargeaud, A. 'Gay-Lussac, Joseph Louis', *La science, journal du progrès des sciences pures et appliquées et des découvertes et inventions. Biographie des savants et des inventeurs*, 2nd year, No. 94, Thursday 27 November 1856, pp. 750–2; No. 95, Sunday 30 November 1856, pp. 756–9; No. 96, Thursday 4 December 1856, pp. 764–6.

Hoefer, F. 'Gay-Lussac', *Nouvelle biographie générale*, vol. 19, 1857, cols. 758–71.

Gay de Vernon 'Gay-Lussac', Congrès scientifique de France, XXVI° session tenue à Limoges. Notice biographique lue à la séance d'ouverture le 12 septembre 1859, Limoges, 1860.

Pouillin, J. and M. *Gay-Lussac*, Limoges, n.d., ix + 133pp.

Fremy, A. 'Gay-Lussac', *Encyclopédie chimique*, vol. 1, part 1, 1882, pp. 61–9.

Dehérain, P. P. Institut de France, Académie des Sciences. Inauguration de la statue de Gay-Lussac à Limoges de 11 aôut 1890. Discours prononcé par M. P. P. Dehérain, Paris, 1890, 17pp.

Lemoine, G. 'Gay-Lussac', *Ecole Polytechnique. Livre du Centenaire 1794–1894*, Paris, 1895, vol. 1, pp. 338–48.

De Vanssay, H. Notes sur ma famille, chapt. 5, pp. 53–68, Famille Gay-Lussac, Paris, 1948 (privately printed).

Blanc, E. & Delhoume, L. *La vie émouvante et noble de Gay-Lussac*, Paris, 1950.

Maurice Daumas (*loc. cit.*, 1952) says of this book that 'its decorative title is a faithful reflection of the spirit in which it was composed. Unfortunately, pious intentions do not themselves justify the publication of a useful book. The tone of the book has a false ring about it; in an irritating way it recalls books of moral reading...' The book makes some use of original documents. Its value is lessened by the authors' extensive use of Arago's *éloge*, from which they quote several pages at a time.

Daumas, M. 'Gay-Lussac (1778–1850)', *Revue d'histoire des sciences*, **3** (1950), 337–42. See also *ibid.*, 5 (1952), 284–5.

Schimank, H. 'J. L. Gay-Lussac und seine Leistung auf dem Gebiete der allgemeinen und physikalischen Chemie', *Die Naturwissenschaften*, **38** (1951), 265–74.

Lecomte, J. 'Quelques documents inédits sur Gay-Lussac. Remarques sur son oeuvre scientifique', *87° Congrès des sociétés savantes*, Limoges, 1962, pp. 123–48.

Crosland, M. P. 'Gay-Lussac', *Dictionary of Scientific Biography*, ed. C. C. Gillispie, New York, vol. 5, 1972, pp. 317–27.

Sadoun-Goupil, M. 'Joseph-Louis Gay-Lussac (1778–1850). Sa formation scientifique et ses premiers travaux', *93ème Congrès de l'Association Française pour l'Avancement des Sciences*, Limoges, 1974, section 25, pp. 1–5.

Crosland, M. P., 'Gay-Lussac', *La Rechehche*, no. 91 (July 1978), 625–33.

## General

(In the many cases where French books were published in Paris or English books in London, the place of publication has been omitted.)

Académie des Sciences *Procès-verbaux des séances de l'Académie des Sciences tenues depuis la fondation de l'Institut jusqu'au mois d'août 1835*. 10 vols., Hendaye, 1910–22.

Achinstein, P. *Law and explanation. An essay in the philosophy of science.* Oxford, 1971.

*Annales de chimie* 96 vols., 1789–1815.

(continued as) *Annales de chimie et de physique*. 75 vols., 1816–40.

3rd series, 69 vols., 1841–63.

Arago, D. F. J. *Oeuvres*. 17 vols., 1854–62.

Arcueil *Mémoires de physique et de chimie de la Société d'Arcueil*. 3 vols., 1807–17.

New edition with Introduction by M. P. Crosland, New York, 1967.

Bache, A. D. *Report on education in Europe*. Philadelphia, 1839.

Bastid, P. *Les institutions politiques de la monarchie parlementaire française, 1814–48*. 1954.

Ben-David, J. 'The profession of science and its powers', *Minerva*, 10 (1972), 362–83.

Ben-David, J. *The scientist's role in society. A comparative study*. Englewood Cliffs, N.J., 1971.

Berthelot, P. E. M. *Science et libre pensée*. 1905.

Berthelot, P. E. M. *Traité pratique de l'analyse des gaz*. 1906.

Berthollet, C. L. *Essai de statique chimique*. 2 vols., 1803.

New edition with Introduction by M. P. Crosland, New York, 1972.

Berthollet, C. L. and A. B. *Eléments de l'art de la teinture*. 2nd edn, 2 vols., 1804.

Berthomé, M. *L'enseignement secondaire dans la Haute Vienne pendant la Révolution (1789–1804)*. Paris and Limoges, 1913.

Berzelius, J. J. *Bref*, ed. H. G. Söderbaum, 15 parts, Uppsala, 1912–35.

Berzelius J. J. *Autobiographical notes*, trans. O. Larsell, Baltimore, 1934.

Biting, A. W. *Appertizing or the art of canning; its history and development*. San Francisco, 1937.

Bodelio, H. *Petite promenade physique contre l'idée de la pesanteur de l'air*...1818.

Bourgin, G. and H. *Le régime de l'industrie en France de 1814 à 1830*. 3 vols., 1912.

Boussingault, J. B. *Mémoires*. 5 vols., 1892–1903.

Bradley, M. 'The facilities for practical instruction in science during the early years of the Ecole Polytechnique', *Annals of Science*, 33 (1976), 425–46.

Brissot de Warville *Un mot à l'oreille des Académiciens de Paris*. [1786].

Brown, T. M. 'The electric current in early nineteenth-century French physics', *Historical Studies in the Physical Sciences*, 1 (1967), 61–103.

Callot, J. P. *Histoire de l'Ecole Polytechnique.* 1958.

Campbell, P. *French electoral systems and elections, 1789–1957.* 1958.

Cardwell, D. S. L. *From Watt to Clausius. The rise of thermodynamics in the early industrial age.* 1971.

Chaptal, J. A. *Elements of chemistry* (trans. W. Nicholson), 2nd edn, 1795.

Choffel, J. *Saint-Gobain. Du miroir à l'atome.* 1960.

Christophe, R. 'L'analyse volumetrique de 1790 à 1860. Caractéristiques et importance industrielle. Evolution des instruments', *Revue d'histoire des sciences,* 24 (1971), 25–44.

Christison, R. *The life of Sir Robert Christison, Bart...edited by his sons.* 2 vols., 1885–6.

Costabel, P. 'Le "calorique du vide" de Clément et Desormes (1812–1819)', *Archives internationales d'histoire des sciences,* 21 année (1968), 1–14.

Crosland, M. P. 'The development of a professional career in science in France', in *The emergence of science in western Europe,* ed. M. P. Crosland, 1975, pp. 139–59.

Crosland, M. P. 'The origins of Gay-Lussac's law of combining volumes of gases', *Annals of Science,* 17 (1961), [1963], 1–26.

Crosland, M. P. *The Society of Arcueil. A view of French science at the time of Napoleon I.* 1967.

Crosland, M. P. *Les héritiers de Lavoisier.* 1968.

Crosland, M. P. (ed.) *Science in France in the revolutionary era described by Thomas Bugge.* Cambridge, Mass., and London, 1969.

Crosland, M. P. 'Theories of acidity in the early nineteenth-century', in *Proceedings of the XIIIth international congress of the history of science,* Moscow, 1971, section 7, pp. 67–74.

Crosland, M. P. 'Lavoisier's theory of acidity', *Isis,* 64 (1972), 306–25.

Crosland, M. P. 'The French Academy of Sciences in the nineteenth century', *Minerva,* 16 (1978), No. 1, in press.

Dalton, J. *New system of chemical philosophy.* Vol. 1, Manchester, 1808, 1810; vol. 2, London, 1827.

Daumard, A. *La bourgeoisie parisienne de 1815 à 1848.* 1963.

Davy, H. *Collected works,* ed. John Davy. 9 vols., 1839–40.

Davy, H. *Researches, chemical and philosophical chiefly concerning nitrous oxide.* 1800.

Davy, H. *Elements of chemical philosophy.* 1812.

Davy, H. *Fragmentary remains,* ed. John Davy. 1858.

Deleuze, M. *Histoire et description du Muséum d'Histoire Naturelle.* 2 vols., 1823.

Despretz, C. M. *Traité élémentaire de physique,* 4th edn. 1836.

Dumas, J. B. *Traité de chimie appliquée aux arts.* 8 vols., 1828.

Dumas, J. B. *Leçons sur la philosophie chimique.* 1837.

Dunham, A. L. *The industrial revolution in France, 1815–1848.* New York, 1955.

Ecole Polytechnique *Livre du centenaire.* 3 vols., 1895.

Faraday, M. *Experimental researches.* 3 vols., 1839–55.

Faraday, M. *Life and letters*, ed. Bence Jones. 2 vols., 1870.

Farrar, W. V. 'Dalton and structural chemistry', in *John Dalton and the progress of science*, ed. D. Cardwell, Manchester, 1968, pp. 290–9.

Figuier, L. *Les applications nouvelles de la science à l'industrie et aux arts en 1855*, 2nd edn. 1857.

Fourcy, A. *Histoire de l'Ecole Polytechnique*. 1828.

Fox, R. *The caloric theory of gases from Lavoisier to Regnault*. Oxford, 1971.

Fox, R. 'The rise and fall of Laplacian physics', *Historical Studies in the Physical Sciences*, 4 (1974–5), 89–136.

Frankland, E. *Sketches from the life of Edward Frankland (1825–1899)* edited...by his two daughters M.N.W. and S.J.C. 1902.

Freund, I. *The study of chemical composition*. 1904. Dover edition, New York, 1968.

Fullmer, J. Z. 'Davy's priority in the iodine dispute: further documentary evidence', *Ambix*, 22 (1975), 39–51.

Gillispie, C. C. (ed.) *Dictionary of scientific biography*. 14 vols., New York, 1970–7.

Gréard, O. *Education et instruction*, vol. 1, *Enseignement supérieur*, 2nd edn. Paris, 1889.

Grimaux, E. and Gerhardt, C. *Charles Gerhardt. Sa vie, son oeuvre, sa correspondance*. 1900.

Guerlac, H. 'Some Daltonian doubts', *Isis*, 52 (1961), 544–54.

Guerlac, H. 'Chemistry as a branch of physics: Laplace's collaboration with Lavoisier', *Historical Studies in the Physical Sciences*, 7 (1976), 193–276.

[Gunpowder] 'Le service des poudres', *Croix de Guerre*, Numéro Spécial. Oct.–Nov. 1961.

Haber, L. F. *The chemical industry during the nineteenth century*. Oxford, 1958.

Hahn, R. *The anatomy of a scientific institution. The Paris Academy of Sciences, 1666–1803*. Berkeley and Los Angeles, 1971.

Hartley, H. *Humphry Davy*. 1966.

Hatin, E. *Bibliographie historique et critique de la presse périodique française*. 1866.

Herivel, J. *Joseph Fourier. The man and the physicist*. Oxford, 1975.

Holmes, F. *Claude Bernard and animal chemistry. The emergence of a scientist*. Cambridge, Mass., 1974.

Kahlbaum, E. W. A. and Schaer, E. *Christian Friedrich Schönbein. 1799–1868*. Leipzig, 1899.

Kent, S. *Electoral procedure under Louis Philippe* (Yale Historical Publications: Studies, X. New Haven, 1937.

Knight, D. M. 'The scientist as sage', *Studies in Romanticism*, 6 (1967), 65–88.

Ladd, A. J. *Ecole Normale Supérieure. An historical sketch*. Grand Falls, North Dakota, 1907.

Lalande, J. de François de *Bibliographie astronomique*. 1803.

Laplace, P. S. *Exposition du système du monde,* 2nd edn. 1799.
3rd edn, 1808.

Laplace, P. S. *Mécanique céleste.* 5 vols., 1798–1827.

Larivière, F. *Expertise d'aréomètres par Jacquelain, Barreswil et Troost. Examen critique du rapport des experts.* 1861.

Launay, L. de *Les Brongniart.* 1940.

Laurens, L. *Observations sur les essais alcalimétriques des sels de soude du commerce et moyen simple pour apprecier le carbonate de soude et le sulfite qui s'y trouvent contenus.* Marseille, 1819.

Lavoisier *Elements of chemistry,* trans. R. Kerr. Edinburgh, 1790.
Dover edition, New York, 1965.

Lavoisier *Oeuvres.* 6 vols., 1862–93.

Lecler, abbé A. *Martyrs et confesseurs de la foi du diocèse de Limoges pendant la révolution française.* 4 vols., Limoges, 1892–1904.

Levasseur, E. *Histoire des classes ouvrières et de l'industrie en France de 1789–1870.* 2 vols., 1903, 1904.

Lhomme, J. *La grande bourgeoisie au pouvoir (1830–1880).* 1960.

Longchamp *Lettre de M. Longchamp, ex commissaire en chef des poudres à M. Gay-Lussac, membre du Comité Consultatif de la Direction des Poudres, l'un des rédacteurs des Annales de Chimie et de Physique, etc., etc.* April, 1827.

Lunge, A. *A theoretical and practical treatise on the manufacture of sulphuric acid and alkali with the collateral branches.* 2 vols., 1879.

Marion, M. *Histoire financière de la France depuis 1715.* 6 vols. [1914], reprint 1927–31.

Merz, J. T. *A history of European thought in the nineteenth century,* vol. 1. London, 1904.

Middleton, W. E. K. *A history of the thermometer.* Baltimore, Maryland, 1966.

Middleton, W. E. K. *The history of the barometer.* Baltimore, Maryland, 1964.

Mitscherlich, A. *Gesammelte Schriften von Eilhard Mitscherlich. Lebensbild und Abhandlungen.* Berlin, 1896.

Mohr, F. *Traité d'analyse chimique à l'ordre de liqueurs titrées,* trans. C. Forthomme. Paris and Nancy, 1857.

Molard, C. P. (ed.) *Description des machines et procédés specifiés dans les brevets d'invention, de perfectionnement et d'importation dont la durée est expirée.* 93 vols., 1811–63.

Moquin-Tandon, A. *Un naturaliste à Paris sous Louis-Philippe. Journal de voyage inédit,* ed. M. Roland. 1944.

Olmsted, J. M. D. *François Magendie.* New York, 1944.

Palmade, G. P. *French capitalism in the nineteenth century,* trans. G. M. Holmes. Newton Abbot, 1972.

Paris, J. A. *The life of Sir Humphry Davy.* 2 vols., 1831.

Partington, J. R. *History of chemistry,* vols. 3 and 4. London, 1962, 1964.

Patterson, E. C. *John Dalton and the atomic theory.* New York, 1970.

Pelouze, T. J. and Fremy, E. *Cours de chimie générale.* 4 vols., 1848–50.

Pierson, S. 'Gay-Lussac and Berthollet's theory', in *Actes du XIIIe congrès internationale d'histoire des sciences*, Paris, 1968, vol. 6, pp. 83–6.

Piobetta, J. B. *Le baccalauréat*. 1937.

Poggiale, A. B. *Traité d'analyse chimique par la méthode des volumes*. 1958.

Rancke Madsen, E. *The development of titrimetric analysis till 1806*. Copenhagen, 1958.

Remusat, C. de *Mémoires de ma vie*, ed. C. H. Pouthas, vol. 1. 1958.

Richardson, N. *The French prefectorial corps, 1814–1830*. Cambridge, 1966.

Sadoun-Goupil, M. *Le chimiste Claude-Louis Berthollet (1748–1822), sa vie, son oeuvre*. 1977.

[Saint-Gobain] *Compagnie de Saint-Gobain, 1665–1965*. [1965].

Sébastiani, Lt-Gen. *Réponse du Général Sébastiani, député de l'Aisne à M. Gay-Lussac, membre de l'Académie des Sciences, etc.* Paris, 30 June [1828].

Smeaton, W. A. *Fourcroy, chemist and revolutionary, 1755–1809*. 1962.

Smith, R. A. *Life and work of Thomas Graham, illustrated by 64 unpublished letters*. Glasgow, 1884.

Sutton, F. *Systematic handbook of volumetric analysis*. 1863.

Szabadvary, F. *History of analytical chemistry*. Trans. E. Svehla, Oxford, 1966.

Taton, R. (ed.) *Enseignement et diffusion des sciences en France au XVIIIe siècle*. 1964.

Thackray, A. *Atoms and powers. An essay on Newtonian matter-theory and the development of chemistry*. Cambridge, Mass., 1970.

Thackray, A. *John Dalton. Critical assessments of his life and science*. Cambridge, Mass., 1972.

Thenard, L. J. *Traité de chimie*, 1st edn. 4 vols., 1813–16. 6th edn, 5 vols., 1834.

Thenard, P. *Un grand Français. Le chimiste Thenard, 1777–1857*, par son fils; avec introduction et notes de Georges Bouchard. Dijon, 1950.

Thomson, T. *Système de chimie*, [trans.]...*précédé d'une introduction de C. L. Berthollet*. 9 vols., 1809.

Thomson, T. *History of chemistry*. 2 vols., 1830, 1831.

Toraude, L. G. *Bernard Courtois (1776–1838) et la découverte de l'iode*. 1921.

Tudesq, A. J. *Les grands notables en France, 1840–1849*. 2 vols., Bordeaux, 1964.

Verdet, M. E. (ed.) *Notes et mémoires*. 1872.

Weeks, M. E. *Discovery of the elements*, 5th edn. Easton, Pennsylvania, 1945.

Williams, L. P. (ed.) *The Selected Correspondence of Michael Faraday*, 2 vols., Cambridge, 1971.

Wurtz, A. 'Sulphurique (acide)', in *Dictionnaire de chimie pure et appliquée*. 3 vols in 5. 1868–78.

## Unpublished theses

Bickerton, D. M. 'M.A. and C. Pictet, the Bibliothèque Britannique (1796–1815), and the dissemination of British science and literature on the Continent' (University of Leeds, Ph.D), 1978/9.

Boughey, K. J. S. 'Studies in the role of positivism in nineteenth-century French chemistry' (University of Leeds, Ph.D), 1972.

Bradley, M. 'The Ecole Polytechnique, 1795–1830: organisational changes and students' (University of Leeds, M.Phil.), 1974.

Cawood, A. J. 'The scientific work of D. F. J. Arago (1786–1853)' (University of Leeds, Ph.D.), 1974.

Court, S. 'The Annales of Chimie, 1789–1815' (University of Leeds, M.Phil.), 1969.

Klosterman, L. J. 'Studies in the life and work of Jean Baptiste André Dumas (1800–1884): the period up to 1850' (University of Kent at Canterbury, Ph.D.), 1976.

Smith, J. Graham 'Studies of certain chemical industries in Revolutionary and Napoleonic France' (University of Leeds, Ph.D.), 1970.

# Name Index

326

# Subject Index

Printed in the United States
By Bookmasters